Frontiers in Atmospheric Pressure Plasma Technology

Frontiers in Atmospheric Pressure Plasma Technology

Editor

Andrei Vasile Nastuta

MDPI • Basel • Beijing • Wuhan • Barcelona • Belgrade • Manchester • Tokyo • Cluj • Tianjin

Editor
Andrei Vasile Nastuta
Physics and Biophysics
Education Research
Laboratory (P&B-EduResLab)
"Grigore T. Popa" University
of Medicine and Pharmacy
Iasi
Romania

Editorial Office
MDPI
St. Alban-Anlage 66
4052 Basel, Switzerland

This is a reprint of articles from the Special Issue published online in the open access journal *Applied Sciences* (ISSN 2076-3417) (available at: www.mdpi.com/journal/applsci/special_issues/ atmospheric_pressure_plasma_technology).

For citation purposes, cite each article independently as indicated on the article page online and as indicated below:

LastName, A.A.; LastName, B.B.; LastName, C.C. Article Title. *Journal Name* **Year**, *Volume Number*, Page Range.

ISBN 978-3-0365-5002-2 (Hbk)
ISBN 978-3-0365-5001-5 (PDF)

Cover image courtesy of Andrei Vasile Nastuta

Contents

About the Editor

Andrei Vasile Nastuta

Andrei Vasile NASTUTA, PhD. Phys., is a university lecturer and a researcher in Physics and Biophysics within the Department of Biomedical Sciences of the Faculty of Medical Bioengineering, University of Medicine and Pharmacy "Grigore T. Popa" Iasi, Romania. He graduated from the Faculty of Physics, Biophysics specialization, from "Alexandru Ioan Cuza" University of Iași, Romania. He also has an M.Sc. degree in Plasma Physics, Spectroscopy and Polymer Physics and a PhD in Physics, with over 15 years of experience in the university environment. He is the coordinator of the Laboratory of Research and Education in Physics and Biophysics (P&B-EduResLab) within the Faculty of Medical Bioengineering. For 10 years, he has been teaching classes and carrying out practical work in the Physics and Biophysics disciplines for the students of the Faculty of Medical Bioengineering.

He has been working in the field of plasma physics and biophysics, covering areas of low-pressure DC and pulsed magnetron discharges; low-pressure DC and pulsed hollow-cathode discharges; plasma thin-film deposition (by means of appj, DC and high-impulse power magnetron and hollow-cathode sputtering discharges); designing and developing plan-parallel and jet-like atmospheric pressure dielectric barrier discharges sources, in a closed or open environment; characterization techniques of plasma discharges through optical-spectroscopy (optical emission spectroscopy, ultra-fast and phase resolved photography, Hook interferometry) and electrical diagnosis methods (voltage–current, Lissajous figures, Pockels effect electro-optic electric field); surface characterization techniques (contact angle, optical microscopy, atomic force microscopy, SEM, ATR-FTIR/RAMAN spectroscopy, X-ray photoelectron spectroscopy); mass spectrometry; and plasma applications for biomedical scenarios.

He has co-authored more than 40 published refereed papers (31 ISI) and over 50 conference papers on these topics.

Preface to "Frontiers in Atmospheric Pressure Plasma Technology"

Atmospheric pressure plasma discharges have grown rapidly in importance in recent decades, due to the ease in handling and operation, plus their eco-friendly applications, for agriculture, food, medicine, materials and even the automotive and aerospace industries. In this context, the need for a collection of results based on plasma technologies is justified. Moreover, at the international level, the increased number of projects that translated to publications and patents in the multidisciplinary field of plasma-based technology gives researchers the opportunity to challenge their knowledge and contribute to a new era of green services and products that society demands. Therefore, this book, based on the Special Issue of "Frontiers in Atmospheric Pressure Plasma Technology" in the "Applied Physics" section of the journal *Applied Sciences*, provides results on some plasma-based methods and technologies for novel and possible future applications of plasmas in life sciences, biomedicine, agriculture, and the automotive industry.

As a Guest Editor of this Special Issue of "Frontiers in Atmospheric Pressure Plasma Technology", in the Editorial, I briefly discuss the results of all the articles published in this Special Issue (8 articles, 2 review articles and 1 editorial), also including some personal results. We know that we are only managing to address a small part of what plasma discharge can be used for, but we hope that the readers will enjoy this book and, therefore, be inspired with new ideas for future research in the field of plasma.

Andrei Vasile Nastuta
Editor

Editorial

Frontiers in Atmospheric Pressure Plasma Technology

Andrei Vasile Nastuta 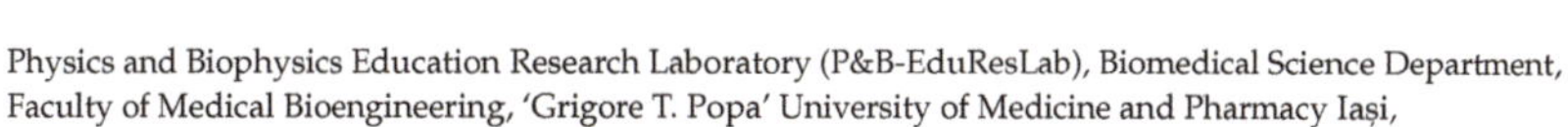

Physics and Biophysics Education Research Laboratory (P&B-EduResLab), Biomedical Science Department, Faculty of Medical Bioengineering, 'Grigore T. Popa' University of Medicine and Pharmacy Iași, Str. M. Kogalniceanu No. 9-13, 700454 Iași, Romania; nastuta.andrei@umfiasi.ro or andrei.nastuta@gmail.com

Citation: Nastuta, A.V. Frontiers in Atmospheric Pressure Plasma Technology. *Appl. Sci.* **2022**, *12*, 6369. https://doi.org/10.3390/app12136369

Received: 23 May 2022
Accepted: 17 June 2022
Published: 22 June 2022

Publisher's Note: MDPI stays neutral with regard to jurisdictional claims in published maps and institutional affiliations.

Atmospheric pressure plasmas represent a feasible and eco-friendly alternative to conventional physicochemical methods used in technology today for facing materials. The complex physical and chemical processes occurring when plasma interacts with mater offer a rich source of short- and long-lived chemical species, mostly reactive nitrogen and oxygen species (RNS/ROS). They are also crucial for many applications ranging from the food industry, environmental related fields, agriculture and healthcare, to material science and even automotive. Exciting novel applications of plasma–surface, plasma–liquid, and plasma–gas interactions are at the focus of many challenging multidisciplinary scientific inquiries.

The range of potential plasma applications is broad, from plasma (bio)medicine (antibacterial/disinfectant/antiseptic agent, wound healing promoter, selective treatment of cancer cells and tumors), plasma pharmacology, plasma and food, plasma bioengineering, plasma agriculture (as seed germination inducer or even as a green fertilizer), to plasma and automotive. Nevertheless, all of these plasma fields are plasma–based technologies.

This Special Issue on 'Frontiers in Atmospheric Pressure Plasma Technology' was focused on, but not limited to, recent findings in novel and possible future applications of plasmas in life sciences, biomedicine, agriculture, and the automotive industry.

Papers providing fundamental insights into the understanding of plasmas and detailed analysis of electrical discharges, pushing forward cutting-edge techniques in plasma science and technology, were especially welcomed and received.

On the basis of the background outlined above, this Special Issue of *Applied Sciences* entitled "Frontiers in Atmospheric Pressure Plasma Technology" includes eight original papers [1–8] and two reviews [9,10] providing new insights into the application of atmospheric pressure plasma technology.

Lee et al. [1] used atmospheric pressure plasma irradiation to facilitate transdermal permeability of aniline blue on porcine skin and increase the cellular permeability of keratinocytes and further demonstrated the production of nitric oxide from keratinocytes. Their findings suggest a promoting effect of low-temperature plasma on transdermal absorption, even for high-molecular-weight molecules, as well as that the plasma-induced nitric oxide from keratinocytes is likely to regulate transdermal permeability in the epidermal layer.

Oliveira et al. [2] showed in their study the inhibitory effect of cold atmospheric plasma (CAP), running in helium, on chronic wound-related multispecies biofilms. The report describes the effect of He-CAP on wound-related multispecies biofilms and confirms the safety of the protocol. It was proven that exposure to He-CAP for 5 min provides inhibitory effects for the wound-related multispecies biofilms formed by methicillin-resistant *Staphylococcus aureus* (MRSA), *Pseudomonas aeruginosa* and also *Enterococcus faecalis*.

Furthermore, Nascimento et al. [3] revealed the effects of O_2 addition on the discharge parameters and production of reactive species of a transferred atmospheric pressure plasma jet produced at the tip of a long and flexible plastic tube. They concluded that the addition of O_2 to the working gas seems to be useful for increasing the effectiveness of the plasma treatment only when the target modification effect is directly dependent on the content of atomic oxygen.

In a report by Janda et al. [4] the role of HNO_2 in the generation of plasma-activated water by air transient spark (TS) discharge, a DC-driven self-pulsing discharge generating a highly reactive atmospheric pressure air plasma, was studied. Moreover, the authors compared TS with water electrospray (ES) in a one-stage system and TS operated in dry or humid air followed by water ES in a two-stage system, and show that gaseous HNO_2, rather than NO or NO_2, plays a major role in the formation of NO_2^-(aq) in plasma-activated water (PAW) that reached the concentration up to 2.7 mM.

Moreover, Huzum and Nastuta [5] were using helium atmospheric pressure plasma jet source in the treatment of white grapes juice for winemaking purposes. Based on principles of dielectric barrier discharge, an atmospheric pressure plasma jet in helium was used to treat two types of grape white fresh musts, after that, the resulting 1 and 2 year old wine was characterized and the correlation of plasma parameters (mean power, current density, RONS) and the physico-chemical proprieties (pH values, BRIX, UV-Vis and FTIR spectroscopy, CIE L*a*b* and RGB) of white must and wine were assessed.

Ivankov et al. [6] investigated the non-thermal plasma jet excitation modes and the optical assessment of its electron concentration. They proposed and described a new method based on digital holography to estimate electronic concentration for a non-thermal plasma source, taking into account its disadvantages and further applications.

Nastuta and Gerling [7] examined the use of a cold atmospheric pressure plasma jet source operated both in Ar and He, optically and electrically, going from basic plasma properties to vacuum ultraviolet, electric field determinations as well as safety thresholds measurements in Plasma Medicine. They reported that the surface temperature and leakage values of both systems showed different slopes, with the biggest surprise being a constant leakage current over distance for argon jet.

Benova et al. [8] followed up on the characteristics of 2.45 GHz surface-wave-sustained argon discharge for bio-medical applications. Their manuscript presents the characterization and optimization of the active region of surface-wave-sustained argon plasma for biological systems treatment. Their result shows that a discharge tube with a bigger inner diameter is able to obtain a surface temperature of 45 °C, as is required for the treatments in biology, medicine and agriculture.

A review by Borges et al. [9] focuses on the applications of cold atmospheric pressure plasma in Dentistry. Their manuscript outlines the application of cold atmospheric pressure plasma (CAPP) in dentistry for the control of several pathogenic microorganisms, as well as for induction of anti-inflammatory, tissue repair effects and apoptosis of cancer cells, with low toxicity to healthy cells. Therefore, CAPP has the potential to be applied in many areas of dentistry such as cardiology, periodontology, endodontics and even oral oncology.

A review by Zaplotnik et al. [10] is directed towards optical emission spectroscopy (OES), as a diagnostic tool for the characterization of atmospheric plasma jets. They revealed the advantages and limitations of the method, in comparison with other information given by other spectroscopic methods, e.g., VUV, that can bring more insights into the energetic species in the plasma.

This collection of reports on different plasma-based technological applications, gathered in this Special Issue brought from the community a total of 28 citations and more than 8600 views, showing the growing interest of society in such green technologies (without chemicals and by-products).

Finally, the Editor is delighted to have had the honor of organizing this Special Issue for *Applied Sciences*, which highlights the research of distinguished scientists in the field of plasma physics. The Editor would like to thank all the contributors to this Special Issue for their commitment and enthusiasm during the compilation of the respective articles. The Editor also wish to thank the members of the editorial staff at Multidisciplinary Digital Publishing Institute (MDPI) for the professionalism and dedication they have shown in completing this Special Issue. We hope that the readers will enjoy this Special Issue and be inspired with new ideas for future research in the field of plasma.

Funding: This research received no external funding.

Conflicts of Interest: The author declare no conflict of interest.

References

1. Lee, S.; Choi, J.; Kim, J.; Jang, Y.; Lim, T. Atmospheric pressure plasma irradiation facilitates transdermal permeability of aniline blue on porcine skin and the cellular permeability of keratinocytes with the production of nitric oxide. *Appl. Sci.* **2021**, *11*, 2390. [CrossRef]
2. Oliveira, M.; Lima, G.; Nishime, T.; Gontijo, A.; Menezes, B.; Caliari, M.; Kostov, K.; Koga-Ito, C. Inhibitory Effect of Cold Atmospheric Plasma on Chronic Wound-Related Multispecies Biofilms. *Appl. Sci.* **2021**, *11*, 5441. [CrossRef]
3. Nascimento, F.; Petroski, K.; Kostov, K. Effects of O_2 Addition on the Discharge Parameters and Production of Reactive Species of a Transferred Atmospheric Pressure Plasma Jet. *Appl. Sci.* **2021**, *11*, 6311. [CrossRef]
4. Janda, M.; Hensel, K.; Tóth, P.; Hassan, M.; Machala, Z. The Role of HNO_2 in the Generation of Plasma-Activated Water by Air Transient Spark Discharge. *Appl. Sci.* **2021**, *11*, 7053. [CrossRef]
5. Huzum, R.; Nastuta, A. Helium Atmospheric Pressure Plasma Jet Source Treatment of White Grapes Juice for Winemaking. *Appl. Sci.* **2021**, *11*, 8498. [CrossRef]
6. Ivankov, A.; Kozhevnikova, A.; Schitz, D.; Alekseenko, I. Investigation of Nonthermal Plasma Jet Excitation Mode and Optical Assessment of Its Electron Concentration. *Appl. Sci.* **2021**, *11*, 9203. [CrossRef]
7. Nastuta, A.; Gerling, T. Cold Atmospheric Pressure Plasma Jet Operated in Ar and He: From Basic Plasma Properties to Vacuum Ultraviolet, Electric Field and Safety Thresholds Measurements in Plasma Medicine. *Appl. Sci.* **2022**, *12*, 644. [CrossRef]
8. Benova, E.; Marinova, P.; Tafradjiiska-Hadjiolova, R.; Sabit, Z.; Bakalov, D.; Valchev, N.; Traikov, L.; Hikov, T.; Tsonev, I.; Bogdanov, T. Characteristics of 2.45 GHz Surface-Wave-Sustained Argon Discharge for Bio-Medical Applications. *Appl. Sci.* **2022**, *12*, 969. [CrossRef]
9. Borges, A.; Kostov, K.; Pessoa, R.; de Abreu, G.; Lima, G.; Figueira, L.; Koga-Ito, C. Applications of cold atmospheric pressure plasma in dentistry. *Appl. Sci.* **2021**, *11*, 1975. [CrossRef]
10. Zaplotnik, R.; Primc, G.; Vesel, A. Optical emission spectroscopy as a diagnostic tool for characterization of atmospheric plasma jets. *Appl. Sci.* **2021**, *11*, 2275. [CrossRef]

applied sciences

MDPI

Review

Applications of Cold Atmospheric Pressure Plasma in Dentistry

Aline C. Borges [1], Konstantin G. Kostov [2], Rodrigo S. Pessoa [3], Geraldo M.A. de Abreu [1], Gabriela de M.G. Lima [1], Leandro W. Figueira [1] and Cristiane Y. Koga-Ito [1,*]

[1] Oral Biopathology Graduate Program and Department of Environmental Engineering, Institute of Science and Technology, São Paulo State University/UNESP, São José dos Campos-SP 12247-016, Brazil; aline_chiodi@hotmail.com (A.C.B.); abreugma@gmail.com (G.M.A.d.A.); gabrielademorais@yahoo.com.br (G.d.M.G.L.); leandrowf@live.com (L.W.F.)

[2] Department of Physics, Guaratinguetá Faculty of Engineering, São Paulo State University/UNESP, Guaratinguetá-SP 12516-410, Brazil; kostov@feg.unesp.br

[3] Plasmas and Processes Laboratory, Aeronautics Institute of Technology/ITA, São José dos Campos-SP 12228-900, Brazil; rspessoa@ita.br

[*] Correspondence: cristiane.koga-ito@unesp.br; Tel.: +55-12-39479708

Abstract: Plasma is an electrically conducting medium that responds to electric and magnetic fields. It consists of large quantities of highly reactive species, such as ions, energetic electrons, exited atoms and molecules, ultraviolet photons, and metastable and active radicals. Non-thermal or cold plasmas are partially ionized gases whose electron temperatures usually exceed several tens of thousand degrees K, while the ions and neutrals have much lower temperatures. Due to the presence of reactive species at low temperature, the biological effects of non-thermal plasmas have been studied for application in the medical area with promising results. This review outlines the application of cold atmospheric pressure plasma (CAPP) in dentistry for the control of several pathogenic microorganisms, induction of anti-inflammatory, tissue repair effects and apoptosis of cancer cells, with low toxicity to healthy cells. Therefore, CAPP has potential to be applied in many areas of dentistry such as cariology, periodontology, endodontics and oral oncology.

Keywords: cold atmospheric pressure plasma; antimicrobial agent; plasma medicine; dentistry

Citation: Borges, A.C.; Kostov, K.G.; Pessoa, R.S.; de Abreu, G.M.A.; Lima, G.d.M.G.; Figueira, L.W.; Koga-Ito, C.Y. Applications of Cold Atmospheric Pressure Plasma in Dentistry. *Appl. Sci.* **2021**, *11*, 1975. https://doi.org/10.3390/app11051975

Academic Editor: Andrei Vasile Nastuta

Received: 19 January 2021
Accepted: 17 February 2021
Published: 24 February 2021

Publisher's Note: MDPI stays neutral with regard to jurisdictional claims in published maps and institutional affiliations.

1. Introduction

Plasma is frequently referred to as the fourth state of the matter and can be described as a gaseous mixture of neutral particles, electrons and ions at different densities and temperatures. Most of the visible matter in the universe (about 99%), such as stars, nebulas and interstellar medium, is in the state of plasma. Plasma can be generated by heating a gas or by subjecting it to strong electromagnetic fields to the point that the gas particles become ionized. Thus, plasma is an electrically conducting medium that responds to electric and magnetic fields, which is also a source of large quantities of highly reactive species, such as ions, energetic electrons, excited atoms and molecules, ultraviolet photons, metastable, and active radicals [1]. In laboratory conditions, plasma is generally produced by an electric discharge in noble or molecular gases, such as argon (Ar), helium (He), oxygen (O_2) and nitrogen (N_2), using different excitation schemes, such as microwaves, radiofrequency and DC or AC electric fields [2].

Normally, the electron and ion densities in plasmas are approximately equal (a condition called quasi-neutrality), but the respective electron and ion temperatures can be quite different. Plasmas are usually classified as thermal and non-thermal plasmas. Thermal plasmas are in thermal equilibrium, which means that their temperatures are relatively homogenous throughout the heavy particles (i.e., atoms, molecules, and ions) and electrons that usually span in the range of thousands of K [1]. The so-called non-thermal or cold plasmas are partially ionized gases and their electron temperatures exceed several tens of thousands K, while the heavy particles (ions and neutrals) have a much lower temperature [2]. Plasma can be also generated under different pressure conditions, including atmospheric pressure [1]. In the last decade, atmospheric pressure plasmas have become a very attractive tool for material processing applications because they are generated in an open environment and can be easily implemented in online processing. However, working at atmospheric pressure has some disadvantages. For instance, gas breakdown at atmospheric pressure occurs at much higher electric field, typically in the order of ten of kV cm^{-1} [1]. Additionally, if special precautions are not taken, the atmospheric plasmas have the tendency to become thermal i.e., hot plasmas that can damage heat sensitive materials or burn living tissues [2].

A gas discharge in which a dielectric layer covers one or both electrodes is called Dielectric Barrier Discharge (DBD) and it was first introduced by Siemens—in 1857—for the production of ozone. Large number of DBD systems has been reported [3] with the planar and cylindrical geometries as the most common employed configurations. Figure 1a depicts a typical planar DBD reactor, while Figure 1b shows the so-called floating electrode DBD (FE-DBD) [4,5]. The plasma, in this case, is formed between an insulated high voltage electrode and a target (human skin or living tissue), which acts as a floating counter electrode.

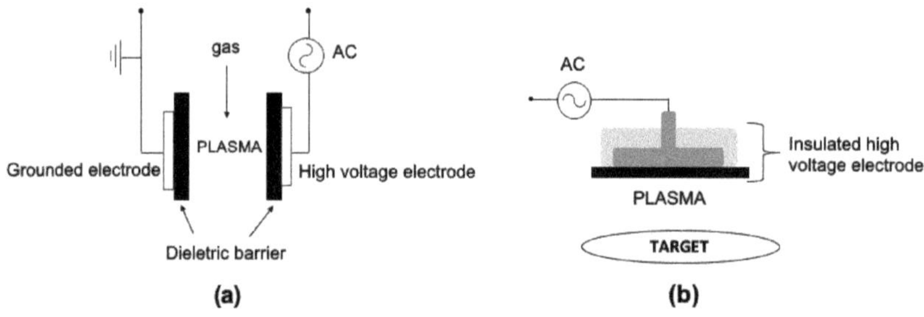

Figure 1. Schematic representation of DBD reactors in planar geometry: (**a**) conventional two electrodes DBD and (**b**) floating electrode DBD (FE-DBD). Adapted from [5].

Due to the current limitation caused by the charge accumulation on the dielectric surface, the gas temperature in DBD can be quite low (about the room temperature), which makes it adequate for biological applications [4]. However, the plasma in DBD devices is confined into small gaps between two electrodes (usually in the order of few mm), which is a disadvantage for some applications. For instance, FE-DBD was successfully used in a number of clinical trials for the treatment of skin diseases, to control melanoma development, blood coagulation and antisepsis of open wounds [5]. However, it is not suitable for plasma application inside the human body's cavities, such as tooth root canals or internal organs.

On the other hand, in the so-called atmospheric pressure plasma jets (APPJs) the plasma generated into a dielectric enclosure (tube or syringe) is expelled through a small orifice into the ambient atmosphere by gas flow (usually a noble gas). The ejected plasma forms a plasma plume that extended several cm into the air and can be easily directed to a target. Figure 2 shows a drawing of the APPJ concept. In the past decade many APPJ configurations, different electrodes arrangements and excitation schemes have been reported in the literature. More details about APPJs and their characteristics can be found in some recent review papers [6–8].

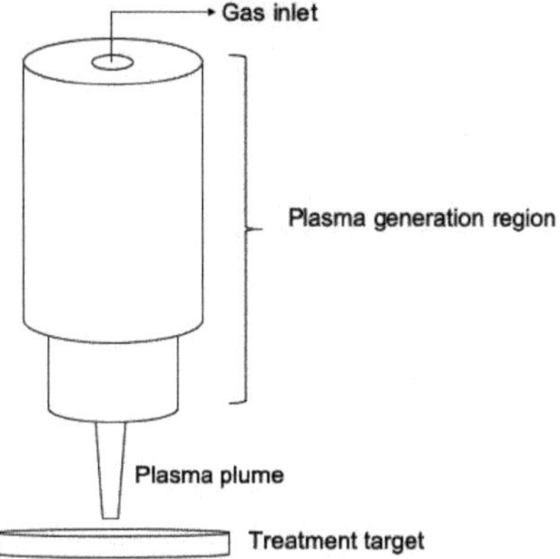

Figure 2. Schematic representation of an APPJ.

Depending on the jet's operating conditions, the plasma plume tip can be maintained below 40 °C, enabling the contact with living tissues without any risk of burns and electric shock. Thus, cold atmospheric pressure plasma (CAPP), such as FE-DBD and APPJ, have been appointed as the most promising tools for biomedical and hospital applications [9–12].

Since CAPPs are generated in ambient air, large quantities of reactive oxygen and nitrogen species (RONS) are produced. Therefore, when CAPP enters in contact with living tissues, the synergistic action of several plasma components, such RONS, energetic (UV) photons, and charge particles should be considered. The biochemical mechanisms involved in the interaction of plasma species with microorganisms and cells, as well as the plasma application for tissue healing and disinfection are extensively studied in a novel interdisciplinary field called Plasma Medicine [9,10]. Recent studies have demonstrated that RONS are the main factor responsible for plasma antimicrobial and tissue healing effects, while the UV photons have only minor effect [13].

CAPP can be applied directly on living tissues and, in this case, RONS reach directly the target. Alternatively, CAPP can be applied indirectly, by previous exposure of liquids (i.e., water, liquid culture media) to the plasma, creating solutions containing RONS, known as plasma-activated media (PAM) or plasma-activated water (PAW) [14,15]. In general, the plasma-liquid interaction generates hydrogen peroxide, nitrites, nitrates and others RONS [16]. However, the composition of PAM depends not only on the plasma-liquid interactions but also on the subsequent chemical reactions in the liquid phase that can cause further changes into PAM composition [15].

2. CAPP Biological Activities

The biological effects of CAPP enable several applications in the medical area [9]. Laroussi [17] was the first author to report on the antibacterial effect of CAPP. After, an expressive number of manuscripts, review articles, contributions to conference and books on the antimicrobial potential of CAPP and on the physicochemical mechanisms for antimicrobial inactivation have been published. The growth control of several pathogenic microorganisms, such as Gram-positive and Gram-negative bacteria, fungal species and bacterial spores have been reported [18–23]. Additionally, an antibiofilm effect has also been observed for bacteria and fungi [24–28].

Interesting data point out to the anti-inflammatory and tissue repair effect induced by CAPP [29,30]. CAPP improved wound healing in mice with induction of type I collagen and MCP-1 protein production in keratinocytes and fibroblasts [31,32]. Brun et al. [33] observed increased migration and proliferation of fibroblasts in response to the production of RONS during CAPP exposure. Similar effects were observed by Bourdens et al. [34] and Haralambiev et al. [35]. Stimulation of keratinocytes by antioxidant pathways was also reported [30,36]. CAPP showed positive effect in the cutaneous microcirculation, increasing the tissue oxygen saturation and radial blood flow [37,38] that can contribute to improved tissue repair.

A remarkable feature of CAPP is its highly selective toxicity that highlights the potential for clinical treatment of infectious diseases [9,39,40]. This differential activity is based on differences in the cellular metabolism in presence of RONS. Eukaryotic cells exhibit protection to RONS, while prokaryotic ones do not demonstrate such protective mechanism [39,41–43]. The disparity in the cell sizes influences response to CAPP. For instance, bacterial cells (typically between 0.2 and 10 μm) have higher surface-volume ratio, which favours the plasma action, while eukaryotic cells are much bigger, from 10 to 100 μm [10,39,42]. Moreover, the organization of eukaryotic cells into tissues additionally increases their resistance to CAPP effects. Therefore, by adjusting the treatment parameters, plasma can be used to eliminate bacteria in planktonic or biofilm forms without damage to the surrounding host tissues [9,44–46].

In this context, treating fungal diseases has an additional challenge, as both fungal and host cells are eukaryotic. Hence, in this particular situation, the simultaneous study of fungal inhibition and toxicity to host cells is extremely important. Interestingly, there have already been some encouraging results in the literature. Borges et al. [24] reported that CAPP treatment for 5 min has antibiofilm effects on *Candida albicans* with low cytotoxicity to Vero cells [47]. In the same study, CAPP was also applied in vivo to treat oral candidiasis in mice without damaging the surrounding tissues.

The ability of CAPP to induce cell death by inducing apoptosis [48] can be also very useful for therapeutic purposes, and it has been applied in the control of cancer cells. In this case, metabolic differences between healthy and malignant cells favour the selectivity of CAPP. Constant cell replications, observed in malignant cells, can expose their DNA to CAPP more frequently, favouring to the cell structural damage [39,42,49–51].

RONS generated by CAPP, including hydroxyl radicals (OH^-), hydrogen peroxide (H_2O_2), singlet oxygen ($^1O_2$), superoxide anion (O^-_2), atomic oxygen (O), atomic nitrogen (N), nitric oxide (NO), nitrogen trioxide (NO_3), influence redox-regulated cell processes [9,11,18,42,52–56]. In particular, reactive oxygen species (ROS) can react with many biological macromolecules, causing oxidative structural modification and the loss of their biological function [57]. At the cellular level, ROS regulate growth, apoptosis and other signalling processes, while at the system level, they contribute to complex functions, including regulation of immune response [58]. In addition, ROS are emerging as the most important agents in the bacterial response to lethal stress. Currently, the effects of superoxide, hydrogen peroxide and hydroxyl radical have been studied. Superoxide and hydrogen peroxide arise when molecular oxygen oxidizes redox enzymes that transfer electrons to other substrates. Hydrogen peroxide that can be produced from the dismutation of superoxide serves as a substrate for the formation of hydroxyl radicals. If this oxidative process is not controlled, an accumulation of hydroxyl radicals can occur. The hydroxyl radical breaks down nucleic acids, carbonylated proteins and peroxidised lipids which can lead to cell death [59].

Reactive nitrogen species (RNS) can be both harmful or beneficial to living systems. At low concentrations, RNS can play an important role as a regulatory mediator in signalling. On the other hand, at moderate or high concentrations, RNS are harmful to living organisms and can inactivate important cellular molecules [60]. Nitric oxide is an important regulator of physiological processes [61] and can mediate the harmful cellular toxicity of metabolic enzymes, generating nitrite peroxide as a final product of reaction with superoxide [62].

The exact mechanism of CAPP and microbial cell interaction is not fully understood yet, but presently it is widely accepted that its antimicrobial activity is associated with synergetic action of two major CAPP components, UV radiation and RONS. They can break covalent bonds of stable compounds, such as the peptidoglycan from the bacterial cell walls and peroxidation of lipids in the cell membrane [17,44,56,63–65]. Also, CAPP interaction with prokaryotic cells can cause cellular rupture by electro erosion with formation of ionic pores and subsequent loss of cellular content [11,63]. CAPP can also break covalent bonds in the polymeric matrix of microbial biofilms, favouring their disruption [28,45,46,56,66].

3. CAPP Application in Periodontology

Periodontal disease (PD) affects the dental support tissues and it is a major cause of tooth loss impacting on individual's function and social behaviour. Nowadays, it is well established that the presence of biofilm formation itself can lead to gingivitis but not to periodontitis. Among the bacteria involved in periodontitis development, *Porphyromonas gingivalis*, *Tannerella forsythia* and *Treponema denticola*, known as red complex, are the most studied and associated to tissue destruction. In particular, *P. gingivalis* can modulate the host response leading to bacterial unbalance into the gingival sulcus [67]. Due to the limited gain of the traditional periodontal treatment, it is necessary to find new adjuvant therapies.

Mahasneh et al. [68] treated blood agar plates previously inoculated with *P. gingivalis* (ATCC 33277) with He-CAPP jet from 5 to 11 min, and found significant difference in the diameter of inhibition zone in a time dependent manner for all periods of application when compared to the control group. Authors attributed the results to cell damage induced by RONS, though no specific analysis was performed to confirm the hypothesis. Liu et al. [69] studied *P. gingivalis* mono species biofilms using confocal microscopy and also evaluated the effect of CAPP on rabbit mucosa. Differently of Mahasneh et al. [68] that used He-CAPP, Liu et al. [69] tested He/O_2 mixture as working gas and observed that 5 min of treatment inhibited the most bacterial cells contrary to negative control. Moreover, after one or five days of CAPP exposure for 10 min on healthy rabbit mucosa, they found no signs of irritation.

Küçük et al. [70], in the first clinical trial using CAPP as an adjuvant therapy for non-surgical treatment in one-time application protocol, found significant gain of clinical attachment length after three months. They also observed the elimination of microorganisms in the red complex and recolonization reduction.

In terms of tissue repair, Kwon et al. [71] evidenced that CAPP treatment for 1 and 2 min could improve cell morphology of human gingival fibroblasts and enhance the mRNA expression of TGF-β and VEGF. Though, when authors tested CAPP for 4 min, both morphology and growth factors expression were better in control group. Interestingly, Eggers et al. [72] observed that one day after a 60-s CAPP application on osteoblast-like cells, the mRNA of proinflammatory cytokines such as IL-6, IL-8 and IL-1 were upregulated as well as TNF-, COX_2, CCL_2 mRNA and COL 1, important genes for wound healing. Plus, proliferation genes such as PCNA and Ki-67 mRNA expression was significantly upregulated.

More recently, most of the studies have been focused on peri-implantitis as the study of Shi et al. [73]. Similar to periodontitis, peri-implantitis is an inflammatory disease dependent on biofilm that causes loss of hard and soft tissue, but it occurs around the dental implant. The authors evaluated the effects of CAPP (3 min) as an adjuvant to clinical treatment in ligature induced peri-implantitis in beagle dogs. The study compared this treatment with the traditional clinical treatment using 0.2% chlorhexidine digluconate (3 min) as decontaminant. Clinical and bone analyses (micro-CT and histology) showed better recovery in CAPP-treated group three months after treatment. Microbial recovery by PCR showed that *P. gingivalis* and *T. forsythia* quantity was significantly reduced when compared to control and baseline. *A. actinomycetemcomitans* had a significant decrease in the first month that was not maintained for the next two months of follow up.

Carreiro et al. [74] showed that CAPP can reduce viability and, supposedly, quantity of *P. gingival* biofilm in titanium discs after 1 and 3 min with no epithelial harm for gingival tissues, in vitro. The same results were obtained by Lee et al. [75] using He-CAPP for 3 and 5 min. Authors noted that not only the exposed, but also the peripheral area of the titanium discs were decontaminated by CAPP. The authors also observed an increase in anti-VEGF expression after CAPP treatment using an in vitro gingival epithelium model.

Not directly focused on periodontitis, some studies on CAPP therapy let us intrigued on its potential on periodontal therapy. Arndt et al. [31,76] observed an increase in the β-defensins levels in fibroblasts and keratinocytes, and type I collagen induction after 2 min of cell activation in vitro using a microwave plasma torch (MicroPlaSter ß plasma torch system) with argon. These findings are extremally important since ß-defensins is a component of the innate immune response that avoid the increase in the count of pathogenic bacteria in the gingival sulcus during the first stages of periodontal disease, and expression of human ß-defensins mRNA is decreased in patients with chronic periodontitis when compared to healthy individuals [77]. Additionally, induction of type I collagen indicates wound healing stimulation.

Meanwhile, Brun et al. [33] demonstrated increase on migration and proliferation of fibroblasts in response to ROS produced by He-CAPP. In a recent study, the same group [78] studied *P. aeruginosa* and *S. aureus* biofilms and demonstrated that CAPP damaged the cell membrane in a way that prevents antimicrobial resistance, and has a synergistic effect with other antimicrobial drugs.

These antimicrobials, antibiofilm, and tissue stimulating evidences of CAPP therapy taken together show the importance of studying this technology as an adjuvant therapy for periodontitis that remains as a chronic disease where current gold standard therapy (scaling and root planning) is based only on damage control, in a sense that pathogenic microbial colonization can be reduced but recolonization commonly occurs in the short-term, and tissue reattachment can be achieved, but still there is no gain of lost tissue.

4. CAPP Application in Endodontics

Endodontic infection occurs in the tooth root canal system that gets exposed to the oral environment. The microorganisms can reach the intra-radicular region from a carious lesion or after a traumatic injury to the coronal tooth structure [79]. Primary endodontic infections are caused by polymicrobial biofilms dominated by aerobes and facultative anaerobes microorganisms [80]. The persistence of *Enterococcus faecalis* in the root canal system is the main cause of post-treatment infection followed by *Fusobacterium* and *Propionibacterium* [80,81].

Endodontic infection treatment involves removing of remnants vital or necrotic tissues, the elimination of microorganisms within the root canal system, and removal of hard tissue debris that is formed during root canal instrumentation. The root canal disinfection is considered the pivot of this therapy [81]. NaOCl solution is commonly used to clean and disinfect the root canal due to its antimicrobial activity, however, it can also affect vital pulp and decreases the mechanical resistance of dentin [82].

Some studies have shown that CAPP is able to inhibit the persistent root canal microorganism *Enterococcus faecalis*. Chang et al. [83] evaluated the effect of CAPP against *E. faecalis* suspensions spread on the surface of sterile glass slides. They observed that CAPP was able to reduce the number of colonies forming unities after 2 min of exposure. They also observed that the antimicrobial effects are time-dependent and the exposure for 3 min showed the best results. In situ studies using dental tissue as the surface for biofilm formation could better represent the root canal system. Armand et al. [84] reported that the viability of *E. faecalis* biofilms formed on the surface of tooth fragments were significantly reduced after exposure to He-CAPP or He/O_2-CAPP for 4, 6 and 8 min.

The association of He-CAPP with other substances can also be a strategy for clinical application. Li et al. [85] observed that Ar/O_2-CAPP treatment for 12 min eliminated 3-week *E. faecalis* biofilms without changes of root canal dentin. Zhou et al. [86] observed that He-CAPP jet flowing through 3% hydrogen peroxide (H_2O_2) showed better effects against *E. faecalis* biofilms formed inside extracted teeth root canals when compared to He-CAPP alone. Optical emission spectroscopy analysis showed that stronger emission lines of atomic oxygen and hydroxyl radical in the He/H_2O_2-CAPP when compared with He-CAPP.

More recently, the indirect CAPP treatment also appeared as an alternative for the disinfection of root canals system, although few information is available so far. Yamamoto et al. [87] showed that plasma-treated water (PTW) can be used as an endodontic irrigant.

The in-situ experiment showed that PTW was able to disinfect the root canal experimentally infected with *E. faecalis* without adverse effects to oral mucosa.

The study on the CAPP effects against *Propionibacterium* is scarce in the literature. According to Ali et al. [88] CAPP exposure inactivated *P. acnes* cells in suspensions and biofilm formed on glass slides. They also observed that the antimicrobial effect increases with time.

No study reporting the antimicrobial effects of CAPP against *Fusobacterium* could be found in the literature. However, CAPP has already showed antimicrobial activity against other Gram-negative and anaerobes oral pathogens as *P. gingivalis* [68].

The aim of endodontic treatment is to eliminate the pathogenic microorganisms and prevent reinfection, avoiding clinical failures. The studies shown that CAPP can be an alternative to the conventional irrigant solutions which can damage healthy tissues. The antimicrobial effect of plasma jet seems to be correlated with the time of exposure and the working gas. Then, an efficient protocol should be developed to the use of CAPP in the endodontic clinical routine.

5. CAPP Application in Cariology

Carious dental tissue infected by cariogenic microorganisms is usually removed by using rotating instruments that can cause pain and discomfort to the patient. However, the treatment of carious lesions has been changing in the last years. Minimal intervention dentistry (MID) is one of the proposed alternatives that aims to prevent or paralyze disease's activity [89]. In this context, atraumatic restorative treatment (ART) is one example of MID and it has been used with success [90]. This method shows positive results in the treatment of caries in childhood and injuries in children and elderly [91–94]. In ART, carious tissue is removed by hand instruments and the restoration is carried out [94]. It has been described that the performance of ART can be improved by the elimination of the cariogenic microorganisms during the cavity's cleaning process carried out before the restorative procedure [95].

CAPP has been suggested as a promising therapeutic tool in cariology due to its antimicrobial efficiency [96]. Promising results have been obtained against cariogenic bacteria [97]. Sladek et al. [98] were the first researchers to suggest the use of CAPP for the disinfection of caries cavities. The researchers concluded that it represents an efficient technique that promote the disinfection of irregular structures and canaliculi inside the affected tooth. Plasma needles generated by specific devices allow the penetration of reactive RONS inside the dental canaliculi inhibiting the cariogenic microorganisms [98]. In addition, plasma has the advantage to inhibit microbial biofilms, without damaging the normal tissue [28,97] and without causing hyperaemia, swelling, ulcer or anabrosis, resulted from absence of thermal damage [28,69].

Subsequent studies also reported the effect of CAPP against cariogenic microorganisms. Hirano et al. [99] analysed the effect of CAPP on free-floating planktonic cells. The treatment with CAPP reduced the counts of *Streptococcus mutans* in 4-logs after 3 min of exposure. Park et al. [100] described the inhibitory effect of atmospheric pressure plasma associated with gold nanoparticles on *S. mutans*. The group observed that gold nanoparticles conjugated to the bacterial surface and stimulated by the CAPP affected the bacterial cell wall, suggesting that this association may be a future alternative for caries treatment.

The inhibitory effect on *S. mutans* and *Lactobacillus acidophilus* biofilms grown on hydroxyapatite discs has been observed a few seconds after the CAPP treatment [101]. The study suggested that CAPP deactivation of microorganisms was caused by bombardment of charged or neutral species or an accumulation of electric charge a few seconds after treatment. Recently, the inhibition of a cariogenic multispecies biofilm formed by *S. mutans*, *S. sanguinis* and *S. gordonii* by Ar-CAPP was reported [102]. Despite of these positive evidences, studies about the effectiveness of CAPP treatment on polymicrobial cariogenic biofilms is still needed [96].

CAPP can also improve the conventional restorative methods. It can optimize the adhesion between the tooth and the restorative material [103]. Another study demonstrated that the CAPP induces the polymerization of a dental adhesive by direct and indirect energy transfer [104]. In this sense, CAPP can control the moisture of demineralized dentin surfaces, improve adhesive penetration and the mechanical properties of the adhesive/dentin interface [105].

Considering all the characteristics and effects of CAPP, the application of this innovative technique can be an alternative for the treatment of caries diseases in the near future, allowing more conservative procedures and improving restorative methods.

6. CAPP Application in Oral Oncology

The incidence and prevalence of oral cancer have been continuously increasing in numbers, as well as the mortality rates, especially among younger patients [106]. Oral cancer stands as an international public health problem and one of most common cancer with more than 177,000 deaths and 354,500 new cases year worldwide [107].

Recent advances and future directions have been proposed on oral oncology [108,109]. In addition to surgical approaches, the stereotactic body radiotherapy associated with smart drug delivery systems (SDDSs) have been proposed to oral cancer therapy [109–112]. The immunotherapy also advanced considerably in the last years [111,113–115]. However, those new discoveries involve expensive treatments and ultra-expensive drugs [116–118]. For this reason, the search for alternative is needed.

The potential of CAPP in oral oncotherapy is based on its selectivity towards malignant cells, capacity to induce cell death, immune response, and controlled discharge of RONS that can interfere on the molecular mechanisms of the disease [51,119,120]. The role of oxidation/reduction potential is already understood as key factor for the progression and establishment of the disease based on the HOCl or the $\cdot$NO/ONOO$^-$ signalling pathway [119,121]. In this way, NO and nitrite therapies have already been applied as anti-cancer agent based on their effects on cancer cells and on catalase-dependent apoptotic pathways, which are implied in development and regression of the disease [51,121,122].

CAPP treatment as a tool to control oral cancer cells has been studied in the last years. Han et al. [123] reported that N_2-CAPP induced DNA damage in SCC-25 oral cancer cells. The effect on head and neck squamous cell carcinoma (HNSCC) was also detected [124]. CAPP can also be an alternative for the treatment of oral lichen planus, a precancerous lesion [125]. Interestingly, cancer cells such as SCC-15 and HNSCC were more sensitive to CAPP when compared to non-cancer cells lines [124,126].

Clinical researches showed that CAPP can reduce the microbial load in the lesions and the pain, as well as partial remission on head and neck cancer patients [127,128]. The proposal to use CAPP for oral cancer treatment is recent and the understanding of in situ effects requires more studies as a promising pro-oxidant therapy [120,121].

7. CAPP for the Treatment of Oral Candidiasis

Oral candidiasis is an opportunistic disease with high prevalence among immunocompromised patients [129]. Lately, reports on refractory cases of oropharyngeal candidiasis are increasing [130,131] and the treatment of these cases has faced considerable challenges due to the increasing occurrence of antifungal resistance and low number of new antifungal molecules [132].

Proton ATPases, efflux pumps, adherence, morphogenesis, and resistance to oxidative stress have posed as new targets to the development of novel antifungal agents [133,134]. In this context, CAPP antifungal effect has been studied.

Anti-*Candida albicans* effect was reported by some studies [21,135–137]. Additionally, CAPP showed modulatory effects on *C. albicans* virulence factors, such as adhesion and filamentation [24,136]. Suppression of ergosterol biosynthesis has been observed [138].

Antibiofilm effect is considered a key factor to superficial candidiasis treatment. Exposure to He-CAPP for 150 s reduced significantly the viability of *C. albicans* biofilms, with low cytotoxicity to Vero cells [24]. The same effect was detected when Ar-CAPP [28,139,140] and a microwave-Induced Plasma Torch were used [141]. Recently, Singh et al. [142] reported that RONS from CAPP inhibited *C. albicans* biofilms by affecting the fungal cell wall.

Murine experimental model of oral candidiasis treated with CAPP displayed marked reduction in inflammation and fungal tissue invasion [47].

8. Conclusions

Cold atmospheric pressure plasma (CAPP) has antimicrobial and anti-inflammatory effects that are useful in several areas of Dentistry, such as in Cariology, Periodontology and Endodontics. Additionally, CAPP has been showing potential to be used in the treatment of oral fungal diseases and to control oral cancer. However, additional in vivo studies to standardize clinical protocols are still needed.

9. Future Perspectives

CAPP has a great potential to be used in dentistry in the near future, with applications in several dental specialities. CAPP might contribute to the treatment of refractory infectious diseases and control of oral cancers. Currently, the major challenge to be overcome is the determination of standardized therapeutic protocols to each disease that can be validated by controlled clinical trials.

Author Contributions: Conceptualization: C.Y.K.-I., R.S.P. and K.G.K.; methodology: C.Y.K.-I., R.S.P., G.d.M.G.L., A.C.B., L.W.F., G.M.A.d.A., K.G.K.; validation: C.Y.K.-I., R.S.P., G.d.M.G.L., A.C.B., L.W.F., G.M.A.d.A., K.G.K.; formal analysis: C.Y.K.-I., R.S.P., G.d.M.G.L., A.C.B., L.W.F., G.M.A.d.A., K.G.K.; investigation: C.Y.K.-I., R.S.P., G.d.M.G.L., A.C.B., L.W.F., G.M.A.d.A., K.G.K.; resources: C.Y.K.-I., R.S.P., G.d.M.G.L., A.C.B., L.W.F., G.M.A.d.A., K.G.K.; data curation: C.Y.K.-I., R.S.P., G.d.M.G.L., A.C.B., L.W.F., G.M.A.d.A., K.G.K.; writing—original draft preparation: C.Y.K.-I., R.S.P., G.d.M.G.L., A.C.B., L.W.F., G.M.A.d.A., K.G.K.; writing—review and editing: C.Y.K.-I., R.S.P., G.d.M.G.L., A.C.B., L.W.F., G.M.A.d.A., K.G.K.; visualization: C.Y.K.-I., R.S.P., G.d.M.G.L., A.C.B., L.W.F., G.M.A.d.A., K.G.K.; supervision: C.Y.K.-I., R.S.P. and K.G.K.; project administration: C.Y.K.-I., R.S.P. and K.G.K.; funding acquisition: C.Y.K.-I., R.S.P. and K.G.K. All authors have read and agreed to the published version of the manuscript.

Funding: São Paulo Research Foundation (FAPESP) (grant number 19/05856-7, 18/17707-3, 15/03470-3) and CNPq (Grant number 308127/2018-8). This study was financed in part by the Coordenação de Aperfeiçoamento de Pessoal de Nível Superior—Brasil (CAPES)—Finance Code 001.

Institutional Review Board Statement: Not applicable.

Informed Consent Statement: Not applicable.

Data Availability Statement: Data sharing is not applicable to this article.

Conflicts of Interest: The authors declare no conflict of interest.

References

1. von Keudell, A.; Schulz-von der Gathen, V. Foundations of low-temperature plasma physics—An introduction. *Plasma Sources Sci. Technol.* **2017**, *26*, 113001. [CrossRef]
2. Bruggeman, P.J.; Iza, F.; Brandenburg, R. Foundations of atmospheric pressure non-equilibrium plasmas. *Plasma Sources Sci. Technol.* **2017**, *26*, 123002. [CrossRef]
3. Gibalov, V.I.; Pietsch, G.J. The development of dielectric barrier discharges in gas gaps and on surfaces. *J. Phys. D Appl. Phys.* **2000**, *33*, 2618–2636. [CrossRef]
4. Chirokov, A.; Gutsol, A.; Fridman, A. Atmospheric pressure plasma of dielectric barrier discharges. *Pure Appl. Chem.* **2005**, *77*, 487–495. [CrossRef]
5. Hoffmann, C.; Berganza, C.; Zhang, J. Cold Atmospheric Plasma: Methods of production and application in dentistry and oncology. *Medical Gas Res.* **2013**, *3*, 21. [CrossRef]
6. Winter, J.; Brandenburg, R.; Weltmann, K.D. Atmospheric pressure plasma jets: An overview of devices and new directions. *Plasma Sources Sci. Technol.* **2015**, *24*, 064001. [CrossRef]
7. Lu, X.; Reuter, S.; Laroussi, M.; Liu, D. *Nonequlibrium Atmospheric Pressure Plasma Jets. Fundamentals, Diagnostics and Medical Applications*; CRC Press, Taylor & Francis Group: Boca Raton, FL, USA, 2019.

8.   Fanelli, F.; Fracassi, F. Atmospheric pressure non-equilibrium plasma jet technology: General features, specificities and applications in surface processing of materials. *Surf. Coat. Technol.* **2017**, *322*, 174–201. [CrossRef]

9.   Laroussi, M. Cold Plasma in Medicine and Healthcare: The New Frontier in Low Temperature Plasma Applications. *Front. Phys.* **2020**, *8*. [CrossRef]

10. Kong, M.G.; Kroesen, G.; Morfill, G.; Nosenko, T.; Shimizu, T.; van Dijk, J.; Zimmermann, J.L. Plasma medicine: An introductory review. *New J. Physi.* **2009**, *11*, 115012. [CrossRef]

11. Bourke, P.; Ziuzina, D.; Han, L.; Cullen, P.J.; Gilmore, B.F. Microbiological interactions with cold plasma. *J. Appl. Microbiol.* **2017**, *123*, 308–324. [CrossRef] [PubMed]

12. Neyts, E.C.; Brault, P. Molecular Dynamics Simulations for Plasma-Surface Interactions. *Plasma Process. Polym.* **2017**, *14*, 1600145. [CrossRef]

13. Nicol, M.J.; Brubaker, T.R.; Honish, B.J.; Simmons, A.N.; Kazemi, A.; Geissel, M.A.; Whalen, C.T.; Siedlecki, C.A.; Bilén, S.G.; Knecht, S.D.; et al. Antibacterial effects of low-temperature plasma generated by atmospheric-pressure plasma jet are mediated by reactive oxygen species. *Sci. Rep.* **2020**, *10*, 3066. [CrossRef]

14. Kaushik, N.K.; Ghimire, B.; Li, Y.; Adhikari, M.; Veerana, M.; Kaushik, N.; Jha, N.; Adhikari, B.; Lee, S.J.; Masur, K.; et al. Biological and medical applications of plasma-activated media, water and solutions. *Biol. Chem.* **2018**, *400*, 39–62. [CrossRef] [PubMed]

15. Bradu, C.; Kutasi, K.; Magureanu, M.; Puač, N.; Živković, S. Reactive nitrogen species in plasma-activated water: Generation, chemistry and application in agriculture. *J. Phys. D Appl. Phys.* **2020**, *53*, 223001. [CrossRef]

16. Tanaka, H.; Nakamura, K.; Mizuno, M.; Ishikawa, K.; Takeda, K.; Kajiyama, H.; Utsumi, F.; Kikkawa, F.; Hori, M. Non-thermal atmospheric pressure plasma activates lactate in Ringer's solution for anti-tumor effects. *Sci. Rep.* **2016**, *6*, 36282. [CrossRef]

17. Laroussi, M. Sterilization of contaminated matter with an atmospheric pressure plasma. *IEEE Trans. Plasma Sci.* **1996**, *24*, 1188–1191. [CrossRef]

18. Laroussi, M.; Mendis, D.A.; Rosenberg, M. Plasma interaction with microbes. *New J. Phys.* **2003**, *5*, 41. [CrossRef]

19. Kostov, K.G.; Rocha, V.; Koga-Ito, C.Y.; Matos, B.M.; Algatti, M.A.; Honda, R.Y.; Kayama, M.E.; Mota, R.P. Bacterial sterilization by a dielectric barrier discharge (DBD) in air. *Surf. Coat. Technol.* **2010**, *204*, 2954–2959. [CrossRef]

20. Klämpfl, T.G.; Isbary, G.; Shimizu, T.; Li, Y.F.; Zimmermann, J.L.; Stolz, W.; Schlegel, J.; Morfill, G.E.; Schmidt, H.U. Cold atmospheric air plasma sterilization against spores and other microorganisms of clinical interest. *Appl. Environ. Microbiol.* **2012**, *78*, 5077–5082. [CrossRef] [PubMed]

21. Nishime, T.M.C.; Borges, A.C.; Koga-Ito, C.Y.; Machida, M.; Hein, L.R.O.; Kostov, K.G. Non-thermal atmospheric pressure plasma jet applied to inactivation of different microorganisms. *Surf. Coat. Technol.* **2017**, *312*, 19–24. [CrossRef]

22. Liao, X.; Muhammad, A.I.; Chen, S.; Hu, Y.; Ye, X.; Liu, D.; Ding, T. Bacterial spore inactivation induced by cold plasma. *Crit. Rev. Food Sci. Nutr.* **2019**, *59*, 2562–2572. [CrossRef]

23. Borges, A.C.; Nishime, T.M.C.; de Moura Rovetta, S.; Lima, G.M.G.; Kostov, K.G.; Thim, G.P.; de Menezes, B.R.C.; Machado, J.P.B.; Koga-Ito, C.Y. Cold Atmospheric Pressure Plasma Jet Reduces Trichophyton rubrum Adherence and Infection Capacity. *Mycopathologia* **2019**, *184*, 585–595. [CrossRef] [PubMed]

24. Borges, A.C.; Castaldelli Nishime, T.M.; Kostov, K.G.; de Morais Gouvêa Lima, G.; Lacerda Gontijo, A.V.; de Carvalho, J.N.M.M.; Yzumi Honda, R.; Yumi Koga-Ito, C. Cold atmospheric pressure plasma jet modulates Candida albicans virulence traits. *Clin. Plasma Med.* **2017**, *7*, 9–15. [CrossRef]

25. Rao, Y.; Shang, W.; Yang, Y.; Zhou, R.; Rao, X. Fighting Mixed-Species Microbial Biofilms With Cold Atmospheric Plasma. *Front. Microbiol.* **2020**, *11*, 1000. [CrossRef]

26. Jiang, C.; Schaudinn, C.; Jaramillo, D.E.; Webster, P.; Costerton, J.W. In Vitro Antimicrobial Effect of a Cold Plasma Jet against Enterococcus faecalis Biofilms. *ISRN Dent.* **2012**, *2012*, 295736. [CrossRef] [PubMed]

27. Idlibi, A.N.; Al-Marrawi, F.; Hannig, M.; Lehmann, A.; Rueppell, A.; Schindler, A.; Jentsch, H.; Rupf, S. Destruction of oral biofilms formed in situ on machined titanium (Ti) surfaces by cold atmospheric plasma. *Biofouling* **2013**, *29*, 369–379. [CrossRef]

28. Delben, J.A.; Zago, C.E.; Tyhovych, N.; Duarte, S.; Vergani, C.E. Effect of Atmospheric-Pressure Cold Plasma on Pathogenic Oral Biofilms and In Vitro Reconstituted Oral Epithelium. *PLoS ONE* **2016**, *11*, e0155427. [CrossRef] [PubMed]

29. Rutkowski, R.; Schuster, M.; Unger, J.; Seebauer, C.; Metelmann, H.R.; Woedtke, T.V.; Weltmann, K.D.; Daeschlein, G. Hyperspectral imaging for in vivo monitoring of cold atmospheric plasma effects on microcirculation in treatment of head and neck cancer and wound healing. *Clin. Plasma Med.* **2017**, *7–8*, 52–57. [CrossRef]

30. Shome, D.; von Woedtke, T.; Riedel, K.; Masur, K. The HIPPO Transducer YAP and Its Targets CTGF and Cyr61 Drive a Paracrine Signalling in Cold Atmospheric Plasma-Mediated Wound Healing. *Oxid. Med. Cell Longev.* **2020**, *2020*, 4910280. [CrossRef] [PubMed]

31. Arndt, S.; Unger, P.; Wacker, E.; Shimizu, T.; Heinlin, J.; Li, Y.F.; Thomas, H.M.; Morfill, G.E.; Zimmermann, J.L.; Bosserhoff, A.K.; et al. Cold atmospheric plasma (CAP) changes gene expression of key molecules of the wound healing machinery and improves wound healing in vitro and in vivo. *PLoS ONE* **2013**, *8*, e79325. [CrossRef]

32. Xu, D.; Wang, S.; Li, B.; Qi, M.; Feng, R.; Li, Q.; Zhang, H.; Chen, H.; Kong, M.G. Effects of Plasma-Activated Water on Skin Wound Healing in Mice. *Microorganisms* **2020**, *8*, 1091. [CrossRef] [PubMed]

33. Brun, P.; Pathak, S.; Castagliuolo, I.; Palù, G.; Zuin, M.; Cavazzana, R.; Martines, E. Helium generated cold plasma finely regulates activation of human fibroblast-like primary cells. *PLoS ONE* **2014**, *9*, e104397. [CrossRef]

34. Bourdens, M.; Jeanson, Y.; Taurand, M.; Juin, N.; Carrière, A.; Clément, F.; Casteilla, L.; Bulteau, A.-L.; Planat-Bénard, V. Short exposure to cold atmospheric plasma induces senescence in human skin fibroblasts and adipose mesenchymal stromal cells. *Sci. Rep.* **2019**, *9*, 8671. [CrossRef] [PubMed]
35. Haralambiev, L.; Bandyophadyay, A.; Suchy, B.; Weiss, M.; Kramer, A.; Bekeschus, S.; Ekkernkamp, A.; Mustea, A.; Kaderali, L.; Stope, M.B. Determination of Immediate. *Anticancer Res.* **2020**, *40*, 3743–3749. [CrossRef] [PubMed]
36. Schmidt, A.; Dietrich, S.; Steuer, A.; Weltmann, K.D.; von Woedtke, T.; Masur, K.; Wende, K. Non-thermal plasma activates human keratinocytes by stimulation of antioxidant and phase II pathways. *J. Biol. Chem.* **2015**, *290*, 6731–6750. [CrossRef] [PubMed]
37. Kisch, T.; Helmke, A.; Schleusser, S.; Song, J.; Liodaki, E.; Stang, F.H.; Mailaender, P.; Kraemer, R. Improvement of cutaneous microcirculation by cold atmospheric plasma (CAP): Results of a controlled, prospective cohort study. *Microvasc. Res.* **2016**, *104*, 55–62. [CrossRef]
38. Busco, G.; Robert, E.; Chettouh-Hammas, N.; Pouvesle, J.M.; Grillon, C. The emerging potential of cold atmospheric plasma in skin biology. *Free Radic. Biol. Med.* **2020**, *161*, 290–304. [CrossRef] [PubMed]
39. Dobrynin, D.; Fridman, G.; Friedman, G.; Fridman, A. Physical and biological mechanisms of direct plasma interaction with living tissue. *New J. Phys.* **2009**, *11*, 115020. [CrossRef]
40. Weltmann, K.D.; von Woedtke, T. Plasma medicine—Current state of research and medical application. *Plasma Phys. Control. Fusion* **2016**, *59*, 014031. [CrossRef]
41. Kumar, N.; Attri, P.; Yadav, D.K.; Choi, J.; Choi, E.H.; Uhm, H.S. Induced apoptosis in melanocytes cancer cell and oxidation in biomolecules through deuterium oxide generated from atmospheric pressure non-thermal plasma jet. *Sci. Rep.* **2014**, *4*, 7589. [CrossRef]
42. Xu, D.; Liu, D.; Wang, B.; Chen, C.; Chen, Z.; Li, D.; Yang, Y.; Chen, H.; Kong, M.G. In Situ OH Generation from O_2- and H_2O_2 Plays a Critical Role in Plasma-Induced Cell Death. *PLoS ONE* **2015**, *10*, e0128205. [CrossRef]
43. Lin, A.; Truong, B.; Patel, S.; Kaushik, N.; Choi, E.H.; Fridman, G.; Fridman, A.; Miller, V. Nanosecond-Pulsed DBD Plasma-Generated Reactive Oxygen Species Trigger Immunogenic Cell Death in A549 Lung Carcinoma Cells through Intracellular Oxidative Stress. *Int. J. Mol. Sci.* **2017**, *18*, 966. [CrossRef] [PubMed]
44. Lunov, O.; Zablotskii, V.; Churpita, O.; Lunova, M.; Jirsa, M.; Dejneka, A.; Kubinová, Š. Chemically different non-thermal plasmas target distinct cell death pathways. *Sci. Rep.* **2017**, *7*, 600. [CrossRef] [PubMed]
45. Alkawareek, M.Y.; Algwari, Q.T.; Gorman, S.P.; Graham, W.G.; O'Connell, D.; Gilmore, B.F. Application of atmospheric pressure nonthermal plasma for the in vitro eradication of bacterial biofilms. *FEMS Immunol. Med. Microbiol.* **2012**, *65*, 381–384. [CrossRef] [PubMed]
46. Puligundla, P.; Mok, C. Potential applications of nonthermal plasmas against biofilm-associated micro-organisms in vitro. *J. Appl. Microbiol.* **2017**, *122*, 1134–1148. [CrossRef]
47. Borges, A.C.; Lima, G.M.G.; Nishime, T.M.C.; Gontijo, A.V.L.; Kostov, K.G.; Koga-Ito, C.Y. Amplitude-modulated cold atmospheric pressure plasma jet for treatment of oral candidiasis: In vivo study. *PLoS ONE* **2018**, *13*, e0199832. [CrossRef] [PubMed]
48. Boehm, D.; Bourke, P. Safety implications of plasma-induced effects in living cells—A review of in vitro and in vivo findings. *Biol. Chem.* **2018**, *400*, 3–17. [CrossRef]
49. Han, X.; Kapaldo, J.; Liu, Y.; Stack, M.S.; Alizadeh, E.; Ptasinska, S. Large-Scale Image Analysis for Investigating Spatio-Temporal Changes in Nuclear DNA Damage Caused by Nitrogen Atmospheric Pressure Plasma Jets. *Int. J. Mol. Sci.* **2020**, *21*, 4127. [CrossRef] [PubMed]
50. Van der Paal, J.; Hong, S.-H.; Yusupov, M.; Gaur, N.; Oh, J.-S.; Short, R.D.; Szili, E.J.; Bogaerts, A. How membrane lipids influence plasma delivery of reactive oxygen species into cells and subsequent DNA damage: An experimental and computational study. *Phys. Chem. Chem. Phys.* **2019**, *21*, 19327–19341. [CrossRef]
51. Bengtson, C.; Bogaerts, A. On the Anti-Cancer Effect of Cold Atmospheric Plasma and the Possible Role of Catalase-Dependent Apoptotic Pathways. *Cells* **2020**, *9*, 2330. [CrossRef]
52. VON Woedtke, T.; Schmidt, A.; Bekeschus, S.; Wende, K.; Weltmann, K.D. Plasma Medicine: A Field of Applied Redox Biology. *In Vivo* **2019**, *33*, 1011–1026. [CrossRef] [PubMed]
53. Zhao, S.; Xiong, Z.; Mao, X.; Meng, D.; Lei, Q.; Li, Y.; Deng, P.; Chen, M.; Tu, M.; Lu, X.; et al. Atmospheric pressure room temperature plasma jets facilitate oxidative and nitrative stress and lead to endoplasmic reticulum stress dependent apoptosis in HepG2 cells. *PLoS ONE* **2013**, *8*, e73665. [CrossRef]
54. Zhao, J.; Nie, L. Five gaseous reactive oxygen and nitrogen species (RONS) density generated by microwave plasma jet. *Phys. Plasmas* **2019**, *26*, 073503. [CrossRef]
55. Kim, S.J.; Chung, T.H. Cold atmospheric plasma jet-generated RONS and their selective effects on normal and carcinoma cells. *Sci. Rep.* **2016**, *6*, 20332. [CrossRef]
56. Kurita, H.; Haruta, N.; Uchihashi, Y.; Seto, T.; Takashima, K. Strand breaks and chemical modification of intracellular DNA induced by cold atmospheric pressure plasma irradiation. *PLoS ONE* **2020**, *15*, e0232724. [CrossRef] [PubMed]
57. Bartosz, G. Reactive oxygen species: Destroyers or messengers? *Biochem. Pharmacol.* **2009**, *77*, 1303–1315. [CrossRef]
58. Brieger, K.; Schiavone, S.; Miller, F.J.; Krause, K.H. Reactive oxygen species: From health to disease. *Swiss Med. Wkly* **2012**, *142*, w13659. [CrossRef] [PubMed]
59. Imlay, J.A. Pathways of oxidative damage. *Annu. Rev. Microbiol.* **2003**, *57*, 395–418. [CrossRef]
60. Finkel, T.; Holbrook, N.J. Oxidants, oxidative stress and the biology of ageing. *Nature* **2000**, *408*, 239–247. [CrossRef]

61. Bogdan, C. Nitric oxide and the immune response. *Nat. Immunol.* **2001**, *2*, 907–916. [CrossRef] [PubMed]
62. Pacher, P.; Beckman, J.S.; Liaudet, L. Nitric oxide and peroxynitrite in health and disease. *Physiol. Rev.* **2007**, *87*, 315–424. [CrossRef]
63. Mai-Prochnow, A.; Clauson, M.; Hong, J.; Murphy, A.B. Gram positive and Gram negative bacteria differ in their sensitivity to cold plasma. *Sci. Rep.* **2016**, *6*, 38610. [CrossRef]
64. Laroussi, M.; Leipold, F. Evaluation of the roles of reactive species, heat, and UV radiation in the inactivation of bacterial cells by air plasmas at atmospheric pressure. *Int. J. Mass Spectrom.* **2004**, *233*, 81–86. [CrossRef]
65. Laroussi, M. Plasma Medicine: A Brief Introduction. *Plasma* **2018**, *1*, 5. [CrossRef]
66. Brelles-Mariño, G. Challenges in biofilm inactivation: The use of cold plasma as a new approach. *J. Bioprocess. Biotech.* **2012**, *2*, 4. [CrossRef]
67. Bostanci, N.; Bao, K.; Wahlander, A.; Grossmann, J.; Thurnheer, T.; Belibasakis, G.N. Secretome of gingival epithelium in response to subgingival biofilms. *Mol. Oral Microbiol.* **2015**, *30*, 323–335. [CrossRef] [PubMed]
68. Mahasneh, A.; Darby, M.; Tolle, S.L.; Hynes, W.; Laroussi, M.; Karakas, E. Inactivation of Porphyromonas gingivalis by Low-Temperature Atmospheric Pressure Plasma. *Plasma Medicine* **2011**, *1*, 191–204. [CrossRef]
69. Liu, D.; Xiong, Z.; Du, T.; Zhou, X.; Cao, Y.; Lu, X. Bacterial-killing effect of atmospheric pressure non-equilibrium plasma jet and oral mucosa response. *J. Huazhong Univ. Sci. Technol.* **2011**, *31*, 852–856. [CrossRef]
70. Küçük, D.; Savran, L.; Ercan, U.K.; Yarali, Z.B.; Karaman, O.; Kantarci, A.; Sağlam, M.; Köseoğlu, S. Evaluation of efficacy of non-thermal atmospheric pressure plasma in treatment of periodontitis: A randomized controlled clinical trial. *Clin. Oral Investig.* **2020**, *24*, 3133–3145. [CrossRef]
71. Kwon, J.S.; Kim, Y.H.; Choi, E.H.; Kim, C.K.; Kim, K.N.; Kim, K.M. Non-thermal atmospheric pressure plasma increased mRNA expression of growth factors in human gingival fibroblasts. *Clin. Oral Investig.* **2016**, *20*, 1801–1808. [CrossRef] [PubMed]
72. Eggers, B.; Marciniak, J.; Memmert, S.; Kramer, F.J.; Deschner, J.; Nokhbehsaim, M. The beneficial effect of cold atmospheric plasma on parameters of molecules and cell function involved in wound healing in human osteoblast-like cells in vitro. *Odontology* **2020**, *108*, 607–616. [CrossRef] [PubMed]
73. Shi, Q.; Song, K.; Zhou, X.; Xiong, Z.; Du, T.; Lu, X.; Cao, Y. Effects of non-equilibrium plasma in the treatment of ligature-induced peri-implantitis. *J. Clin. Periodontol.* **2015**, *42*, 478–487. [CrossRef] [PubMed]
74. Carreiro, A.F.P.; Delben, J.A.; Guedes, S.; Silveira, E.J.D.; Janal, M.N.; Vergani, C.E.; Pushalkar, S.; Duarte, S. Low-temperature plasma on peri-implant-related biofilm and gingival tissue. *J. Periodontol.* **2019**, *90*, 507–515. [CrossRef] [PubMed]
75. Lee, J.Y.; Kim, K.H.; Park, S.Y.; Yoon, S.Y.; Kim, G.H.; Lee, Y.M.; Rhyu, I.C.; Seol, Y.J. The bactericidal effect of an atmospheric-pressure plasma jet on. *J. Periodontal. Implant. Sci.* **2019**, *49*, 319–329. [CrossRef] [PubMed]
76. Arndt, S.; Landthaler, M.; Zimmermann, J.L.; Unger, P.; Wacker, E.; Shimizu, T.; Li, Y.F.; Morfill, G.E.; Bosserhoff, A.K.; Karrer, S. Effects of cold atmospheric plasma (CAP) on ß-defensins, inflammatory cytokines, and apoptosis-related molecules in keratinocytes in vitro and in vivo. *PLoS ONE* **2015**, *10*, e0120041. [CrossRef]
77. Wang, P.; Duan, D.; Zhou, X.; Li, X.; Yang, J.; Deng, M.; Xu, Y. Relationship between expression of human gingival beta-defensins and levels of periodontopathogens in subgingival plaque. *J. Periodontal. Res.* **2015**, *50*, 113–122. [CrossRef] [PubMed]
78. Brun, P.; Bernabè, G.; Marchiori, C.; Scarpa, M.; Zuin, M.; Cavazzana, R.; Zaniol, B.; Martines, E. Antibacterial efficacy and mechanisms of action of low power atmospheric pressure cold plasma: Membrane permeability, biofilm penetration and antimicrobial sensitization. *J. Appl. Microbiol.* **2018**, *125*, 398–408. [CrossRef]
79. Jhajharia, K. Microbiology of endodontic diseases: A review article. *Int. J. Appl. Dent. Sci.* **2019**, *5*, 4.
80. Neelakantan, P.; Romero, M.; Vera, J.; Daood, U.; Khan, A.U.; Yan, A.; Cheung, G.S.P. Biofilms in Endodontics-Current Status and Future Directions. *Int. J. Mol. Sci.* **2017**, *18*, 1748. [CrossRef] [PubMed]
81. Prada, I.; Micó-Muñoz, P.; Giner-Lluesma, T.; Micó-Martínez, P.; Collado-Castellano, N.; Manzano-Saiz, A. Influence of microbiology on endodontic failure. Literature review. *Med. Oral Patol. Oral Cir. Bucal.* **2019**, *24*, e364–e372. [CrossRef] [PubMed]
82. Borzini, L.; Condò, R.; De Dominicis, P.; Casaglia, A.; Cerroni, L. Root Canal Irrigation: Chemical Agents and Plant Extracts Against. *Open Dent. J.* **2016**, *10*, 692–703. [CrossRef]
83. Chang, Y.T.; Chen, G. Oral bacterial inactivation using a novel low-temperature atmospheric-pressure plasma device. *J. Dent. Sci.* **2016**, *11*, 65–71. [CrossRef]
84. Armand, A.; Khani, M.; Asnaashari, M.; AliAhmadi, A.; Shokri, B. Comparison study of root canal disinfection by cold plasma jet and photodynamic therapy. *Photodiagnosis Photodyn. Ther.* **2019**, *26*, 327–333. [CrossRef]
85. Li, Y.; Sun, K.; Ye, G.; Liang, Y.; Pan, H.; Wang, G.; Zhao, Y.; Pan, J.; Zhang, J.; Fang, J. Evaluation of Cold Plasma Treatment and Safety in Disinfecting 3-week Root Canal Enterococcus faecalis Biofilm In Vitro. *J. Endod.* **2015**, *41*, 1325–1330. [CrossRef]
86. Zhou, X.C.; Li, Y.L.; Liu, D.X.; Cao, Y.G.; Lu, X.P. Bactericidal effect of plasma jet with helium flowing through 3% hydrogen peroxide against. *Exp. Ther. Med.* **2016**, *12*, 3073–3077. [CrossRef] [PubMed]
87. Yamamoto, K.; Ohshima, T.; Kitano, K.; Ikawa, S.; Yamazaki, H.; Maeda, N.; Hosoya, N. The efficacy of plasma-treated water as a root canal irrigant. *Asian Pac. J. Dent.* **2017**, *17*, 8.
88. Ali, A.; Kim, Y.H.; Lee, J.Y.; Lee, S.; Uhm, H.S.; Cho, G.; Park, B.J.; Choi, E.H. Inactivation of Propionibacterium acnes and its biofilm by non-thermal plasma. *Curr. Appl. Phys.* **2014**, *14*, S142–S148. [CrossRef]
89. Mackenzie, L.; Banerjee, A. Minimally invasive direct restorations: A practical guide. *Br. Dent. J.* **2017**, *223*, 163–171. [CrossRef] [PubMed]

90. Frencken, J.E. Atraumatic restorative treatment and minimal intervention dentistry. *Br. Dent. J.* **2017**, *223*, 183–189. [CrossRef]
91. Arrow, P. Restorative Outcomes of a Minimally Invasive Restorative Approach Based on Atraumatic Restorative Treatment to Manage Early Childhood Caries: A Randomised Controlled Trial. *Caries Res.* **2016**, *50*, 1–8. [CrossRef] [PubMed]
92. Duangthip, D.; Chen, K.J.; Gao, S.S.; Lo, E.C.M.; Chu, C.H. Managing Early Childhood Caries with Atraumatic Restorative Treatment and Topical Silver and Fluoride Agents. *Int. J. Environ. Res. Public Health* **2017**, *14*, 1204. [CrossRef]
93. Frencken, J.E.; Leal, S.C.; Navarro, M.F. Twenty-five-year atraumatic restorative treatment (ART) approach: A comprehensive overview. *Clin. Oral Investig.* **2012**, *16*, 1337–1346. [CrossRef]
94. Cruz Gonzalez, A.C.; Marín Zuluaga, D.J. Clinical outcome of root caries restorations using ART and rotary techniques in institutionalized elders. *Braz. Oral Res.* **2016**, *30*. [CrossRef] [PubMed]
95. Mohan, P.V.; Uloopi, K.S.; Vinay, C.; Rao, R.C. In vivo comparison of cavity disinfection efficacy with APF gel, Propolis, Diode Laser, and 2% chlorhexidine in primary teeth. *Contemp. Clin. Dent.* **2016**, *7*, 45–50. [CrossRef]
96. Ranjan, R.; Krishnamraju, P.V.; Shankar, T.; Gowd, S. Nonthermal Plasma in Dentistry: An Update. *J. Int. Soc. Prev. Community Dent.* **2017**, *7*, 71–75. [CrossRef]
97. Yang, B.; Chen, J.; Yu, Q.; Li, H.; Lin, M.; Mustapha, A.; Hong, L.; Wang, Y. Oral bacterial deactivation using a low-temperature atmospheric argon plasma brush. *J. Dent.* **2011**, *39*, 48–56. [CrossRef] [PubMed]
98. Sladek, R.E.J.; Stoffels, E.; Walraven, R.; Tielbeek, P.J.A.; Koolhoven, R.A. Plasma treatment of dental cavities: A feasibility study. *IEEE Trans. Plasma Sci.* **2004**, *32*, 1540–1543. [CrossRef]
99. Hirano, Y.; Hayashi, M.; Tamura, M.; Yoshino, F.; Yoshida, A.; Masubuchi, M.; Imai, K.; Ogiso, B. Singlet oxygen generated by a new nonthermal atmospheric pressure air plasma device exerts a bactericidal effect on oral pathogens. *J. Oral Sci.* **2019**, *61*, 521–525. [CrossRef] [PubMed]
100. Park, S.R.; Lee, H.W.; Hong, J.W.; Lee, H.J.; Kim, J.Y.; Choi, B.B.; Kim, G.C.; Jeon, Y.C. Enhancement of the killing effect of low-temperature plasma on Streptococcus mutans by combined treatment with gold nanoparticles. *J. Nanobiotechnol.* **2014**, *12*, 29. [CrossRef]
101. Blumhagen, A.; Singh, P.; Mustapha, A.; Chen, M.; Wang, Y.; Yu, Q. Plasma deactivation of oral bacteria seeded on hydroxyapatite disks as tooth enamel analogue. *Am. J. Dent.* **2014**, *27*, 84–90. [PubMed]
102. Figueira, L.W.; Panariello, B.H.D.; Koga-Ito, C.Y.; Duarte, S. Low-Temperature Plasma as an Approach for Inhibiting a Multi-Species Cariogenic Biofilm. *Appl. Sci.* **2021**, *11*, 570. [CrossRef]
103. Zhang, Y.; Yu, Q.; Wang, Y. Non-thermal atmospheric plasmas in dental restoration: Improved resin adhesive penetration. *J. Dent.* **2014**, *42*, 1033–1042. [CrossRef] [PubMed]
104. Chen, M.; Zhang, Y.; Yao, X.; Li, H.; Yu, Q.; Wang, Y. Effect of a non-thermal, atmospheric-pressure, plasma brush on conversion of model self-etch adhesive formulations compared to conventional photo-polymerization. *Dent. Mater.* **2012**, *28*, 1232–1239. [CrossRef] [PubMed]
105. Kim, J.H.; Han, G.J.; Kim, C.K.; Oh, K.H.; Chung, S.N.; Chun, B.H.; Cho, B.H. Promotion of adhesive penetration and resin bond strength to dentin using non-thermal atmospheric pressure plasma. *Eur. J. Oral Sci.* **2016**, *124*, 89–95. [CrossRef]
106. Sarode, G.; Maniyar, N.; Sarode, S.C.; Jafer, M.; Patil, S.; Awan, K.H. Epidemiologic aspects of oral cancer. *Dis. Mon.* **2020**, *66*, 100988. [CrossRef]
107. Pilleron, S.; Soto-Perez-de-Celis, E.; Vignat, J.; Ferlay, J.; Soerjomataram, I.; Bray, F.; Sarfati, D. Estimated global cancer incidence in the oldest adults in 2018 and projections to 2050. *Int. J. Cancer* **2021**, *148*, 601–608. [CrossRef] [PubMed]
108. Speight, P.M.; Farthing, P.M. The pathology of oral cancer. *Br. Dent. J.* **2018**, *225*, 841–847. [CrossRef] [PubMed]
109. Muzaffar, J.; Bari, S.; Kirtane, K.; Chung, C.H. Recent Advances and Future Directions in Clinical Management of Head and Neck Squamous Cell Carcinoma. *Cancers* **2021**, *13*, 338. [CrossRef]
110. Iqbal, M.S.; West, N.; Richmond, N.; Kovarik, J.; Gray, I.; Willis, N.; Morgan, D.; Yazici, G.; Cengiz, M.; Paleri, V.; et al. A systematic review and practical considerations of stereotactic body radiotherapy in the treatment of head and neck cancer. *Br. J. Radiol.* **2021**, *94*, 20200332. [CrossRef]
111. Hossen, S.; Hossain, M.K.; Basher, M.K.; Mia, M.N.H.; Rahman, M.T.; Uddin, M.J. Smart nanocarrier-based drug delivery systems for cancer therapy and toxicity studies: A review. *J. Adv. Res.* **2019**, *15*, 1–18. [CrossRef]
112. Lorscheider, M.; Gaudin, A.; Nakhlé, J.; Veiman, K.L.; Richard, J.; Chassaing, C. Challenges and opportunities in the delivery of cancer therapeutics: Update on recent progress. *Ther. Deliv.* **2021**, *12*, 55–76. [CrossRef]
113. Liu, C.; Han, C.; Liu, J. The Role of Toll-Like Receptors in Oncotherapy. *Oncol. Res.* **2019**, *27*, 965–978. [CrossRef] [PubMed]
114. Wijetunga, N.A.; Yu, Y.; Morris, L.G.; Lee, N.; Riaz, N. The head and neck cancer genome in the era of immunotherapy. *Oral Oncol.* **2021**, *112*, 105040. [CrossRef] [PubMed]
115. Carvalho, B.G.; Vit, F.F.; Carvalho, H.F.; Han, S.W.; de la Torre, L.G. Recent advances in co-delivery nanosystems for synergistic action in cancer treatment. *J. Mater. Chem. B* **2021**, *9*, 1208–1237. [CrossRef] [PubMed]
116. DiStefano, M.J.; Kang, S.-Y.; Yehia, F.; Morales, C.; Anderson, G.F. Assessing the Added Therapeutic Benefit of Ultra-Expensive Drugs. *Value Health* **2021**. [CrossRef]
117. Vrdoljak, E.; Sekerija, M.; Plestina, S.; Belac Lovasic, I.; Katalinic Jankovic, V.; Garattini, L.; Bobinac, A.; Voncina, L. Is it too expensive to fight cancer? Analysis of incremental costs and benefits of the Croatian National Plan Against Cancer. *Eur. J. Health Econ.* **2021**, 1–11. [CrossRef]

118. Tringale, K.R.; Gennarelli, R.L.; Gillespie, E.F.; Mitchell, A.P.; Zelefsky, M.J. Association between Site-of-Care and the Cost and Modality of Radiotherapy for Prostate Cancer: Analysis of Medicare Beneficiaries from 2015 to 2017. *Cancer Investig.* **2021**, 1–9. [CrossRef]

119. Bauer, G.; Sersenová, D.; Graves, D.B.; Machala, Z. Cold Atmospheric Plasma and Plasma-Activated Medium Trigger RONS-Based Tumor Cell Apoptosis. *Sci. Rep.* **2019**, *9*, 14210. [CrossRef]

120. Semmler, M.L.; Bekeschus, S.; Schäfer, M.; Bernhardt, T.; Fischer, T.; Witzke, K.; Seebauer, C.; Rebl, H.; Grambow, E.; Vollmar, B.; et al. Molecular Mechanisms of the Efficacy of Cold Atmospheric Pressure Plasma (CAP) in Cancer Treatment. *Cancers* **2020**, *12*, 269. [CrossRef]

121. Malyavko, A.; Yan, D.; Wang, Q.; Klein, A.L.; Patel, K.C.; Sherman, J.H.; Keidar, M. Cold atmospheric plasma cancer treatment, direct versus indirect approaches. *Mater. Adv.* **2020**, *1*, 12. [CrossRef]

122. Graves, D.B. Reactive Species from Cold Atmospheric Plasma: Implications for Cancer Therapy. *Plasma Process. Polym.* **2014**, *11*, 1120–1127. [CrossRef]

123. Han, X.; Klas, M.; Liu, Y.; Sharon Stack, M.; Ptasinska, S. DNA damage in oral cancer cells induced by nitrogen atmospheric pressure plasma jets. *Appl. Phys. Lett.* **2013**, *102*, 5. [CrossRef]

124. Guerrero-Preston, R.; Ogawa, T.; Uemura, M.; Shumulinsky, G.; Valle, B.L.; Pirini, F.; Ravi, R.; Sidransky, D.; Keidar, M.; Trink, B. Cold atmospheric plasma treatment selectively targets head and neck squamous cell carcinoma cells. *Int. J. Mol. Med.* **2014**, *34*, 941–946. [CrossRef] [PubMed]

125. Seebauer, C.; Hasse, S.; Segebarth, M.; Bekeschus, S.; von Woedtke, T.; Weltmann, K.-D.; Schuster, M.; Rutkowski, R.; Metelmann, H.-R. Cold Atmospheric Plasma for the treatment of Oral Lichen Planus as intraoral precancerous lesion. *Clin. Plasma Med.* **2018**, *9*, 44–45. [CrossRef]

126. Pereira, S.; Pinto, E.; Ribeiro, P.A.; Sério, S. Study of a Cold Atmospheric Pressure Plasma jet device for indirect treatment of Squamous Cell Carcinoma. *Clin. Plasma Med.* **2019**, *13*, 9–14. [CrossRef]

127. Metelmann, H.-R.; Nedrelow, D.S.; Seebauer, C.; Schuster, M.; von Woedtke, T.; Weltmann, K.-D.; Kindler, S.; Metelmann, P.H.; Finkelstein, S.E.; Von Hoff, D.D.; et al. Head and neck cancer treatment and physical plasma. *Clin. Plasma Med.* **2015**, *3*, 17–23. [CrossRef]

128. Metelmann, H.-R.; Seebauer, C.; Miller, V.; Fridman, A.; Bauer, G.; Graves, D.B.; Pouvesle, J.-M.; Rutkowski, R.; Schuster, M.; Bekeschus, S.; et al. Clinical experience with cold plasma in the treatment of locally advanced head and neck cancer. *Clin. Plasma Med.* **2018**, *9*, 6–13. [CrossRef]

129. Lombardi, A.; Ouanounou, A. Fungal infections in dentistry: Clinical presentations, diagnosis, and treatment alternatives. *Oral Surg. Oral Med. Oral Pathol. Oral Radiol.* **2020**, *130*, 533–546. [CrossRef]

130. Baumgardner, D.J. Oral Fungal Microbiota: To Thrush and Beyond. *J. Patient Cent. Res. Rev.* **2019**, *6*, 252–261. [CrossRef] [PubMed]

131. StatPearls. *Esophageal Candidiasis*; NBK537268; StatPearls: Treasure Island, FL, USA, 2020.

132. Bhattacharya, S.; Sae-Tia, S.; Fries, B.C. Candidiasis and Mechanisms of Antifungal Resistance. *Antibiotics* **2020**, *9*, 312. [CrossRef]

133. Ahmad, A.; Molepo, J.; Patel, M. Challenges in the Development of Antifungal Agents Against Candida: Scope of Phytochemical Research. *Curr. Pharm. Des.* **2016**, *22*, 4135–4150. [CrossRef]

134. Caldara, M.; Marmiroli, N. Known Antimicrobials Versus Nortriptyline in. *Microorganisms* **2020**, *8*, 742. [CrossRef] [PubMed]

135. Yamazaki, H.; Ohshima, T.; Tsubota, Y.; Yamaguchi, H.; Jayawardena, J.A.; Nishimura, Y. Microbicidal activities of low frequency atmospheric pressure plasma jets on oral pathogens. *Dent. Mater. J.* **2011**, *30*, 384–391. [CrossRef] [PubMed]

136. Sun, Y.; Yu, S.; Sun, P.; Wu, H.; Zhu, W.; Liu, W.; Zhang, J.; Fang, J.; Li, R. Inactivation of Candida biofilms by non-thermal plasma and its enhancement for fungistatic effect of antifungal drugs. *PLoS ONE* **2012**, *7*, e40629. [CrossRef] [PubMed]

137. Kostov, K.G.; Borges, A.C.; Koga-Ito, C.Y.; Nishime, T.M.C.; Prysiazhnyi, V.; Honda, R.Y. Inactivation of Candida albicans by Cold Atmospheric Pressure Plasma Jet. *IEEE Trans. Plasma Sci.* **2015**, *43*, 770–775. [CrossRef]

138. Rahimi-Verki, N.; Shapoorzadeh, A.; Razzaghi-Abyaneh, M.; Atyabi, S.M.; Shams-Ghahfarokhi, M.; Jahanshiri, Z.; Gholami-Shabani, M. Cold atmospheric plasma inhibits the growth of Candida albicans by affecting ergosterol biosynthesis and suppresses the fungal virulence factors in vitro. *Photodiagnosis Photodyn. Ther.* **2016**, *13*, 66–72. [CrossRef] [PubMed]

139. Doria, A.C.O.C.; Sorge, C.D.P.C.; Santos, T.B.; Brandão, J.; Gonçalves, P.A.R.; Maciel, H.S.; Khouri, S.; Pessoa, R.S. Application of post-discharge region of atmospheric pressure argon and air plasma jet in the contamination control of Candida albicans biofilms. *Res. Biomed. Eng.* **2015**, *31*, 358–362. [CrossRef]

140. Brany, D.; Dvorská, D.; Halašová, E.; Škovierová, H. Cold Atmospheric Plasma: A Powerful Tool for Modern Medicine. *Int. J. Mol. Sci.* **2020**, *21*, 2932. [CrossRef] [PubMed]

141. Handorf, O.; Schnabel, U.; Bösel, A.; Weihe, T.; Bekeschus, S.; Graf, A.C.; Riedel, K.; Ehlbeck, J. Antimicrobial effects of microwave-induced plasma torch (MiniMIP) treatment on Candida albicans biofilms. *Microb. Biotechnol.* **2019**, *12*, 1034–1048. [CrossRef]

142. Singh, S.; Halder, A.; Mohid, S.A.; Bagchi, D.; Sinha, O.; Banerjee, A.; Sarkar, P.K.; Bhunia, A.; Ghosh, S.K.; Mitra, A.; et al. Nonthermal Atmospheric Plasma-Induced Cellular Envelope Damage of Staphylococcus aureus and Candida albicans Biofilms: Spectroscopic and Biochemical Investigations. *IEEE Trans. Plasma Sci.* **2020**, *48*, 2768–2776. [CrossRef]

Review

Optical Emission Spectroscopy as a Diagnostic Tool for Characterization of Atmospheric Plasma Jets

Rok Zaplotnik, Gregor Primc and Alenka Vesel *

Department of Surface Engineering, Jozef Stefan Institute, Jamova Cesta 39, 1000 Ljubljana, Slovenia; rok.zaplotnik@ijs.si (R.Z.); gregor.primc@ijs.si (G.P.)
* Correspondence: alenka.vesel@guest.arnes.si

Abstract: A suitable technique for localized surface treatment of solid materials is an atmospheric pressure plasma jet (APPJ). The properties of the APPJ plasma often depend on small details like the concentration of gaseous impurities what influences the surface kinetics. The simplest and often most useful configuration of the APPJ is presented, characterized by optical emission spectroscopy (OES), and results are discussed in view of various papers. Furthermore, results of additional recent papers on the characterization of the APPJ by OES are presented as well. Because the APPJ is operating at atmospheric pressure, even the water vapor traces may significantly alter the type and concentration of reactive species. The APPJ sustained in noble gases represents a source of vacuum ultraviolet (VUV) radiation that is absorbed in the surface of the treated material, thus causing bond scission. The addition of minute amounts of reactive gases causes significant suppression of VUV radiation and the formation of reactive radicals. These radicals such as OH, O, N, NO, O_3, and alike interact chemically with the surface causing its functionalization. Huge gradients of these radicals have been reported, so the surface finish is limited to the area reached by the radicals. Particularly OH radicals significantly prevail in the OES spectra, even when using very pure noble gas. They may cause suppression of other spectral features. OH radicals are especially pronounced in Ar plasmas. Their density decreases exponentially with a distance from the APPJ orifice.

Keywords: atmospheric pressure plasma jet (APPJ); optical emission spectroscopy (OES); plasma-surface interactions; local surface modification; polymers; functionalization

Citation: Zaplotnik, R.; Primc, G.; Vesel, A. Optical Emission Spectroscopy as a Diagnostic Tool for Characterization of Atmospheric Plasma Jets. *Appl. Sci.* **2021**, *11*, 2275. https://doi.org/10.3390/app11052275

Academic Editor: Andrei Vasile Nastuta

Received: 30 January 2021
Accepted: 2 March 2021
Published: 4 March 2021

Publisher's Note: MDPI stays neutral with regard to jurisdictional claims in published maps and institutional affiliations.

1. Introduction

A standard method for tailoring surface properties is a brief treatment of a solid material by non-equilibrium gaseous plasma. Plasma sustained in molecular gases is rich in reactive chemical species that interact with surfaces of different materials, causing its modification. A variety of gaseous plasmas have been used for the treatment of different materials. Non-equilibrium plasmas are divided into low-pressure [1] and atmospheric-pressure plasmas [2]. Low-pressure plasmas are usually sustained in the range of pressures between about 1 and 1000 Pa. In-between 1000 and 100,000 Pa (1 bar), there is a black zone rarely tackled by researchers. Gaseous plasma can also be sustained at pressures well above 1 bar, but the practical applications are limited. Low-pressure plasmas often occupy large volumes, so any material is treated rather uniformly over the entire surface facing the plasma. The application of such low-pressure plasmas for modification of a specific area on the surface of products is, however, limited because of the complicated handling of the products that should enter a vacuum chamber. Treatment of numerous products in the continuous mode thus represents a technological challenge when low-pressure plasma is used for surface modification. This obstacle is the main reason for using atmospheric pressure plasmas which can be focused onto the desired surface area, especially when localized treatment is the goal. Atmospheric-pressure plasma jet (APPJ) enables, for example, polymer activation over an area of the order of the cross-section of the visible plasma jet [3–5]. The APPJ is also very popular for medical applications such as wound treatment [6,7], cancer treatment [8,9], bacterial inactivation [10,11], and even for the deposition of thin films [12,13] and coatings [14,15]. In the latter case, the localized surface treatment may not be an advantage any more, especially if larger substrates need to be coated. This can be overcome by scanning the surface of the substrate material [16]. A review of recent applications of the APPJ was published by Penkov et al. [17].

Atmospheric-pressure plasmas are sustained either by a rather low-frequency alternative current (AC) discharges, or radiofrequency (RF), or even microwave (MW) discharges. Depending on the frequency of the power supply, the plasma may be continuous or in short pulses. When the frequency of the power supply is large enough so that the period is shorter than the typical life-time of gaseous plasma, the plasma is sustained in the continuous mode. The continuous mode often occurs at frequencies larger than MHz. Low-frequency discharges will typically create individual streamers of electrons, which will, in turn, cause the formation of a rather dense plasma with the life-time much shorter than the period of the voltage supply.

Numerous methods for atmospheric pressure plasma characterization have been invented [18–21]. Most are based on optical emission or absorption. As long as the discharge is in the continuous mode, methods will give the correct values of plasma parameters. In the case of pulsed plasmas, only a few methods will enable measuring plasma parameters in real-time. Most methods will just give values averaged over the acquisition time. Some methods have a high spatial resolution, while many others will average the signal over the probed volume. Most of the methods also require a deep understanding of the physical phenomena, so the interpretation of the measured signal is far from being trivial. This is a reason why they are not routinely used by researchers who work on the modification of materials' surface properties by plasmas but are not specialists in plasma physics.

Because users of plasma techniques are often not familiar with physical processes in non-equilibrium gaseous plasmas, they are seeking an appropriate method that will give basic information about the properties of plasmas. Especially characterization of the APPJ plasmas [22,23] is of particular interest because they are simple to construct and do not require expensive, complicated instruments. APPJ plasmas may be of different configurations, but the simplest one is single-electrode APPJ. A review of APPJ devices can be found in [22,23]. Plasmas at atmospheric pressure are much more demanding for characterization than low-pressure plasmas because of the simple fact that large spatial or temporal gradients are typically present. Any improper interpretation of the measured signal will lead to a misunderstanding of the phenomena taking place on the surface of a sample during plasma treatment. To this end, the users are looking for a simple, cheap, and reliable technique for the characterization of atmospheric-pressure plasmas. One such technique is optical emission spectroscopy (OES), which can give information on the presence of various species in the discharge.

Let us briefly present this technique. Because plasma emits a light, the basic principle of OES is to collect and detect the emitted light photons. The gaseous atoms and molecules in plasma are excited to various levels upon plasma conditions. The excitation is often a consequence of an inelastic collision of a free electron with a molecule or atom. The electron energy has to be above the threshold. Many excited states are at the potential energy of approximately 10 eV, so much higher than the electron temperature. Therefore, only the electrons within the high-energy tail of the energy distribution function will be capable of exciting the radiative states. An alternative is a step-wise excitation: An electron first excites an atom or a molecule to a metastable state, and another electron causes excitation to the radiative level. Such two-step processes are effective as long as the excited metastable is not relaxed by any method (quenched). An alternative is a collision of at least one atom or molecule in a metastable state of high potential energy with another particle. The resultant molecule is of a short life-time and relaxes by radiation (excimer, exciplex molecules). A three-body collision is often required in the latter case. Such collisions are likely to occur at atmospheric pressure because the three-body collision frequency is roughly 1 MHz at 1 bar.

Whatever mechanism, the shelf-time of the radiative state is very short—of the order of nanoseconds. The extremely short shelf-time assures for relaxation of such states by radiation rather than any other method (such as quenching). Different atoms and molecules will radiate in a specific range of wavelengths that correspond to the potential energy difference between the upper and lower excited states. The lower state may or may not be the ground state. The relaxation of excited atoms is usually exhibited in separated lines. High-resolution spectrometers will also reveal the fine structure.

The two-atom molecules will radiate in bands. For a given electronic transition, there will be numerous vibrational transitions, and for a given vibrational, there will be rotational transitions. In some cases, the lower electronic state of a two-atom molecule is not stable but dissociates. In such cases, the molecule will relax by continuum radiation. More complex molecules will normally relax by radiating a continuum.

In any case, the wavelengths indicate the upper and lower excited levels; therefore, the spectral features can be attributed to specific atoms or molecules. The radiation intensity depends on numerous factors, so optical emission spectroscopy is a qualitative technique. It may be semi-quantitative if a known concentration of another atom or molecule is added intentionally. The technique is often called actinometry or titration. Great care should be taken when interpreting the results of actinometry or titration [24].

Inexpensive spectrometers will cover a range of wavelengths between approximately 200 and 1200 nm, i.e., in the visible, near-infrared, and ultraviolet ranges. The spectral response is far from being constant, so calibration with a standard source is recommended at any attempt to use OES for plasma characterization. The standard spectrometers will therefore detect radiation arising in a range of photon energies between the upper and lower energy levels (1 and 6 eV). Most atomic transitions such as Ar, He, O, N, H, to the ground state are below 200 nm. These transitions are, therefore, invisible by standard spectrometers. The transitions to the ground state are normally much more extensive than to other states, so one misses most radiation when using standard spectrometers. This limitation should be taken into account at any attempt to interpret the plasma spectra acquired by OES. The VUV spectrometers (wavelength 100–200 nm) are increasingly popular in plasma science, but one should keep in mind that the radiation of photon energy above the dissociation energy of oxygen molecules (5.2 eV) is absorbed in air, what complicates the experimental setup. Still, knowing the above limitations, one can find OES a useful, simple, and inexpensive technique, so it is widely used for the basic characterization of atmospheric plasmas.

OES can also enable the determination of the gas temperature, electron density, and electron excitation temperature; however, this already requires more in-depth knowledge. The methods for deduction of these plasma properties are not trivial and are beyond the scope of this paper. Gas temperature is assumed to be similar to rotational temperature ($T_g \approx T_{rot}$). Gas temperature T_g is thus determined from the calculation of rotational temperature T_{rot} of diatomic molecules considering Boltzmann distribution of rotational states [13,25,26]. For the APPJ plasmas, OH(A) and N_2(C) excited states are usually used for the determination of the rotational states. However, the obtained T_{rot} may differ if calculated from OH or N_2, and it often leads to an overestimation of T_g [13].

The spatial resolution depends on the lenses mounted at the end of the optical fiber. Normally, the radiation is collected from a solid angle of few degrees. The acquisition time depends on the radiation intensity. Widely used inexpensive spectrometers will have an adjustable acquisition period between approximately 1 ms and several s. Obviously, simple spectrometers will be particularly useful for characterization of plasma in a continuous mode. When the plasma is in the form of streamers of shelf-time well below a ms, the OES will give values averaged over numerous streamers.

The present paper explains some basic properties of the OES technique and gives a review of published literature, which is summarized in Table 1 and further discussed in the main part (Sections 2 and 3). Moreover, additional details regarding the APPJ configuration, flow rate, voltage, power, etc., are given in Table 1, making the comparison of various results easier. In the main part, some authors' own experiments are presented and discussed in a view of the cited literature, which gives a more comprehensive insight in the phenomena likely to occur in APPJs.

Table 1. T Summary of characterization of APPJ by OES.

REF	Gas Composition	APPJ Configuration	Findings by OES Diagnostics
[13]	Ar/TEOS (tetraethyl orthosilicate)	AlmaJET, single electrode corona jet, AC power supply, 12 kHz, 12 kV amplitude, power 11.5 W, Ar flow 2 slm, 17 ppm TEOS	Besides peaks characteristic for Ar plasma with the presence of water vapor, the addition of TEOS resulted in the appearance of additional excited species: CH, CN, and C_2 bands. T_{rot} of OH(A), N_2(C), and CH(A) were determined using two different simulation software. T_{rot} calculated from OH was much lower (410 K) than T_{rot} calculated from N_2 (550 K) and CH (590 K).
[27]	Ar + H_2O	2.45 GHz MW plasma, 104 W, gas flow 1.7 slm, H_2O content = 0–1.9%	The length of the plasma jet was decreasing with increasing water content. At very high water content, the plasma was unstable. In pure Ar plasma, N_2 dominated in OES spectra measured 2 mm from the nozzle. With the addition of water vapor, N_2 diminished, and OH became dominant. T_{rot} was determined, and it increased with water addition.
[28]	Ar with/o CO_2	DBD plasma jet, 71 kHz, 6 kV (peak-to-peak), 3 slm of pure Ar and 2.9/0.1 slm of Ar/CO_2	Besides Ar, molecular N_2, atomic O and OH were observed in pure Ar plasma. The addition of CO_2 caused a decrease in emission lines and the disappearance of oxygen, whereas OH and N_2 were still observed. Additionally, the CO Angstrom band was observed in Ar/CO_2 mixture.
[29]	Ar + H_2O (up to 7600 ppm)	Second ring-shaped ground electrode, sinusoidal voltage frequency 71 kHz, 10–20 kV (peak to-peak), power 12.88 W, gas flow 0.695–4.82 slm	OES was measured 15 mm from the ground electrode. OH and lines of Ar were found. T_{rot} was determined, and it was increasing with the increase in the water content. The addition of water to Ar caused the decrease in all atomic line intensities. The emission of OH radicals was not linearly dependent on the water content, but it had a maximum at 350 ppm.
[30]	He or Ar (with/o addition of O_2)	Additional outer ground electrode, pulsed power supply (voltage up to 3 kV, 60 kHz), sinusoidal or RF power supply, Ar or He flow rate 1–8 L/min	The electron temperature was measured (0.4 eV). The plume temperature was also measured versus voltage and flow rate. Depending on the frequency and gas flow, it was in the range 25–40 °C. Ar, as well as atomic O, OH, and N_2 peaks, were found in Ar plasma. OH peak was much stronger than O line even if O_2 was added to Ar. The origin of water was the desorption of water molecules from the quartz discharge tube.
[31]	Ar with/o H_2O	Additional grounded outer electrode, 71 kHz, power 12.8 W, peak-to-peak voltage 12.2–17 kV, 0.7–4.8 slm with 0.05–0.76% of water	Besides Ar lines, significant OH radiation was observed as well as N_2 lines and O-atom line. The increase in water content caused a decrease in all atomic lines. The addition of water also caused an increase in the gas temperature from 620 (pure Ar) to 1130 K (0.76% H_2O). OES profiles were measured along with the jet every 5 mm up to a distance of 35 mm. The OH concentration was maximal at the minimum addition of water to the feed gas (0.05%).

Table 1. *Cont.*

REF	Gas Composition	APPJ Configuration	Findings by OES Diagnostics
[32]	Ar, He (with/o addition of CH_4 or C_2H_2)	Coaxial configuration with the inner and outer electrode, 13.56 MHz, rms voltage 200–250 V, rms current 0.4–0.6 A, 3 slm of Ar/He and 160 sccm Ar(He)/CH_4	The collimating lens was positioned 140 mm from the end of the jet. The special construction of APPJ allowed the insertion of Ar(He)/CH_4 gas by a capillary tube into the main Ar/He flow, which prevents the mixing of the surrounding gas. In pure Ar plasma, the Ar and OH emission dominated. Additionally, in the Ar/CH_4 mixture, CH and C_2 bands were observed. In Ar/CH_4, CH was more dominant than C_2; whereas, in Ar/C_2H_2, C_2 band was more intense. Opposite to other authors, no N_2 emission was detected by OES because of special construction that prevented the mixing of the surrounding air. Electron density (8×10^{20} m^{-3}) and gas temperature (<400 K) were also measured from the fine structure of the rotational bands of OH.
[33]	Ar	kINPen, DC supply, 3–6 W, Ar flow 5 slm	OES was measured at various axial positions along with the plasma jet up to 12 mm (sampled every 2 mm). A maximum in the intensity of Ar lines was at 4 mm from the exit. The intensity of N_2 lines was increasing along with the jet because of the mixing of the surrounding gas. Significant OH emission was also detected. They found that APPJ is a source of UV-B (309 nm) and UV-A radiation (350, 380 nm), whereas no UV-C was detected in the range 250–280 nm. Irradiance in the UV range was determined to be 5 mW/cm^2 at the maximum power and the lowest axial distance. It can be concluded that several seconds of exposure of skin does not exceed the allowed exposure limit of 3 mJ/cm^2.
[34]	He + H_2O	AC power supply, 18 kV, 15 kHz, He flow 20 L/min, H_2O flow 3 mL/min	An increase in OH was observed when water droplets were added to He APPJ.
[35]	Ar or He	DBD type APPJ, AC power supply, 5 kV, 50 kHz, He flow: 2–4 slm, Ar flow: 1–3.5 slm	For Ar plasma, OH and N_2 species were dominant. For He plasma, N_2^+ and O were observed. Emission profiles were measured versus a distance from the tube exit.
[36]	Air/Ar/H_2O	2% H_2O in Ar/Air mixture, Ar flow 3 lpm, air flow: 3–15 lpm, voltage 4–7 kV, 250 Hz square wave power source	Production of OH and O radicals was investigated. The production of OH and O was increasing with increasing Ar content in the gas mixture and with the applied voltage, but decreasing with increasing air flow.
[37]	He	DBD plasma jet, pulsed DC power supply, 10 kHz, duty cycle 20%, peak-to-peak voltage 2–3.2 kV, min. power 5 W, He flow 2.5–5.5 slm	NO, OH, N_2 molecular, atomic N, He and O, and O_2^+ species observed. Gas temperature was found to be ~310 K, and the electron excitation temperature was ~5420 K. OES was also used to investigate the generation of various species when the plasma jet was impinging onto the liquid.
[38]	He + O_2	DBD-type APPJ, AC power supply, 60 kHz, peak to peak voltage 6–14 kV, 18 W, gas flow 3 slm, 0.1–0.5% of O_2	Besides He lines, also OH, O, N_2, and N_2^+ species were detected. Rotational and vibrational temperatures from OH radical were determined in the active and afterglow region. Gas temperature slightly increased when O_2 was added or when moved from active to afterglow region. In the active plasma region, it was 310 and 340 K for He or He/O_2, respectively. Whereas T_{vib} was 2200 or 2500 K, respectively.

Table 1. *Cont.*

REF	Gas Composition	APPJ Configuration	Findings by OES Diagnostics
[39]	He	DBD type APPJ, square wave AC supply, peak-to-peak voltage 6 kV, flow rate: 2–7 L/min	The spatial distribution of radicals along the plasma plume was investigated versus the applied voltage and gas flow rate. Various He emission lines, O, OH, N_2 and N_2^+ were observed. A strong decrease in emission intensity of He, O, and OH with a distance from the APPJ exit was observed. Opposite, N_2 and N_2^+ reached a maximum at a distance of 5 mm from the exit and then gradually decreased.
[40]	He/O_2	Powered hollow electrode, 2–10 kV_{pp}, 18 kHz, up to 0.33% O_2 addition, flow 3 slm	The radiation intensity of RN and ROS radicals was strongly dependent on the O_2 concentration. For ROS radicals, different behavior of the emission intensity versus the O_2 concentration was observed than for RN radicals.
[41]	Ar	kINPen, 1.4 MHz, gas flow: 5–15 slm	Production of NO radicals versus air admixture was investigated. The NO emission had a maximum at 0.1–0.2% of air admixture in Ar.
[42]	Ar, He, Ar + He	MW power 5–70 W, gas flow 100–1000 sccm	The influence of the MW power and gas flow rate on the emission intensity of plasma radicals was investigated. A strong dependence of emission intensities on the power was observed.
[43]	He/O_2	Micro-APPJ, 13.56 MHz, up to 0.6% O_2 admixture	OES measurements in UV and VUV for the various admixture of O_2 from 0 to 0.6%. The following species were observed in the range 100–350 nm: O, O_2, NO, and OH.
[44]	Ar/C_2H_5OH	33 MHz RF power source, various powers	C, CN, CH, and C_2 species were observed in ethanol plasma. An increase in ethanol content in Ar caused a decrease in Ar lines and an increase in the relative amount of C_2. No significant dependence of the intensity ratio of excited species with the increase in RF power.

2. OES Characterization of the Reactive Species along the APPJ Plasma Jet

APPJs have nowadays become very important in many applications, especially for localized treatment of polymers and for various biomedical applications. Therefore, the knowledge on the generation of excited species and the control of their intensities during the treatment of materials with the APPJ is important. OES is a method that enables the real-time monitoring of the reactive plasma species during the operation of the APPJ.

Here we report an example of the OES measurement along the plasma jet of one of the most commonly used and simple configurations of the APPJ (Figure 1). It is a single electrode APPJ having only one electrode. Alternatively, an additional electrode can be placed outside the dielectric tube, but in many practical cases, this is not necessary. A glowing plasma is stretching from the tip of the metallic electrode (a copper wire) mounted coaxially into the dielectric tube (made of Pyrex glass). The luminosity of gaseous plasma is the largest at the electrode tip and decreases with increasing distance. More details of this APPJ can be found in [3]. The discharge tube was made from Pyrex. The length of the discharge tube was 15 cm, and the inner diameters was 3 mm. A copper wire electrode was mounted inside the tube. A diameter of the copper electrode was 0.3 mm and the length of the electrode was the same as the length of the dielectric tube, i.e., 15 cm. The length of luminous plasma depends on the type of gas flown through the dielectric tube, the gas flow, and the properties of the power supply. In our case, a visible part of the plasma jet extended up to 3 cm from the exit of the discharge tube. The plasma jet shown in Figure 1 was fed with Ar gas with a purity of 5.0. The flow rate was adjusted to 1 slm. The powered copper electrode that was mounted inside the Pyrex dielectric tube was connected to an AC power supply. The voltage of the power supply used in this study was 7 kV (peak-to-peak), whereas the excitation (sinusoidal) frequency was 25 kHz. The power was estimated to a few Watts. The electrical field is the highest at the tip of the electrode. The high enough electric field causes the formation of electron streamers. Therefore, our plasma was in the form of streamers rather than continuous plasma. The electron streamers may also appear along the electrode, but they are much weaker because of the lower electric field. The configuration of streamers, as shown in Figure 2, is typical for the case when the plasma device is in ambient air. If the device is placed in a container filled with a noble gas, the intense streamers will not be observed only along the axis but will also propagate radially.

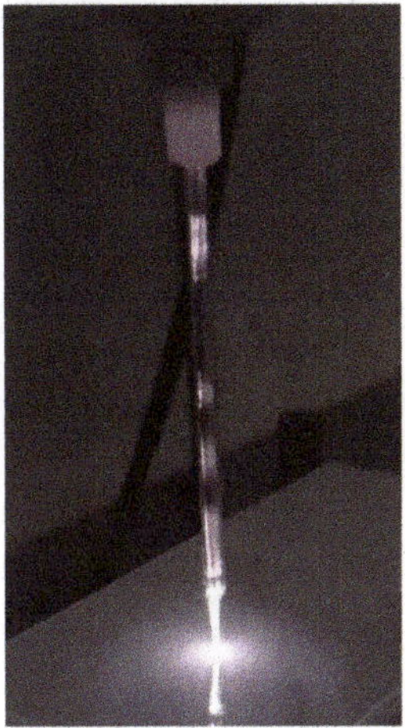

Figure 1. A photo of the atmospheric-pressure plasma jet. The substrate is a wooden plate at a floating potential.

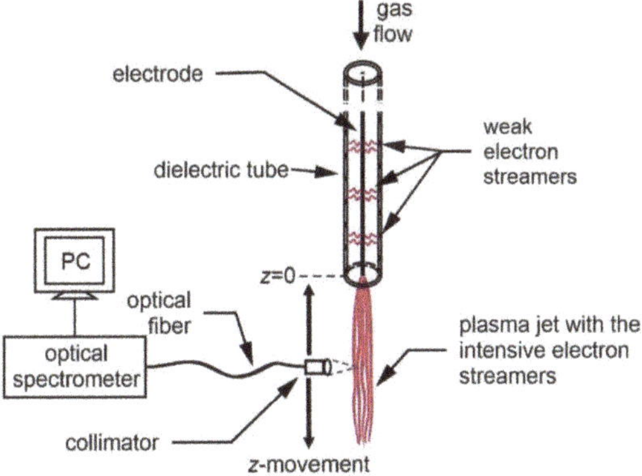

Figure 2. Schematic presentation of the axial plasma jet characterization by moving the optical fiber.

OES was applied for basic plasma characterization of the glowing plasma as shown in Figure 2. The radiation was acquired through an optical fiber, which was mounted on a movable holder. An appropriate lens was placed on the tip of the optical fiber to enable space-resolved measurements. The spectrum acquisition time depends on the luminosity of the plasma jet and the properties of the spectrometer, but the typical value was 10 ms. The time resolution of the simple spectrometer was therefore good enough to perform a gradual characterization with a reasonable vertical resolution by moving the optical fiber vertically along the jet, as shown in Figure 2.

A typical OES spectrum of Ar plasma acquired close to the electrode tip of the APPJ device (i.e., 2 mm from the exit) is shown in Figure 3 (lower curve). Moreover, the OES spectra acquired 10 and 25 mm from the electrode tip are shown in Figure 3 as well. Figure 3 reveals transitions of neutral Ar atoms that correspond to the relaxation of highly excited states to the metastable state. The transitions are in the red part of the spectrum. There are also transitions to the ground state, but they appear in the far UV range, which cannot be probed with the simple optical spectrometer. Interesting enough, Ar lines are not the only spectral features in plasma 2 mm from the exit of the dielectric tube. The most intensive feature appears at the bandhead of approximately 309 nm, and it corresponds to the transition of the OH radicals from the electronically excited to the ground state. Furthermore, there is a band of lines in the near UV range that corresponds to the transition of highly excited neutral nitrogen molecules in $C^3\Pi_u$ state to N_2 molecules in $B^3\Pi_g$ state. This transition is called 2nd positive system. Details on the most common transitions of various atoms and molecules likely to occur in the APPJ are very well presented in the paper of Golda et al. [45]. Spectral features as observed in Figure 3 in the spectrum acquired close to the APPJ exit also persist in the spectra recorded at longer distances from the exit of the discharge tube; however, their intensity is significantly lower, and also their ratio strongly depends on the distance from the dielectric tube. The axial distribution of the major spectral features versus the distance is shown in Figure 4.

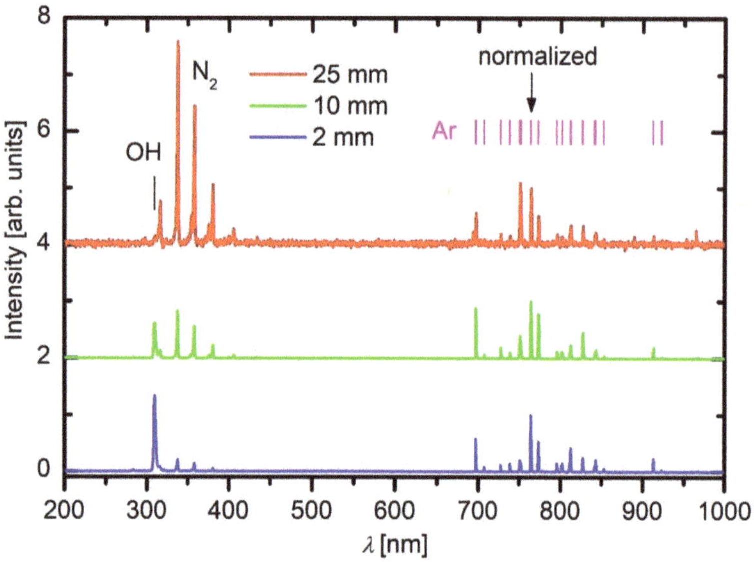

Figure 3. The OES spectra recorded at various distances from the exit of the discharge tube. The spectra were normalized to Ar line at 764 nm.

Park et al. [30] used a device similar to what is shown in Figure 1 but powered it with a DC pulsed discharge. In such discharges, the streamers, as shown in Figure 2, shrink to bullets of a spatially confined dense plasma, which propagate away from the powered electrode at a velocity of the order of 10^4 m s^{-1}. The visible appearance of gaseous plasma is not much different from the one shown in Figure 1, but the electrical properties are completely different. Despite the difference, the optical spectra acquired by Park et al. are similar to those in Figure 3. In the optical spectra acquired by Park et al., the OH line at 309 nm was dominant, and the N_2 band in the near UV range was clearly visible, too (Figure 5). They also provided images of the discharges taken with the fast camera to show the evolution of the plasma bullets. The optical spectra taken in such cases obviously give values averaged over numerous such localized plasma bullets.

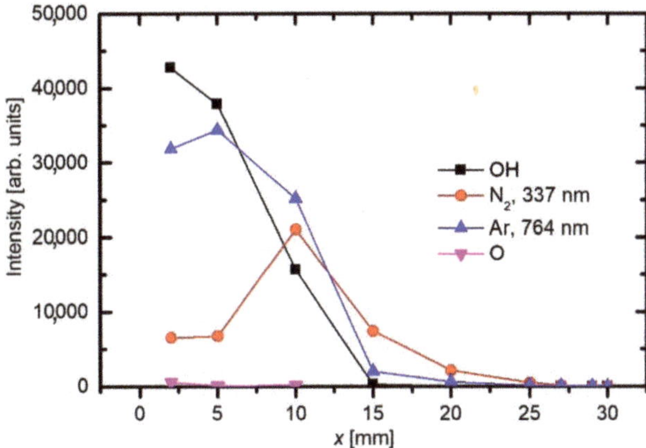

Figure 4. The axial distribution of the major OES spectral features versus the distance from the exit of the discharge tube.

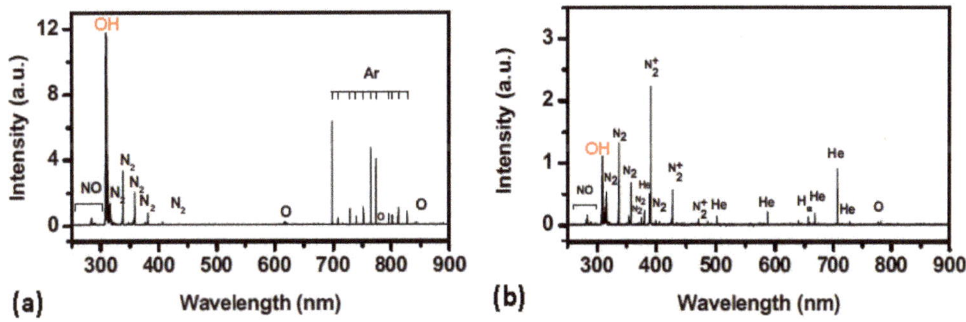

Figure 5. Emission spectra of: (**a**) Ar pulsed 2 kV plasma, and (**b**) He pulsed 2 kV plasma showing the presence of a significant amount of OH and N_2 species. Reprinted with permission from [30] (Park et al.). Copyright 2021 American Institute of Physics.

Additionally, also Sarani et al. [28] and Seo et al. [35] observed similar spectra as shown in Figure 3 for the dielectric-barrier discharge (DBD) type APPJ feed with pure Ar as well as Weltman et al. [33]. Sarani et al. [28] used the APPJ very similar to the device shown in Figure 1. Intensive OH radiation was observed even in pure Ar, and suppressed by the intentional addition of other gases. They provided possible channels for the formation of hydroxyl radicals. The OH radical can be generated by a collision of the water molecule and an electron (electron-impact dissociation), or a metastable Ar atom, or by the reaction of an excited oxygen atom with a water molecule, which was found the dominant mechanism by Sarani et al. Their arguments were based on the papers of Herron et al. [46] and Kim et al. [47]. In addition, Roy et al. [36] used OES to investigate the production of OH and O radicals in air/Ar/H_2O plasma and found that Ar plays a significant role in their production. The production of OH and O radicals was increasing with increasing Ar content.

As already explained before, the appearance of OH and N_2 spectral features in Figure 3 is not because of the inadequate purity of Ar gas, but rather because of the interaction of gaseous plasma with the neighboring atmosphere. As shown schematically in Figure 2, the electron streamers are formed stochastically in the APPJ, which is powered with an AC discharge at a typical frequency of 10 kHz. The streamers propagate along the axis of the Ar jet. They will not propagate in air because of the numerous channels for the loss of electrons' kinetic energy. Still, some streamers reach the edge of the Ar gas jet, where diffusion of air from the surrounding atmosphere occurs. Furthermore, the streamers inside the Ar jet cause formation of Ar metastables with a large radiation life-time. The metastables are therefore quenched rather than relaxed by the radiation. The quenching is unlikely to occur in pure Ar, but very probable at a collision between an Ar metastable and a gaseous molecule. The Ar metastables diffuse within the jet and may finally reach the edge of the pure Ar jet, where they transfer their potential energy to air molecules. The optical spectrometer will collect any radiation arising from a certain solid angle, as determined by the lenses (Figure 2). That is why the nitrogen emission lines appear in the spectrum collected close to the electrode (Figure 3). The relative intensity of nitrogen emission depends on the kinetics of gas mixing and on the plasma excitation of gaseous species. Figure 5 shows a maximum in the nitrogen emission along the axis. A maximum is usually a consequence of two opposite effects. In this particular case, the diffusion of nitrogen inside the plasma jet increases with increasing distance from the electrode. From this point of view, the N_2 emission should increase with increasing distance from the electrode. On the other hand, the absolute luminosity of gaseous plasma decreases with increasing distance. The combination of these two opposite effects causes the maximum in nitrogen radiation, as revealed in Figure 5. Figure 6 shows the intensities of nitrogen and OH radicals radiation normalized to the main Ar line at 764 nm. Now we observe that the relative intensity of nitrogen radiation keeps increasing with increasing distance and remains large even at the distance of 25 mm where the absolute intensity is poor. The effect is explained by the mixing of the gases. A maximum in N_2 emission versus a distance was also observed by Chauvet et al. [39]; however, they observed a maximum at a somehow shorter distance at approximately 5 mm.

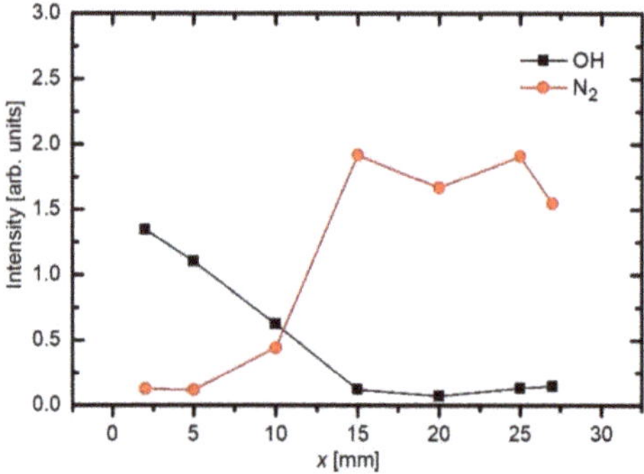

Figure 6. The intensities of nitrogen (337 nm) and OH normalized to the main Ar line at 764 nm.

Such a maximum as observed for nitrogen species in Figure 5 is not observed for OH radiation. For OH radicals, we observe intensive radiation in the first few mm from the exit of the discharge tube and then rather fast decay. Water vapor abounds in air in any laboratory. The saturated water vapor pressure at room temperature (25 °C) is about 3.2 kPa. This value corresponds to 100% humidity. The air humidity in the lab is never that high, but it is rarely below 10%, so the partial pressure of water vapor is still several tenths of kPa. The water vapor from ambient air cannot explain the completely different behavior of the OH as compared to N_2. A more feasible explanation is the presence of water vapor within the dielectric tube. Unlike nitrogen, which is quickly removed from the dielectric tube by passing Ar at a rather large flow rate, water molecules stick to surfaces of solid materials, where they form a layer that is only slowly desorbed at room temperature. The consequence of this water film is the appearance of strong OH radiation even at the electrode tip. The OH radiation depends on the amount of water released from the surfaces, which in turn depends on the amount of water absorbed. Upon continuous flushing of the dielectric tube with Ar, the water concentration decreases with time but remains significant for hours. Figure 7 represents the evolution of Ar and OH lines for one and a half hours [48]. As the discharge is turned on, the OH signal is larger than the Ar line. Within the first minutes, the Ar line increases, and the OH remains fairly intact. After prolonged treatment, the Ar line finally reaches a constant value, but OH keeps decreasing. This important fact should be taken into account at any attempt to explain the plasma-surface interaction upon activation of a polymer product by the APPJ sustained in a noble gas. We should be aware that any APPJ sustained in pure Ar at ambient conditions is therefore rich in OH radicals.

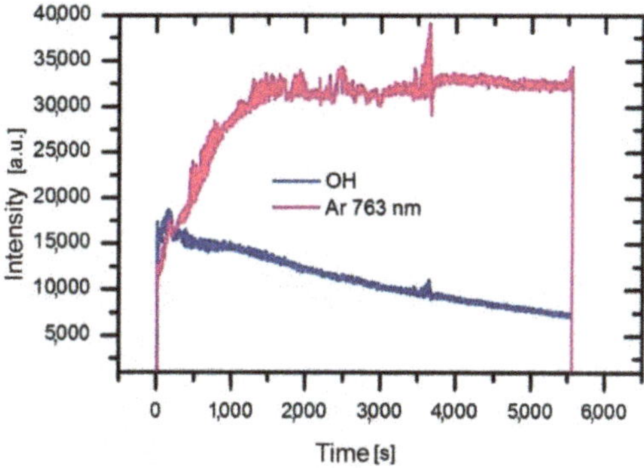

Figure 7. Stability and time evolution of Ar and OH lines after ignition of the APPJ [48].

Similar results were observed by other authors too. Seo et al. [35] used Ar of purity 99.999% for sustaining the APPJ at the frequency of 50 kHz and reported optical spectra similar to those in Figure 3. They employed a somehow more sophisticated discharge configuration with an additional grounded ring mounted on the other side of the dielectric tube but observed practically identical behavior of the OH signal as in Figures 5 and 6 when using Ar feed gas. In contrary, the peak of the nitrogen radiation was not so obvious as in Figure 5, but the relative intensity was almost identical to the behavior, as shown in Figure 6. Here we should also mention, that Seo et al. found a dominant OH emission only in the case of Ar plasma, whereas in the case of He plasma, OH emission was only minor. Because of a very high OH radical concentration in Ar plasma, they obtained a much better improvement of surface hydrophilicity of polydimethylsiloxane when treated with Ar gas than with He.

Figures 5 and 6 are also in agreement with the reports by Srivastava et al., who found that the decay length of OH radicals was much shorter than the length of the visible plasma jet [27]. As shown in Figure 5, the intensity of OH radiation remains roughly constant for the first few mm and then decreases rapidly. In Figures 5 and 6, the OH signal becomes marginal for lengths more than approximately 15 mm, although the radiation from nitrogen bands persists much longer. One can assume that the OH radicals are lost along the jet, but it should be stressed that optical emission spectroscopy is a qualitative technique, and it does not reveal the concentration of radicals in the ground state. A comparison with similar configurations, where absolute techniques were used, is recommended. Srivastava et al. [27] performed a complete characterization of the APPJ sustained in pure Ar with a small admixture of water vapor. They measured the absolute density of OH radicals by a reliable technique (cavity ring-down spectroscopy) and found the OH density as large as 3×10^{22} m^{-3} for high purity Ar. The OH density increased up to approximately 5×10^{22} m^{-3} for the admixture of approximately 1 vol.% of water. Therefore, an important conclusion is that a common APPJ, as shown in Figure 1, always contains a large number of OH radicals, which typically arise from the desorption of water molecules from the dielectric tube or even connecting tubes. Srivastava et al. also employed optical absorption spectroscopy to measure the variation of the OH density along the axis of the APPJ and found a rather exponential decay with a characteristic length of a few mm, as shown in Figure 8 [27]. The decay length depended on the concentration of water vapor in Ar. The decay was the fastest at the maximal concentration of OH radicals, i.e., at the water admixture of approximately 1%. Srivastava et al. also provided explanations for the loss of OH radicals, which included recombination reactions, three-body collision quenching, and reactions of OH with H, O, or even NH radicals. They found the most extensive loss by the reaction $H + OH \rightarrow H_2O$ with a coefficient of approximately 2×10^{-16} m^3 s^{-1} [27].

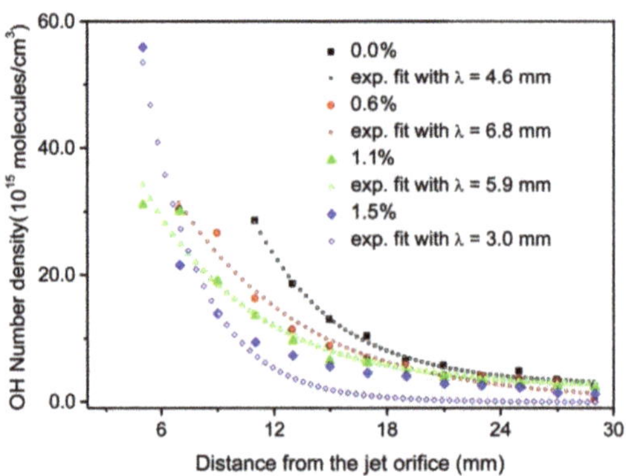

Figure 8. Decay length of the density of OH radicals in Ar plasma with various H_2O contents from 0 to 1.5%. Reprinted with permission from [27] (Srivastava et al.). Copyright 2021 American Institute of Physics.

The results reported above clearly show that a relative gas humidity of the atmosphere may significantly alter the density of reactive species what may lead to unrepeatable results of the APPJ treatment. APPJs are usually feed by a noble gas (Ar or He), because in this case, the gaseous plasma expands rather far from the powered electrode, whereas in the case of molecular gases, the plasma plume is much shorter. Any APPJ sustained in pure noble gas at ambient conditions is thus rich in OH radicals. In fact, there is hardly any report in the literature on OES characterization of the APPJ without a significant signal arising from OH radicals. Some authors used high purity Ar and carefully pre-treated the APPJ devices, but none reported the OH-free optical spectrum [27–33]. Many authors also intentionally added a small admixture of water vapor to the feed gas to investigate the kinetics of the OH formation and its concentration profile [27–29,31,34].

Water vapor was added to the APPJ intentionally by Ilik et al. [34]. They used He as a carrier gas. The flow rate was as large as 20 slm. They managed to spray water droplets into the APPJ and studied the evolution of optical spectra. The discharge was powered with a sinusoidal AC supply operating at 18 kV and 15 kHz. The spectra were averaged over a rather long jet, but were similar to the one presented in Figure 3 except that a weak He line was observed instead of Ar lines. The authors demonstrated that the addition of water droplets did not influence the spectra much. In particular, the relative intensity of nitrogen prevailed at all experimental conditions. The water spraying only caused suppression of all spectral features, thus indicating the negative effect on the electron density or temperature.

The influence of water vapor on the electrical and spectral properties of the APPJ was further investigated by Nikiforov et al. [29]. They used a device similar to that presented in Figure 1, except that the electrode was deep inside the dielectric tube, and a grounded electrode was also applied. They provided precious information about the electrical behavior of an AC discharge because they measured both the voltage and current, as well as the Lissajous curves, which give information on the exact power of the discharge. The intensive electron streamers, as shown schematically in Figure 2, are clearly presented in the paper by Nikiforov et al. The measurements allowed for estimation of the real discharge power absorbed by gaseous plasma versus the adjustable voltage of the AC power supply, which operated at the frequency of 70 kHz. At a given voltage, the maximum power was absorbed by plasma at the lowest concentration of the water vapor. The maximum emission arising from OH radicals was found at a water vapor concentration of 350 ppm. Although the content of H_2O was varied in a broad range up to several 1000 ppm, the maximum effect of polymer sample surface modification was observed at the water admixture in the range between 200 and 500 ppm. Based on this observation, Nikiforov et al. concluded that the yield of OH radicals was the highest at such a low water admixture.

Weltmann et al. elaborated the evolution of optical spectra along the APPJ operating in the semi-continuous mode [33]. They used an RF supply operating at a frequency of 1.1 MHz. They observed gradual decay of the OH and N_2 spectral features with the distance from the powered electrode. The OH radiation reached the peak value 2 mm away from the electrode, while nitrogen about 10 mm. The observations are, therefore, similar to those presented in Figures 3, 5 and 6. They also managed to measure the axial and radial temperature profiles using a glass fiber temperature probe. As expected, the highest gas temperature was observed where the luminosity was the highest, but the plasma tip temperature at the axis was about 48 °C. Rather strong radial gradients were observed, and the gas remained at room temperature a few mm from the visible plasma jet. These results indicate that the rotational temperatures, as reported by the above-cited authors, may not be representative for the estimation of the neutral gas kinetic temperature.

Production of OH radicals in the bullet-like APPJ was recently elaborated also by Gott and Xu [49]. The APPJ was sustained in He gas of purity 99.9999% with a pulsed discharge powered with a 10 kV supply. The voltage rise-time of 60 ns only assured for almost perfectly-shaped bullets propagating from a tungsten electrode which was placed into a quartz glass tube, similar to Figure 1. The emission from OH radicals was much larger inside than outside the tube, confirming that the water vapor originated from the walls inside the discharge tube. The authors explained the extensive radiation from OH radicals by electron-impact dissociation and subsequent excitation rather than the influence of He metastables.

The formation of the bullet-like discharge was further elaborated by Cordaro et al. [50]. They performed a detailed characterization of a plasma jet useful for blood coagulation. Plasma was sustained in He or Ne using a low-frequency discharge, which provided current pulses of a duration of approximately 1 µs. No additional gas was added intentionally, but the OES spectra revealed nitrogen bands even within the dielectric tube despite the rather large gas flow of 2 slm. At a distance of 5 mm from the exhaust, the spectral features corresponding to N_2 and O transitions prevailed over the radiation from He atoms. Unfortunately, the spectrometer used by Cordaro et al. [50] did not allow for measuring the OH line at 309 nm to compare the intensities like in Figure 3. In another paper, the same group reported the thermal effects caused by the treatment of biological matter with such an APPJ [51]. They found experiments with biological materials unfeasible because of the unknown thermal properties, so they used metallic samples and found significant differences for grounded or floating objects. The heating rate depended on the distance from the electrode, indicating the loss of the discharge power along the APPJ.

Chauvet et al. [39] performed an investigation of the spatial distribution of excited and ionized species along the bullet-like APPJ sustained in He at different flow rates. The emission of H, O, and OH strongly decreased with the distance from the APPJ exit, whereas N_2^+ and N_2 * reached a maximum at few mm from the tube exit.

Gerling et al. [40] investigated the production of reactive oxygen species (ROS) such as OH and O and reactive nitrogen species (RN) such as N_2 and N_2^+ in He plasma with various oxygen admixtures up to 0.33%. They found a strong dependence of ROS and RN species versus O_2 admixture. The different behavior of the emission intensity of ROS and RN radicals versus O_2 concentration was observed. In the case of RN radicals, N_2^+

emission exhibited a maximum at 0.05%, whereas N_2 at 0.1% of O_2 addition. For ROS radicals, the emission intensity of O was first increasing with the O_2 addition and reached a saturation when 0.15% of O_2 was added to He. For OH radicals, the opposite behavior was observed. After the initial saturation, it decreased if the addition of O_2 was more than 0.1%. Similar work was also performed by Pipa et al. [41]. The authors investigated the production of NO versus air admixture in Ar plasma and found a maximum in the NO emission intensity at 0.1–0.2% of the air in Ar.

Jia et al. [42] used OES to investigate the influence of microwave power and gas flow on the emission intensity of specific species (Figure 9). The authors used the APPJ created in Ar, He, or Ar + He mixture. The intensity of the emission was monotonously increasing with power. The increase was explained by the increase in the electron density and electron temperature. The variation of the emission intensity with the flow was less dramatic. After a slight initial increase, it slightly decreased at higher flow rates.

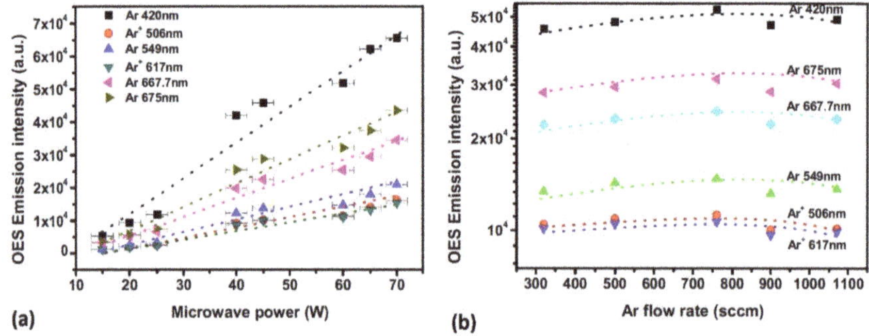

Figure 9. Influence of: (**a**) microwave power and (**b**) gas flow rate on the emission intensities of Ar spectral features. Reprinted with permission from [42] (Jia et al.). Copyright 2021 American Institute of Physics.

A simple optical spectrometer usually operates in the spectral range between approximately 200 and 1000 nm. The optical transitions in this range occur from the excited to the ground level only for the case of OH radicals and any metallic impurities, but most other transitions arise from the relaxation from one (higher) to another (lower) excited state. For example, atomic Ar lines shown in Figure 3 appear at transitions to the level at the excitation energy as high as about 12 eV. Apart from these transitions, there are also transitions that appear to the ground state. These transitions are in the far UV range of the spectrum, below the detection limit of a simple spectrometer, which is usually approximately 200 nm [45]. Such short wavelength radiation is quickly absorbed in the air, as well as in the optical fiber and even lenses, so special types of spectrometers should be used to detect transitions of many gaseous species to the ground state. Among such transitions, there is H-atom Lyman series with the predominant line at 121 nm (transition from the first excited state to the ground state), O-atom transitions with the predominant line at 130 nm, Ar-atom transitions with the most intensive line at 107 nm, and N-atom transition at 149 nm. These transitions may be dominant in the spectra of atmospheric-pressure plasma jets.

Schulz-von der Gathen et al. [52] measured OES of He discharge with various oxygen admixtures in UV and VUV regions. For low or zero oxygen admixture, NO and OH emissions were very strong. In addition, the atomic oxygen line was dominant at low oxygen admixture (0.06%).

An excellent comparison of radiation intensity in the far and near UV range was provided by Golda et al. [45]. They measured the spectra of the APPJ sustained by a sinusoidal RF voltage at the frequency of 13.56 MHz. The VUV spectrometer was mounted onto a special aerodynamic window and placed in a high-vacuum chamber. For He discharge, they found the intensity of OH at about 282 nm marginal as compared to radiation in the range between 60 and 100 nm. A similar result was also observed for Ar, except for this case, the major far UV radiation occurred between 110 and 130 nm. The OH peak at 282 nm is a consequence of the radiative transition from the first vibrational excited state of the OH (A) electronic state to the ground state. This transition is normally 10-times less intense than the major transition at 309 nm, but the latter could not be probed by the spectrometer used by Golda. In any case, the major radiation from the pure argon APPJ aroused from Ar_2 continuum. Even small impurities of water vapor in the monochromator caused decreasing of this radiation to half of the original value. The absorption cross-section for H_2O molecules was also provided by Golda et al. and may be as high as 10^{-17} cm^2. For He plasma, the intensive radiation within 60 and 100 nm was suppressed even by a very small addition of a reactive gas. For example, the addition of 0.01% O_2 caused suppression of the He radiation by a factor of two. Unfortunately, no data for water vapor were provided by Golda et al. Based on the discussion above, it is possible to conclude that a simple plasma jet operating in the configuration of Figure 1, may be a source of radiation in the far UV range, but this cannot be probed with the simple spectrometer. The experimental configuration with a VUV spectrometer, as adopted by Golda et al., is highly recommended, but difficult to realize.

According to this literature review and our results, we can conclude that OH radicals are the most important reactive species in the APPJ. As OES is a qualitative technique, it enables investigation of the relative variation of the density of plasma species along the axis. However, for quantitative determination of the densities, other techniques must be applied. The OH density profile along the plasma jet is usually measured by laser-induced fluorescence (LIF) [53–56] or two-photon absorption LIF (TALIF) [57–59]. Other methods, such as cavity ringdown spectroscopy (CRDS) [27,60–62] and absorption spectroscopy [57,63] can be used as well. Schröter et al. measured the concentration of the OH radicals in the humidified He discharge sustained in an RF driven APPJ by using vacuum ultra-violet high-resolution Fourier-transform absorption spectroscopy and ultra-violet broad-band absorption spectroscopy [57]. They found an extensive dissociation of H_2O molecules upon plasma conditions. The density of OH radicals was approximately 3×10^{20} m^{-3}. Such a large density expanded several cm along the discharge channel. An order of magnitude lower OH density was observed by Fuh et al. [62] in a pulsed He discharge with CRDS method. Whereas Yonemori et al. [56] obtained even lower densities of approximately 1×10^{18} m^{-3} with LIF. In another paper, Yonemori et al. [55] measured the OH density in the humidified He pulsed discharge. The H_2O content in He gas was varied between 0 and 800 ppm. A maximum in the OH density peaking at 6×10^{18} m^{-3} was found at the H_2O content of 200 ppm. Additionally, Srivastava et al. measured the variation of the OH density along the axis of the APPJ [27,60] by using CRDS method. Srivastava et al. [60] measured the OH density in Ar, Ar/N_2, and Ar/O_2 APPJ discharge at various powers and flow rates. The OH density was decreasing along the jet. The OH density also strongly depended on the power and it was increasing with the increasing power. For Ar plasma, the OH density ranged from 1.3×10^{18} to 1.1×10^{22} m^{-3}, depending on particular conditions. For Ar/N_2 it was in the range between 4.1×10^{19} to 3.9×10^{21} m^{-3}, whereas for Ar/O_2 it was between 7.0×10^{18} to 4.6×10^{22} m^{-3} [60].

3. Examples of Using OES for Gas Temperature and Electron Density Measurements

The temperatures can be deduced from the optical spectra only when high-resolution spectrometers are used. Furthermore, more knowledge is needed and application of the simulation software. That is why the rotational and vibrational temperatures are rarely reported. Therefore, the spectra shown in Figure 3 will not reveal the temperatures because of the insufficient resolution of the spectrometer. A user of APPJ should, however, beware when concluding about the temperatures involved within the plasma jet. The temperature may differ depending on which emission line (OH(A) or N_2(C)) is used for its calculation [13]. Furthermore, the temperatures reported in the literature may significantly differ, although similar jets and plasma parameters were used.

Nikiforov et al. measured rather high-resolution OES spectra to deduce the rotational (T_{rot}) and vibrational (T_{vib}) temperatures of OH radicals [29]. For "pure" Ar, the temperatures were found independent from the distance from the electrode at the values of T_{rot} = 450 K, and T_{vib} = 1500 K. An addition of 500 ppm of water vapor did not have a significant effect on the temperatures away from the electrode, but near the electrode, the vibrational temperature was approximately 2000 K. When 7600 ppm of H_2O was added, T_{rot} increased to approximately 800 K rather far from the electrode, where T_{vib} was almost 5000 K. The values were even larger next to the electrode.

High T_{rot} and T_{vib} temperatures were also reported by Srivastava et al. for the case of the APPJ sustained with a microwave (MW) discharge (i.e., in the continuous mode) [27]. In almost pure Ar, T_{rot}, and T_{vib} temperatures were practically equal at approximately 900 K. The addition of 1 vol.% H_2O caused an increase in both temperatures to approximately 1100 K. The huge discrepancy between the results of Srivastava et al. [27] and Nikiforov et al. [29] may be explained by different discharges. While Srivastava et al. used a continuous plasma, Nikiforov et al. selected pulsed discharge with electron streamers, as shown schematically in Figure 2. Obviously, the pulsed plasmas created by rather low-frequency discharges cause overpopulation of highly vibrational excited states, whereas continuous plasmas cause almost equal T_{rot} and T_{vib}. In any case, the temperatures are large, in many cases well above the softening or melting point of polymers. Such hot OH radicals interact abruptly with polymer surfaces, causing functionalization and chemical etching at prolonged treatment times. These effects should be taken into account at any attempt to activate a polymer surface by atmospheric pressure plasmas.

Opposite to Srivastava [27] and Nikiforov [29], Sahu et al. [37] reported a very low T_{rot} of only approximately 330 K. Such low T_{rot} was also determined by Thiyagarajan et al. [38] for DBD-type APPJ in He plasma with small O_2 admixture. Plasma was characterized in the active plasma region and in the afterglow region. It was found that gas and vibrational temperatures increased when O_2 was added. The temperature was also slightly higher in the afterglow region compared to the active plasma region. Gas temperature was found to be 310 and 340 K for He and He/O_2, respectively, whereas T_{vib} was 2200 and 2500 K, respectively. Sahu et al. [37] determined rotational temperature using N_2 line at 337 nm, whereas Thiyagarajan et al. [38] used OH. As will be shown later, the calculated temperature may differ depending on the radical used for calculation.

Sarani et al. [31] also performed a detailed characterization of pulsed discharges using the same equipment as Nikiforov et al. [29]. They used pulsed discharges of a duration of approximately a microsecond. They managed to estimate the electron density inside such a plasma streamer to approximately 10^{19} m^3. The rotational and vibrational temperatures, as calculated from the high-resolution OES spectrum of the OH radical, were found dependent on the population of the excited states. A strong overpopulation of highly excited rotational states was observed, so the rotational temperature is not defined for such discharges. Such a rather unexpected population of excited states was explained by the quenching of excited OH radicals which, according to Sarani et al., depended on the excitation level.

High electron densities were also reported for the APPJ similar to the one in Figure 2 sustained in Ar with an admixture of organic gas, in particular, tetraethyl orthosilicate (TEOS) [13]. The electron density peaked at approximately 10^{23} m^{-3} in a short time during the evolution of streamers. This value is much larger than what was reported by other authors. For example, Sarani et al. [31] reported four orders of magnitude lower electron density. As already explained, the discharges sustained by rather low-frequency AC power supplies cause the formation of numerous streamers of short duration where the peak electron density is difficult to measure. One should beware of averaging results over the time scale much longer than a streamer duration. The intensity of spectral features followed the evolution of the electron density. The OH band at 309 nm prevailed despite the fact that no water vapor was added to the gas mixture intentionally. The peak power absorbed by plasma enabled extensive radicalization of the organic precursors, and even C_2 dimers were detected in the optical spectrum. Barletta et al. [13] also determined rotational temperature calculated from CH, N_2, or OH radicals. This was the first time that CH radical was used for the estimation of T_g. Despite the rather large average power of 11.5 W, the rotational temperature was between approximately 410 K, if calculated from OH, and 550 and 590 K if calculated from N_2 and CH radicals, respectively. The authors showed that taking CH radical leads to an overestimation of T_g. Taking into account the results reported by previously cited authors, it is clear that the deduction of the gas temperatures involved in APPJs may not be straightforward. The APPJ device used by Barletta et al. [13] was also applied for the deposition of thin-films from Ar/TEOS, where knowing T_g is one of the key parameters needed to have control over the deposition process.

An RF discharge operated at 13.56 MHz at rather high power was also used for sustaining a micro-plasma jet by Yanguas-Gil et al. [32]. As mentioned earlier, such plasmas operate in continuous mode. The authors found the electron density close to 10^{21} m^{-3}, although the neutral gas temperature was below 400 K. They used a coaxial configuration, and the plasma plume expanded a few mm from the tip of the powered electrode. They managed to assure negligible diffusion of air into the plasma plume, so the optical spectra consisted only of Ar lines and OH band. A small addition of N_2 into the gas caused significant radiation of the N_2 second positive band and suppression of OH radiation, similar to what is observed in Figure 3.

4. Some Examples of Using OES in Practical Applications of APPJs

In the introduction, some possible applications of the APPJ were mentioned to outline its increasing importance. Although the aim of this paper was to give some examples of using OES for basic characterization of the APPJ plasma, we here further report some examples of using OES in combination with practical applications of the APPJ. For example, Cheng et al. [64] used OES for characterization of the APPJ for wound healing. The authors found that relative intensities of reactive species increased 2–3 times when the plasma jet was in contact with the epidermis. The authors concluded that the epidermis reacted as a floating electrode, causing a locally enhanced electric field at the interface. Additionally, Jacofsky et al. [65] used OES to characterize the APPJ used for wound healing. OES was used to find the most optimal parameters in terms of the flow rate, the distance of the APPJ from the surface, and gas composition. No differences in active species generation were found when the metallic or organic substrate were used. The authors conclude that the secondary electron emission plays a small role in the chemistry induced in the jet. Lin et al. [66] used OES to investigate the correlation between the intensity of OH radicals in the APPJ with various O_2 addition to Ar feed gas and inactivation of bacteria. With the help of OES the authors concluded that the trace amount of O_2 addition into working gas enhanced the OH radicals formation what also speeded up the killing of bacteria and endospore. Another example reported the application of OES in the case when the APPJ was used for the deposition of thin films from styrene and methyl methacrylate precursors [67]. OES was used to investigate the degree of monomer dissociation versus the power and a carrier gas flow rate. In addition, Zhao et al. [68] used OES to characterize the APPJ for deposition of Cu thin films on a polymer, where a Cu wire inserted inside the discharge tube acted as the evaporation source. Enhancement in the Cu film deposition rate and increased purity was obtained by H_2 gas addition to Ar feed gas. Furthermore, increased rotational temperature was observed when adding H_2. The authors concluded that atomic hydrogen produced by the plasma plays an important role in heating the gas to promote the evaporation of Cu atoms. The last example is the application of OES in combination with the APPJ for polymer surface modification. Kehrer et al. [69] used OES to investigate the influence of water admixture (OH radicals) on the functionalization of polymer surfaces. The authors found that the influence of water admixture on polymer surface modification strongly depends on the type of the carrier gas.

5. Summary

An atmospheric pressure plasma jet is useful for localized surface treatment of various materials, as well as tissues. Therefore, knowing the interactions of both radiation and chemically reactive plasma species with the surface is important. Characterization of the APPJ is recommended before any application to control surface reactions and reproducibility of the results. The simplest technique is optical emission spectroscopy (OES). This technique enables space-resolved determination of the radiative species in the APPJ. Usually, OES is mostly used to identify species and monitor reaction products when the plasma interacts with the surface being treated. However, the application of OES to measure the gas temperature is rising, although it requires more in-depth knowledge and the use of computer models. Spectrometers are available with different spectral resolution. Spatial and temporal resolution of the spectrometers are continuously increasing with technological development. Recently, Weiss et al. [70] performed spatially resolved OES using integrating (Ulbrecht) sphere for the first time. This ensured many diffusive reflections of the light in the spherical cavity thus enabling uniform and spatial resolved OES. When comparing to the regular use of OES, similar emission and distribution of excited species was detected, however, species in the visible and near infrared (VIS/NIR) region and especially those in UV region were detected with 4–5 times higher intensity. Nevertheless, because plasma is a rich source of VUV photons, the major drawback of OES is still inability to measure in the VUV range, where more expensive VUV spectrometers are needed.

Detailed information about the behavior of different reactive species can be obtained by comparison of the OES spectra with results of more thorough plasma characterization with other techniques reported in the literature. Several authors have reported detailed studies using more expensive and demanding characterization techniques but similar discharges. Such a comparison is useful for the interpretation, but one should beware of huge discrepancies summarized in this paper.

The APPJ is often sustained in Ar without the intentional addition of a reactive gas. It was shown that the traces of water (that have been adsorbed in the discharge or connecting tubes) might play a dominant role in surface reactions. Any APPJ plasma operating in the atmosphere thus contains OH radicals. The emission from OH radicals is much stronger in pure Ar plasmas than in pure He plasma. The OH radicals are among chemicals of the highest oxidation potential; therefore, they interact intensively with a material surface causing rapid activation. Furthermore, the OH radicals have a rather large internal energy (the vibrational temperature often exceeds 1000 K) what contributes to the chemical reactivity of these radicals. Various OH densities and its rotational and vibrational temperatures were reported by different authors, but as a general rule, the OH density decreases quickly with increasing distance from the APPJ powered electrode. Some authors reported the OH densities well over 10^{20} m^{-3} what should be sufficient for saturation of a polymer surface with functional groups. The optical emission spectroscopy will not give the absolute density of the OH radicals in the ground state but will rather represent a useful technique for detection and estimation of the density, as explained in appendix A. For the determination of the absolute values, the optical absorption spectroscopy is the method of choice. The absorption spectroscopy comes in various configurations. Because the absorption cross-section for 4 eV photons from a laser on OH radicals is rather small, multiple passes are needed to obtain reliable results. An alternative is the simultaneous absorption of two or three photons of lower energy and observation of the fluorescence. The technique is known as two-atom LIF (TALIF). These techniques operate very well at low pressures, but at atmospheric pressure, one should take into account the final shelf-time of excited states and quenching. In any case, the OES will indicate the presence of the radicals useful for tailoring surface properties of solid materials.

The APPJ sustained in pure noble gases is also a rich source of VUV radiation. The penetration depth of radiation with a wavelength of approximately 100 nm in a polymer is short (of the order of 10 nm); therefore, extensive bond scission occurs in the surface layer upon the interaction of the APPJ with a polymer sample. The dangling bonds interact with oxygen even at ambient conditions, so the surface of any polymer exposed to APPJ sustained in pure noble gas is activated. As already mentioned, the radiation in the VUV range can be probed only with special instruments, but a comparison of the OES spectra with VUV spectra reported by other authors may be useful for interpretation of the observed surface finish. Even a small addition of reactive gases to Ar or He causes significant suppression of VUV radiation from dimers and the appearance of radiation of O or N atoms. The VUV radiation from plasma sustained in a noble gas with approximately 0.1–1% of a reactive gas (water vapor, air, or alike) is governed by reactive species, not Ar or He dimers.

The exact surface finish depends on the fluences of both radiation and reactive plasma species. Several authors cited in this paper have probed both parameters what represents a solid background for interpretation of the observed results. The OES is a qualitative technique, so it will not reveal the fluxes or fluences, but the comparison of measured results with reports of other authors that have used more sophisticated techniques for plasma characterization helps to understand the complex mechanisms involved in surface treatment of materials by atmospheric plasma jets.

Author Contributions: Methodology, R.Z. and G.P.; validation, R.Z. and G.P.; formal analysis, R.Z. and A.V.; investigation, R.Z. and A.V.; data curation, R.Z.; writing—original draft preparation, A.V.; writing—review and editing, A.V. and R.Z; supervision, A.V.; project administration, G.P. All authors have read and agreed to the published version of the manuscript.

Funding: This research was funded by the Slovenian Research Agency, project No. L2-2616 (Selected area functionalization of polymeric components by gaseous plasma) and P2-0082 (Thin film structures and plasma surface engineering).

Institutional Review Board Statement: Not applicable.

Informed Consent Statement: Not applicable.

Data Availability Statement: Not applicable.

Acknowledgments: Not applicable.

Conflicts of Interest: The authors declare no conflict of interest. The funders had no role in the design of the study; in the collection, analyses, or interpretation of data; in the writing of the manuscript, or in the decision to publish the results.

References

1. Vesel, A.; Mozetic, M. New developments in surface functionalization of polymers using controlled plasma treatments. *J. Phys. D Appl. Phys.* **2017**, *50*, 293001. [CrossRef]
2. Reuter, S.; von Woedtke, T.; Weltmann, K.-D. The kINPen—A review on physics and chemistry of the atmospheric pressure plasma jet and its applications. *J. Phys. D Appl. Phys.* **2018**, *51*, 233001. [CrossRef]
3. Vesel, A.; Zaplotnik, R.; Primc, G.; Mozetič, M. Evolution of the surface wettability of PET polymer upon treatment with an atmospheric-pressure plasma jet. *Polymers* **2020**, *12*, 87. [CrossRef] [PubMed]
4. Zaplotnik, R.; Vesel, A. Effect of VUV radiation on surface modification of polystyrene exposed to atmospheric pressure plasma jet. *Polymers* **2020**, *12*, 1136. [CrossRef]
5. Nishime, T.M.C.; Wagner, R.; Kostov, G.K. Study of modified area of polymer samples exposed to a He atmospheric pressure plasma jet using different treatment conditions. *Polymers* **2020**, *12*, 1028. [CrossRef] [PubMed]
6. Darmawati, S.; Rohmani, A.; Nurani, L.H.; Prastiyanto, M.E.; Dewi, S.S.; Salsabila, N.; Wahyuningtyas, E.S.; Murdiya, F.; Sikumbang, I.M.; Rohmah, R.N.; et al. When plasma jet is effective for chronic wound bacteria inactivation, is it also effective for wound healing? *Clin. Plasma Med.* **2019**, *14*, 100085. [CrossRef]
7. Xu, G.M.; Shi, X.M.; Cai, J.F.; Chen, S.L.; Li, P.; Yao, C.W.; Chang, Z.S.; Zhang, G.J. Dual effects of atmospheric pressure plasma jet on skin wound healing of mice. *Wound Repair Regen.* **2015**, *23*, 878–884. [CrossRef]
8. Keidar, M. Plasma for cancer treatment. *Plasma Sources Sci. Technol.* **2015**, *24*, 033001. [CrossRef]
9. Xu, G.; Liu, J.; Yao, C.; Chen, S.; Lin, F.; Li, P.; Shi, X.; Zhang, G.J. Effects of atmospheric pressure plasma jet with floating electrode on murine melanoma and fibroblast cells. *Phys. Plasmas* **2017**, *24*, 083504. [CrossRef]
10. Sarani, A.; Nicula, C.; Gonzales, X.F.; Thiyagarajan, M. Characterization of kilohertz-ignited nonthermal He and He/O_2 plasma pencil for biomedical applications. *IEEE Trans. Plasma Sci.* **2014**, *42*, 3148–3160. [CrossRef]

11. Schneider, S.; Lackmann, J.-W.; Ellerweg, D.; Denis, B.; Narberhaus, F.; Bandow, J.E.; Benedikt, J. The role of VUV radiation in the inactivation of bacteria with an atmospheric pressure plasma jet. *Plasma Process. Polym.* **2012**, *9*, 561–568. [CrossRef]

12. Reyes, P.G.; Gomez, A.; Martinez, H.; Flores, O.; Torres, C.; Vergara, J. Characterization of ethanol plasma glow discharge, decomposition in several species and solid film formation. *IEEE Trans. Plasma Sci.* **2016**, *44*, 2995–3000. [CrossRef]

13. Barletta, F.; Leys, C.; Colombo, V.; Gherardi, M.; Britun, N.; Snyders, R.; Nikiforov, A. Insights into plasma-assisted polymerization at atmospheric pressure by spectroscopic diagnostics. *Plasma Process. Polym.* **2020**, *17*, 1900174. [CrossRef]

14. Penkov, O.V.; Lee, D.-H.; Kim, H.; Kim, D.-E. Frictional behavior of atmospheric plasma jet deposited carbon–ZnO composite coatings. *Compos. Sci. Technol.* **2013**, *77*, 60–66. [CrossRef]

15. Penkov, O.V.; Lee, D.H.; Kim, D.E. Wear resistant coatings for polymeric substrates deposited by atmospheric pressure plasma jet. *Sci. Adv. Mater.* **2015**, *7*, 113–119. [CrossRef]

16. Chang, K.-M.; Huang, S.-H.; Wu, C.-J.; Lin, W.-L.; Chen, W.-C.; Chi, C.-W.; Lin, J.-W.; Chang, C.-C. Transparent conductive indium-doped zinc oxide films prepared by atmospheric pressure plasma jet. *Thin Solid Films* **2011**, *519*, 5114–5117. [CrossRef]

17. Penkov, O.V.; Khadem, M.; Lim, W.-S.; Kim, D.-E. A review of recent applications of atmospheric pressure plasma jets for materials processing. *J. Coat. Technol. Res.* **2015**, *12*, 225–235. [CrossRef]

18. Reuter, S.; Sousa, J.S.; Stancu, G.D.; Hubertus van Helden, J.-P. Review on VUV to MIR absorption spectroscopy of atmospheric pressure plasma jets. *Plasma Sources Sci. Technol.* **2015**, *24*, 054001. [CrossRef]

19. Zaplotnik, R.; Bišćan, M.; Krstulović, N.; Popović, D.; Milošević, S. Cavity ring-down spectroscopy for atmospheric pressure plasma jet analysis. *Plasma Sources Sci. Technol.* **2015**, *24*, 054004. [CrossRef]

20. Mozetič, M.; Ricard, A.; Babič, D.; Poberaj, I.; Levaton, J.; Monna, V.; Cvelbar, U. Comparison of NO titration and fiber optics catalytic probes for determination of neutral oxygen atom concentration in plasmas and postglows. *J. Vac. Sci. Technol. A* **2003**, *21*, 369–374. [CrossRef]

21. Lu, X.P.; Wu, S.Q. On the active species concentrations of atmospheric pressure nonequilibrium plasma jets. *IEEE Trans. Plasma Sci.* **2013**, *41*, 2313–2326. [CrossRef]

22. Winter, J.; Brandenburg, R.; Weltmann, K.D. Atmospheric pressure plasma jets: An overview of devices and new directions. *Plasma Sources Sci. Technol.* **2015**, *24*, 064001. [CrossRef]

23. Lu, X.; Laroussi, M.; Puech, V. On atmospheric-pressure non-equilibrium plasma jets and plasma bullets. *Plasma Sources Sci. Technol.* **2012**, *21*, 034005. [CrossRef]

24. Niemi, K.; Reuter, S.; Graham, L.M.; Waskoenig, J.; Gans, T. Diagnostic based modeling for determining absolute atomic oxygen densities in atmospheric pressure helium-oxygen plasmas. *Appl. Phys. Lett.* **2009**, *95*, 151504. [CrossRef]

25. Zhang, Q.Y.; Shi, D.Q.; Xu, W.; Miao, C.Y.; Ma, C.Y.; Ren, C.S.; Zhang, C.; Yi, Z. Determination of vibrational and rotational temperatures in highly constricted nitrogen plasmas by fitting the second positive system of N_2 molecules. *AIP Adv.* **2015**, *5*, 057158. [CrossRef]

26. Zhao, T.-L.; Xu, Y.; Song, Y.-H.; Li, X.-S.; Liu, J.-L.; Liu, J.-B.; Zhu, A.-M. Determination of vibrational and rotational temperatures in a gliding arc discharge by using overlapped molecular emission spectra. *J. Phys. D Appl. Phys.* **2013**, *46*, 345201. [CrossRef]

27. Srivastava, N.; Wang, C.J. Effects of water addition on OH radical generation and plasma properties in an atmospheric argon microwave plasma jet. *J. Appl. Phys.* **2011**, *110*, 053304. [CrossRef]

28. Sarani, A.; De Geyter, N.; Nikiforov, A.Y.; Morent, R.; Leys, C.; Hubert, J.; Reniers, F. Surface modification of PTFE using an atmospheric pressure plasma jet in argon and argon+CO_2. *Surf. Coat. Technol.* **2012**, *206*, 2226–2232. [CrossRef]

29. Nikiforov, A.Y.; Sarani, A.; Leys, C. The influence of water vapor content on electrical and spectral properties of an atmospheric pressure plasma jet. *Plasma Sources Sci. Technol.* **2011**, *20*, 015014. [CrossRef]

30. Park, H.S.; Kim, S.J.; Joh, H.M.; Chung, T.H.; Bae, S.H.; Leem, S.H. Optical and electrical characterization of an atmospheric pressure microplasma jet with a capillary electrode. *Phys. Plasmas* **2010**, *17*, 033502. [CrossRef]

31. Sarani, A.; Nikiforov, A.Y.; Leys, C. Atmospheric pressure plasma jet in Ar and Ar/H_2O mixtures: Optical emission spectroscopy and temperature measurements. *Phys. Plasmas* **2010**, *17*, 063504. [CrossRef]

32. Yanguas-Gil, A.; Focke, K.; Benedikt, J.; von Keudell, A. Optical and electrical characterization of an atmospheric pressure microplasma jet for Ar/CH_4 and Ar/C_2H_2 mixtures. *J. Appl. Phys.* **2007**, *101*, 103307. [CrossRef]

33. Weltmann, K.-D.; Kindel, E.; Brandenburg, R.; Meyer, C.; Bussiahn, R.; Wilke, C.; von Woedtke, T. Atmospheric pressure plasma jet for medical therapy: Plasma parameters and risk estimation. *Contrib. Plasma Phys.* **2009**, *49*, 631–640. [CrossRef]

34. Ilik, E.; Durmus, C.; Akan, T. Adding water droplets into atmospheric pressure plasma jet of helium. *IEEE Trans. Plasma Sci.* **2019**, *47*, 5000–5005. [CrossRef]

35. Seo, K.S.; Cha, J.H.; Han, M.K.; Ha, C.S.; Kim, D.H.; Lee, H.J.; Lee, H.J. Surface treatment of glass and poly(dimethylsiloxane) using atmospheric-pressure plasma jet and analysis of discharge characteristics. *Jap. J. Appl. Phys.* **2015**, *54*. [CrossRef]

36. Roy, N.C.; Talukder, M.R.; Chowdhury, A.N. OH and O radicals production in atmospheric pressure air/Ar/H_2O gliding arc discharge plasma jet. *Plasma Sci. Technol.* **2017**, *19*. [CrossRef]

37. Sahu, B.B.; Jin, S.B.; Han, J.G. Development and characterization of a multi-electrode cold atmospheric pressure DBD plasma jet aiming plasma application. *J. Anal. At. Spectrom.* **2017**, *32*, 782–795. [CrossRef]

38. Thiyagarajan, M.; Sarani, A.; Nicula, C. Optical emission spectroscopic diagnostics of a non-thermal atmospheric pressure helium-oxygen plasma jet for biomedical applications. *J. Appl. Phys.* **2013**, *113*, 233302. [CrossRef]

39. Chauvet, L.; Therese, L.; Caillier, B.; Guillot, P. Characterization of an asymmetric DBD plasma jet source at atmospheric pressure. *J. Anal. At. Spectrom.* **2014**, *29*, 2050–2057. [CrossRef]

40. Gerling, T.; Nastuta, A.V.; Bussiahn, R.; Kindel, E.; Weltmann, K.D. Back and forth directed plasma bullets in a helium atmospheric pressure needle-to-plane discharge with oxygen admixtures. *Plasma Sources Sci. Technol.* **2012**, *21*, 034012. [CrossRef]

41. Pipa, A.V.; Reuter, S.; Foest, R.; Weltmann, K.D. Controlling the NO production of an atmospheric pressure plasma jet. *J. Phys. D Appl. Phys.* **2012**, *45*, 085201. [CrossRef]

42. Jia, H.; Fujiwara, H.; Kondo, M.; Kuraseko, H. Optical emission spectroscopy of atmospheric pressure microwave plasmas. *J. Appl. Phys.* **2008**, *104*, 054908. [CrossRef]

43. Sainct, F.P.; Durocher-Jean, A.; Gangwar, R.K.; Mendoza Gonzalez, N.Y.; Coulombe, S.; Stafford, L. Spatially-resolved spectroscopic diagnostics of a miniature RF atmospheric pressure plasma jet in argon open to ambient air. *Plasma* **2020**, *3*, 5. [CrossRef]

44. Yuan, Q.; Ren, P.; Zhou, Y.; Yin, G.; Dong, C. OES diagnostic of radicals in 33 MHz radio-frequency Ar/C_2H_5OH atmospheric pressure plasma jet. *Plasma Sci. Technol.* **2018**, *21*, 025402. [CrossRef]

45. Golda, J.; Biskup, B.; Layes, V.; Winzer, T.; Benedikt, J. Vacuum ultraviolet spectroscopy of cold atmospheric pressure plasma jets. *Plasma Process. Polym.* **2020**, *17*. [CrossRef]

46. Herron, J.T.; Green, D.S. Chemical kinetics database and predictive schemes for nonthermal humid air plasma chemistry. Part II. Neutral species reactions. *Plasma Chem. Plasma Process.* **2001**, *21*, 459–481. [CrossRef]

47. Kim, S.J.; Chung, T.H.; Bae, S.H.; Leem, S.H. Characterization of atmospheric pressure microplasma jet source and its application to bacterial inactivation. *Plasma Process. Polym.* **2009**, *6*, 676–685. [CrossRef]

48. Resnik, M. Plasma-Induced Modifications of Polypropylene Tubes for Biomedical Applications. Ph.D. Thesis, Jozef Stefan International Postgraduate School, Ljubljana, Slovenija, 2018.

49. Gott, R.P.; Xu, K.G. OH production and jet length of an atmospheric-pressure plasma jet for soft and biomaterial treatment. *IEEE Trans. Plasma Sci.* **2019**, *47*, 4988–4999. [CrossRef]

50. Cordaro, L.; De Masi, G.; Fassina, A.; Mancini, D.; Cavazzana, R.; Desideri, D.; Sonato, P.; Zuin, M.; Zaniol, B.; Martines, E. On the electrical and optical features of the plasma coagulation controller low temperature atmospheric plasma jet. *Plasma* **2019**, *2*, 12. [CrossRef]

51. Cordaro, L.; De Masi, G.; Fassina, A.; Gareri, C.; Pimazzoni, A.; Desideri, D.; Indolfi, C.; Martines, E. The role of thermal effects in plasma medical applications: Biological and calorimetric analysis. *Appl. Sci.* **2019**, *9*, 5560. [CrossRef]

52. Schulz-von der Gathen, V.; Schaper, L.; Knake, N.; Reuter, S.; Niemi, K.; Gans, T.; Winter, J. Spatially resolved diagnostics on a microscale atmospheric pressure plasma jet. *J. Phys. D Appl. Phys.* **2008**, *41*, 194004. [CrossRef]

53. Yang, Y.; Zhang, Y.Z.; Liao, Z.L.; Pei, X.K.; Wu, S.Q. OH radicals distribution and discharge dynamics of an atmospheric pressure plasma jet above water surface. *IEEE Trans. Radiat. Plasma Med. Sci.* **2018**, *2*, 223–228. [CrossRef]

54. Voráč, J.; Dvořák, P.; Procházka, V.; Ehlbeck, J.; Reuter, S. Measurement of hydroxyl radical (OH) concentration in an argon RF plasma jet by laser-induced fluorescence. *Plasma Sources Sci. Technol.* **2013**, *22*, 025016. [CrossRef]

55. Yonemori, S.; Nakagawa, Y.; Ono, R.; Oda, T. Measurement of OH density and air–helium mixture ratio in an atmospheric-pressure helium plasma jet. *J. Phys. D Appl. Phys.* **2012**, *45*, 225202. [CrossRef]

56. Yonemori, S.; Ono, R. Flux of OH and O radicals onto a surface by an atmospheric-pressure helium plasma jet measured by laser-induced fluorescence. *J. Phys. D Appl. Phys.* **2014**, *47*, 125401. [CrossRef]

57. Schröter, S.; Wijaikhum, A.; Gibson, A.R.; West, A.; Davies, H.L.; Minesi, N.; Dedrick, J.; Wagenaars, E.; de Oliveira, N.; Nahon, L.; et al. Chemical kinetics in an atmospheric pressure helium plasma containing humidity. *Phys. Chem. Chem. Phys.* **2018**, *20*, 24263–24286. [CrossRef]

58. Van Gessel, A.F.H.; van Grootel, S.C.; Bruggeman, P.J. Atomic oxygen TALIF measurements in an atmospheric-pressure microwave plasma jet within situxenon calibration. *Plasma Sources Sci. Technol.* **2013**, *22*, 055010. [CrossRef]

59. Reuter, S.; Niemi, K.; Schulz-von der Gathen, V.; Döbele, H.F. Generation of atomic oxygen in the effluent of an atmospheric pressure plasma jet. *Plasma Sources Sci. Technol.* **2008**, *18*, 015006. [CrossRef]

60. Wang, C.; Srivastava, N. OH number densities and plasma jet behavior in atmospheric microwave plasma jets operating with different plasma gases (Ar, Ar/N_2, and Ar/O_2). *Eur. Phys. J. D* **2010**, *60*, 465–477. [CrossRef]

61. Srivastava, N.; Wang, C.J. Determination of OH radicals in an atmospheric pressure helium microwave plasma jet. *IEEE Trans. Plasma Sci.* **2011**, *39*, 918–924. [CrossRef]

62. Fuh, C.A.; Clark, S.M.; Wu, W.; Wang, C. Electronic ground state OH(X) radical in a low-temperature atmospheric pressure plasma jet. *J. Appl. Phys.* **2016**, *120*, 163303. [CrossRef]

63. Hibert, C.; Gaurand, I.; Motret, O.; Pouvesle, J.M. OH(X) measurements by resonant absorption spectroscopy in a pulsed dielectric barrier discharge. *J. Appl. Phys.* **1999**, *85*, 7070–7075. [CrossRef]

64. Cheng, K.-Y.; Lin, Z.-H.; Cheng, Y.-P.; Chiu, H.-Y.; Yeh, N.-L.; Wu, T.-K.; Wu, J.-S. Wound healing in streptozotocin-induced diabetic rats using atmospheric-pressure argon plasma jet. *Sci. Rep.* **2018**, *8*, 12214. [CrossRef]

65. Jacofsky, M.; Lubahn, C.; McDonnell, C.; Seepersad, Y.; Fridman, G.; Fridman, A.; Dobrynin, D. Spatially resolved optical emission spectroscopy of a helium plasma jet and its effects on wound healing rate in a diabetic murine model. *Plasma Med.* **2014**, *4*, 177–191. [CrossRef]

66. Lin, Z.H.; Tschang, C.Y.T.; Liao, K.C.; Su, C.F.; Wu, J.S.; Ho, M.T. Ar/O_2 argon-based round atmospheric-pressure plasma jet on sterilizing bacteria and endospores. *IEEE Trans. Plasma Sci.* **2016**, *44*, 3140–3147. [CrossRef]

67. Narimisa, M.; Krčma, F.; Onyshchenko, Y.; Kozáková, Z.; Morent, R.; De Geyter, N. Atmospheric pressure microwave plasma jet for organic thin film deposition. *Polymers* **2020**, *12*, 354. [CrossRef] [PubMed]

68. Zhao, P.; Zheng, W.; Meng, Y.D.; Nagatsu, M. Characteristics of high-purity Cu thin films deposited on polyimide by radio-frequency Ar/H_2 atmospheric-pressure plasma jet. *J. Appl. Phys.* **2013**, *113*, 123301. [CrossRef]

69. Kehrer, M.; Duchoslav, J.; Hinterreiter, A.; Mehic, A.; Stehrer, T.; Stifter, D. Surface functionalization of polypropylene using a cold atmospheric pressure plasma jet with gas water mixtures. *Surf. Coat. Technol.* **2020**, *384*, 125170. [CrossRef]

70. Weiss, M.; Barz, J.; Ackermann, M.; Utz, R.; Ghoul, A.; Weltmann, K.D.; Stope, M.B.; Wallwiener, D.; Schenke-Layland, K.; Oehr, C.; et al. Dose-Dependent Tissue-Level Characterization of a Medical Atmospheric Pressure Argon Plasma Jet. *ACS Appl. Mater. Interfaces* **2019**, *11*, 19841–19853. [CrossRef]

Article

Atmospheric Pressure Plasma Irradiation Facilitates Transdermal Permeability of Aniline Blue on Porcine Skin and the Cellular Permeability of Keratinocytes with the Production of Nitric Oxide

Sunmi Lee [1,†], Jongbong Choi [2,†], Junghyun Kim [3], Yongwoo Jang [2,*] and Tae Ho Lim [1,3,*]

1   Department of Emergency Medicine, Hanyang University Hospital, Seoul 04763, Korea; leesunmi1035@gmail.com
2   Department of Biomedical Engineering, Hanyang University, Seoul 04763, Korea; cjbonghyu@gmail.com
3   Research Center, CODESTERI Inc., Seoul 04763, Korea; optimist.jhkim@gmail.com
*   Correspondence: ywjang@hanyang.ac.kr (Y.J.); erthim@hanyang.ac.kr (T.H.L.)
†   These authors contributed equally to this work.

Abstract: The transdermal delivery system of nutrients, cosmetics, and drugs is particularly attractive for painless, noninvasive delivery and sustainable release. Recently, atmospheric pressure plasma techniques have been of great interest to improve the drug absorption rate in transdermal delivery. Currently, plasma-mediated changes in the lipid composition of the stratum corneum are considered a possible mechanism to increase transdermal permeability. Nevertheless, its molecular and cellular mechanisms in transdermal delivery have been largely confined and still veiled. Herein, we present the effects of cold plasma on transdermal transmission on porcine skin and the cellular permeability of keratinocytes and further demonstrate the production of nitric oxide from keratinocytes. Consequently, argon plasma irradiation for 60 s resulted in 2.5-fold higher transdermal absorption of aniline blue dye on porcine skin compared to the nontreated control. In addition, the plasma-treated keratinocytes showed an increased transmission of high-molecular-weight molecules (70 and 150 kDa) with the production of nitric oxide. Therefore, these findings suggest a promoting effect of low-temperature plasma on transdermal absorption, even for high-molecular-weight molecules. Moreover, plasma-induced nitric oxide from keratinocytes is likely to regulate transdermal permeability in the epidermal layer.

Keywords: atmospheric pressure plasma; transdermal permeability; transdermal delivery; nitric oxide; plasma medicine

Citation: Lee, S.; Choi, J.; Kim, J.; Jang, Y.; Lim, T.H. Atmospheric Pressure Plasma Irradiation Facilitates Transdermal Permeability of Aniline Blue on Porcine Skin and the Cellular Permeability of Keratinocytes with the Production of Nitric Oxide. *Appl. Sci.* **2021**, *11*, 2390. https://doi.org/10.3390/app11052390

Academic Editors: Matteo Zuin and Andrei Vasile Nastuta

Received: 13 January 2021
Accepted: 4 March 2021
Published: 8 March 2021

Publisher's Note: MDPI stays neutral with regard to jurisdictional claims in published maps and institutional affiliations.

1. Introduction

The percutaneous transmission of nutrients, cosmetics, and drugs is a painless and noninvasive delivery method wherein they are absorbed by crossing through the skin layers to the systemic circulation [1,2]. In particular, transdermal drug delivery systems using various types of skin patches [3,4] and other transdermal methods, such as ion penetration [5,6] and ultrasound [7,8], have improved absorption rates for local and/or systemic delivery of therapeutic agents through the skin layers. In addition, the atmospheric pressure plasma technique is also of great interest to improve the drug absorption rate in transdermal delivery [9].

In the past couple of decades, cold atmospheric pressure plasma opened up a new frontier in medicine and healthcare as a term of plasma medicine [10–13]. Remarkably, numerous studies have demonstrated biomedical applications for alleviating dermatological problems, such as skin wounds and infections, that are relatively easy to process with plasma [14–17]. In addition, some studies have focused on the effects of cold plasma on the skin barrier, not as a disease treatment but as a drug-delivery technology. For instance, it was found that atmospheric plasma jet irradiation for 3 min promotes transdermal delivery of hydrophilic rhodamine B dye on rat skin due to a reversible small pore induced by the collision of charged particles [18]. Moreover, further investigations showed a practical transdermal delivery of galantamine hydrobromide (368 Da) and lidocaine (234 Da) used for Alzheimer's disease treatment and local anesthetic, respectively [19,20]. Interestingly, plasma irradiation through atmospheric-pressure plasma jets or dielectric barrier discharge exhibited an increased transdermal delivery of high-molecular-weight lipophilic cyclosporine A (1203 Da), whereas there was no absorption without plasma treatment [21]. Despite evident findings of improved transdermal delivery by plasma, its molecular and cellular mechanism(s) are still elusive.

The stratum corneum is the outermost layer of the epidermis, which is an important component of the skin barrier [22,23]. As the stratum corneum primarily prevents the transdermal absorption of drugs, several studies have suggested increased permeability due to changes in lipid composition in the plasma-treated stratum corneum layer [18,24]. In addition, one remarkable study most recently showed transcriptomic changes in cell junction proteins, cytoskeletal proteins, and extracellular matrix proteins in the stratum corneum of plasma-treated skin [25]. However, it is still unclear how these transcriptomic changes can be induced in the epidermis layer.

The stratum corneum layer is composed of keratinized and flattened corneocytes that are differentiated dead keratinocytes. In the lower stratum corneum, living keratinocytes are connected through cell–cell junctions in the epidermal layer called the stratum granulosum. In the present study, we attempted to investigate the transdermal permeability of our plasma device on porcine skin and further determined the effect of drug delivery and the production of nitric oxide in cultured keratinocytes, which is an important regulator of skin barrier function and includes junctional proteins, cytoskeletal proteins, and extracellular matrix proteins.

2. Material and Method

2.1. Plasma Device

For the present study, we installed an atmospheric-pressure plasma jet device. The outer and inner diameters of the nozzle tube of the device were 3 mm and 2 mm, respectively (Figure 1A). As shown in Figure 1B, dielectric barrier discharge (DBD) jet consists of a high voltage applied hollow electrode, an alumina dielectric with 0.5 mm thickness, and an external ground electrode with a width of 5 mm. Flow control units (RK1600R, Kofloc, Japan) were independently connected to control argon and helium gas, and the flow rate was maintained at 2 standard liters per minute (slm). The DC input source inverted to AC power by push-pull type inverter. The applied AC voltage and discharge current were measured using an oscilloscope (MDO3000, Tektronix, Beaverton, OR, USA), a high voltage probe (P6015A, Tektronix, USA), and a current probe (P6022, Tektronix, USA) to compare the electrical properties of argon and helium plasma jets. The emission spectra of argon and helium plasma jets were measured by varying the wavelength range from 200 to 900 nm through an OES spectral analyzer (HR4000CG, Ocean Optics, Orlando, FL, USA). The optic fiber was connected to the OES spectral analyzer and fixed at a distance of 5 mm from the plasma device nozzle.

2.2. Materials

Aniline blue cream was prepared from cetaphil, glycerin, petrolatum, dicaprylyl ether, dimethicone, glyceryl stearate, cetyl alcohol, seed oil, PEG-30 stearate, tocopheryl acetate, dimethiconol, acrylates/C10-30 alkyl acrylate crosspolymer, benzyl alcohol, phenoxyerhanol, prunus amygdalus duicis oil, prophylene glycol, disodium EDTA, carnomer, sodium hydroxide, aniline blue (1 mg/mL), and purified water to create a cream with an aniline blue concentration of 737.73 µg/g.

A

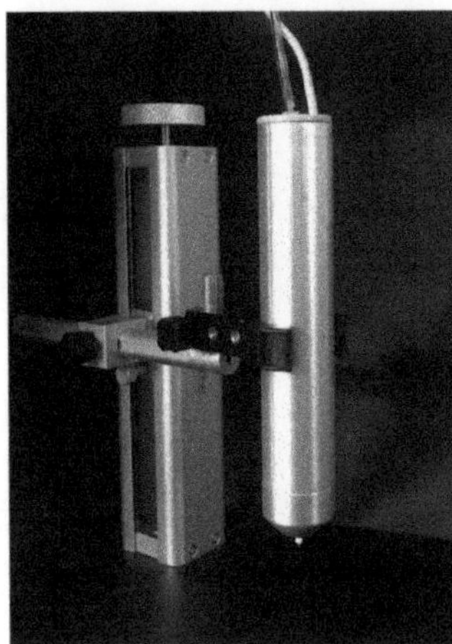

B

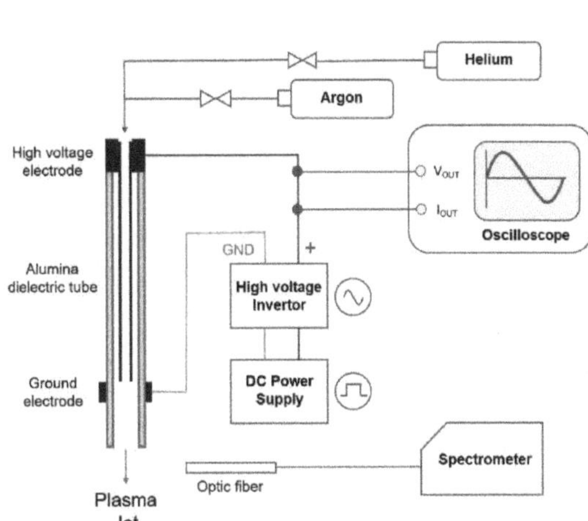

Figure 1. Schematic diagram of the atmospheric pressure plasma jet device. (**A**) Atmospheric pressure plasma jet device used in the present study. (**B**) Schematic diagram of an argon and helium plasma treatment system.

2.3. Permeability of Porcine Skin

Porcine skins were prepared in a size of 2.5 × 2.0 cm. They were divided into four regions of 0.7 × 0.7 cm: no plasma treatment and high-intensity plasma treatment for 10, 30, and 60 s. After the plasma treatment, 1 g of aniline blue cream was applied to the quartered area of porcine skin, and the reaming cream was washed with running water after 5 min. Finally, the aniline blue remaining on the porcine skin after 10 h was quantified by the ImageJ program. The captured picture was executed in the imageJ program. It converted the completed the picture into 8 bit type. To designate the part to be quantified, we used rectangular selection tool of the imageJ program. The rectangular size remained constant for all measurements at each set. The measured values in each set were compared by dividing them into the values of the untreated site (0 s). Finally, statistical analysis was compared based on the experimental results of a total of three sets. Additionally, temperature of porcine skin surface measured using thermal imager (DT 9885, CEM, Shenzhen, China), IR resolution is 384 × 288 pixels and temperature range is from 20 °C to 150 °C. IR image acquisition that we measured in the experiment is 1 spot temperature mode with no zoom.

2.4. Cell Culture

HaCaT cells were cultured in Dulbecco's modified Eagle's medium (DMEM) supplemented with 10% fetal bovine serum (FBS) at 37 °C and 5% CO_2 under previously reported conditions [26].

2.5. Dextran Fluorescein

HaCaT cells (2×10^5) were cultured for 24 h and treated with our plasma device for 30 s in phenol red-free DMEM with 10% FBS. After washing with phosphate-buffered saline (PBS), the cells were further incubated with DMEM containing fluorescein-dextran of 70 or 150 kDa at 37 °C for 1 h. Consequently, the fluorescent dextran in the cells was observed by fluorescence microscopy.

2.6. Measurement of Intracellular Nitric Oxide Level

Intracellular nitric oxide generation was assessed using a QuantiChromTM nitric oxide assay kit (Bioassay Systems, USA). HaCaT cells (2×10^5) were plated on 35-mm dishes and cultured for 24 h. The cells were deprived of serum for 1 h and treated with our plasma device at multiple time points in phenol red-free medium. After plasma irradiation, cell samples were homogenized in PBS. After centrifugation at 12,000 rpm at 4 °C, the supernatant was used for the nitric oxide assay, and the quantitative colorimetric changes were determined at 540 nm.

2.7. Statistical Analysis

Data analysis was performed with Student's *t*-test for comparisons between two groups or ANOVA with Tukey's hoc test for multiple groups (SPSS 12.0 K for Windows, SPSS, Chicago, IL, USA) to determine the statistical significance (p value). Statistical significance was considered for p values of <0.05 (*), <0.01 (**), or 0.001 (***).

3. Results and Discussion

3.1. Electrical Properties and Emission Spectra of Argon and Helium Plasma

To analyze the electrical properties, we first measured DC input voltage and DC discharged current, the breakdown voltage of argon and helium gas were 6 V and 4 V, respectively, and discharge currents as breakdown voltage were 0.29 A and 0.25 A at a 2 slm flow rate, respectively (Figure 2A). The DC input voltage conditions for transdermal permeability experiment that length of plasma plume was the longest without plasma arcing and burn in the porcine skin were 18 V and 12 V in argon and helium gas, respectively. As shown in Figure 2B, we measured two cycle waveforms of the inverted AC applied voltage and discharge current of the argon and helium plasma jet at DC input voltage 18 V and 12 V, respectively, and their frequency were 47.28 kHz and 46.67 kHz, respectively. The RMS (Root Mean Square) voltages of argon and helium plasma jets were 2.54 kV and 1.73 kV, respectively, and their RMS discharge currents were 11.01 mA and 6.21 mA at a 2 slm flow rate, respectively. Additionally, the calculated AC power of the argon and helium plasma jet were 4.85 W and 1.86 W, respectively, while DC power was 5.22 W and 3 W, respectively. As previously reported, this difference seems to be caused by argon being relatively easy to ionize because the ionization threshold energies of argon and helium are 15.76 eV and 24.59 eV, respectively [27], even with the increased voltage, the generation of radicals is much higher in argon than in helium plasma [28]. We examined various active species that are observed in argon and helium plasma. The emission spectra of argon and helium plasma indicate that OH (309 nm) and N_2 (310 to 370 nm) were observed in both argon and helium plasma, but NO (283 nm) and N_2^+ first negative bands (391 nm) were observed only in helium plasma (Figure 3). As the ionization threshold energy of N_2 is 14.53 eV, N_2 is likely to be ionized by helium plasma with a metastable energy level of 19.82 eV compared to that of argon plasma (11.5 eV) [29].

3.2. Argon and Helium Plasma Jet Irradiation Increases the Transdermal Permeability of Porcine Skin

The plasma-induced transdermal absorption rate was determined on pig skin, which is known to be most similar to human skin [30]. To optimize the most effective absorption rate condition, we first investigated the transdermal permeability according to the exposure duration of the atmospheric-pressure helium or argon plasma. As shown in Figure 4, the helium or argon plasma was irradiated for 0, 10, 30, and 60 s in the quartered regions of the porcine skin at a distance of 4 mm from the skin surface. Subsequently, a dye reagent (aniline blue cream) was equivalently applied to the quartered porcine skin. After 10 h, the transdermal permeability was analyzed by colorimetric changes depending on the absorbed amount of aniline blue dye. As shown in Figure 4B, argon plasma remarkably facilitated the transdermal absorption of aniline blue on porcine skin in an exposure time-dependent manner. This consequently resulted in approximately 2.5-fold higher permeation after argon plasma irradiation for 60 s. In addition, the time-dependent irradiation of helium plasma also showed an increased absorption ratio of aniline blue dye and finally saturated from the exposure time of 30 s (Figure 4C). When the plasma irradiation distance moved to 8 mm, there was no significant influence on the absorption rate after irradiation of the argon and helium plasma (Figure 5). Taken together, argon and helium plasma exhibited the most transdermal permeability of aniline blue dye on porcine skin when irradiated for approximately 30 s at a plasma irradiation distance of 4 mm. In addition, argon plasma-treated porcine skin showed higher transdermal permeability than helium plasma. To further investigate the change in temperature by plasma irradiation, we measured the skin temperature after plasma exposure for 10, 30, 60 s at distance of 4 mm. As shown in Figure 6, the skin temperature after exposure of argon plasma for 0, 10, 30, and 60 s was 21.4, 22, 4, 25.9 and 28.7, respectively, and helium plasma was 21.4, 22, 4, 25.9 and 28.7, respectively. The argon and helium plasma treatment for 60 s increased the surface temperature up to approximately 7 °C and 4 °C, respectively.

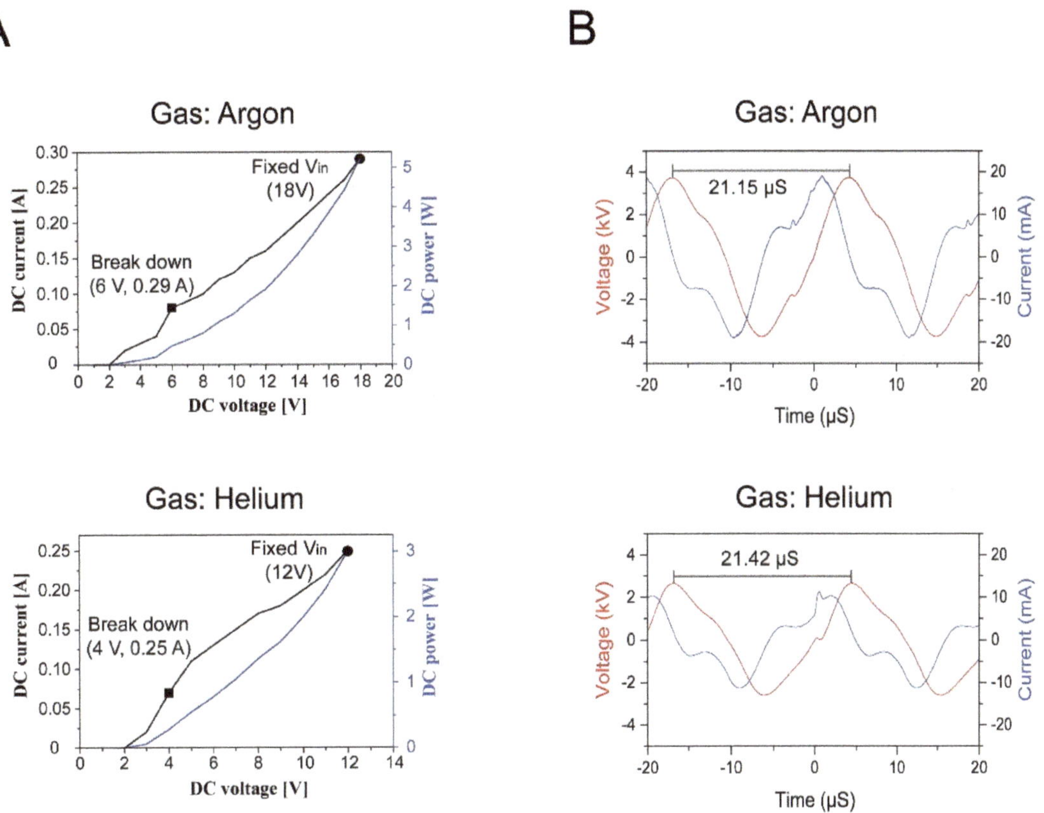

Figure 2. IV curve and electrical waveform of argon and helium plasma jets. The gas flow rates of argon and helium were fixed at 2 slm. (**A**) DC input voltage and DC Discharged current for argon and helium plasma. The breakdown voltage were 6 V for argon and 4 V for helium, The DC input voltage for transdermal permeability experiment were fixed 18 V and 12 V in argon and helium gas, respectively. (**B**) Waveforms of the inverted AC applied voltage and discharge current for the argon and helium plasma at DC input voltage 18 V and 12 V, respectively. The applied RMS voltages were 2.54 kV for argon and 1.73 kV for helium, and their RMS discharge currents were 11.01 mA and 6.21 mA, respectively.

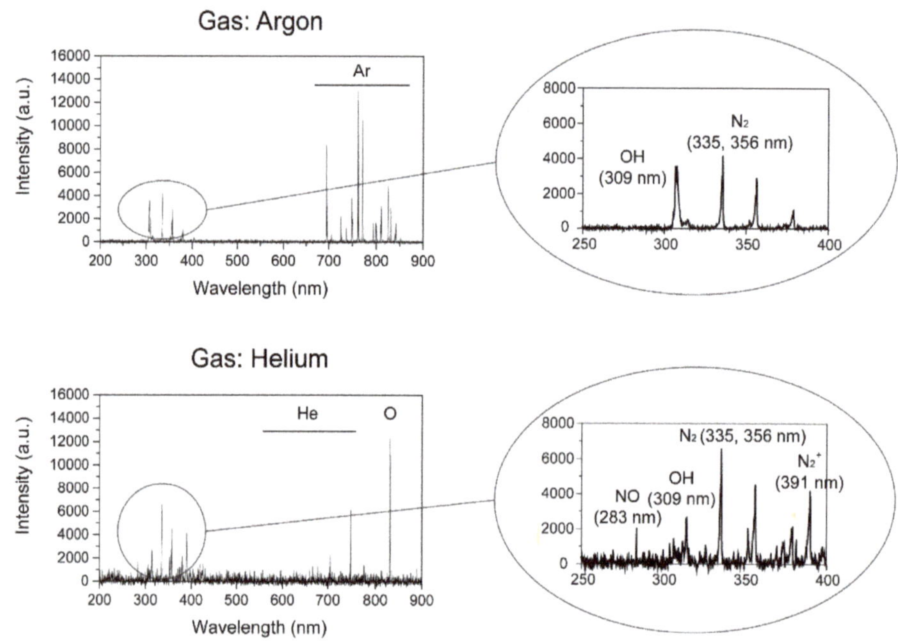

Figure 3. Optical emission spectroscopy of argon and helium plasma jets with a wavelength range from 200–900 nm.

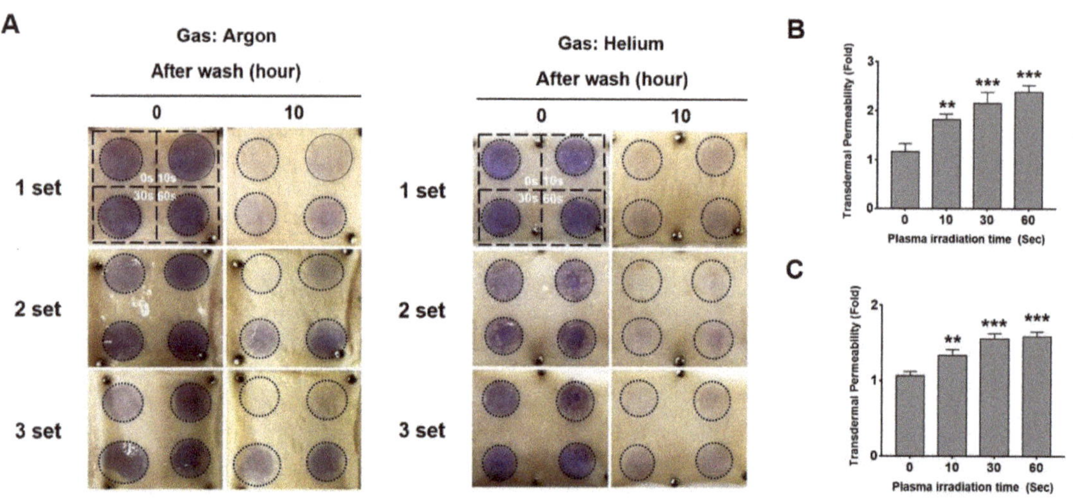

Figure 4. Transdermal absorption of aniline blue dye on porcine skin by argon and helium plasma irradiation at a distance of 4 mm (**A**) The quartered area of the rectangular area in porcine skin was treated with argon or helium plasma for 0, 10, 30, or 60 s at a distance of 4 mm between the plasma jet and the surface of the skin and thereafter stained with aniline blue dye. The transdermal permeability of aniline blue dye on porcine skin was quantified and normalized to that of the nontreated control. (**B**,**C**) Data are presented as means ± S.D. ** $p < 0.01$, *** $p < 0.001$ compared to the nontreated control (one-way ANOVA, Tukey's post hoc test) with argon (**B**) or helium (**C**). n = 3 independent samples per group.

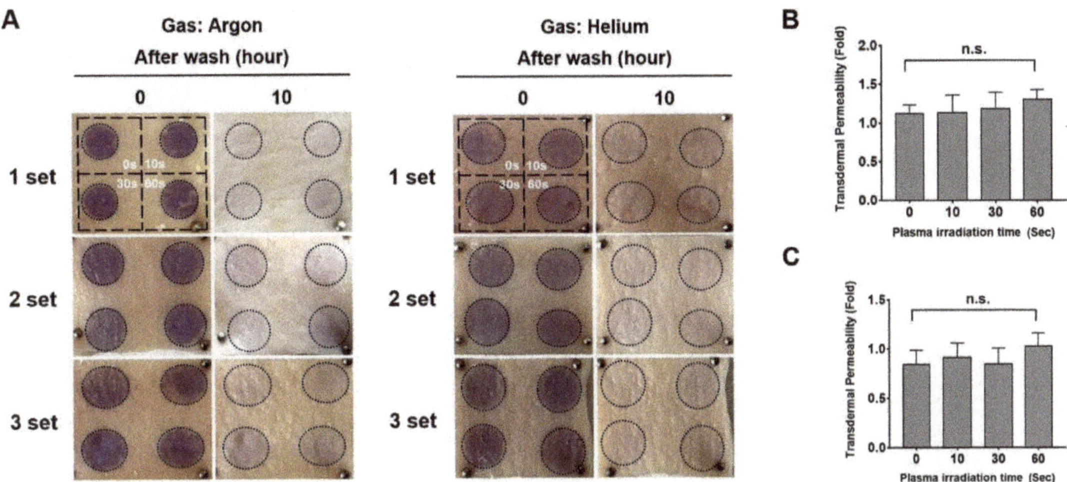

Figure 5. Transdermal absorption of aniline blue dye on porcine skin by argon and helium plasma irradiation at a distance of 8 mm (**A**) The quartered area of the rectangular area in porcine skin was treated with argon or helium plasma for 0, 10, 30, or 60 s at a distance of 8 mm between the plasma jet and the surface of the skin and thereafter stained with aniline blue dye. The transdermal permeability of aniline blue dye on porcine skin was quantified and normalized to that of the nontreated control. (**B,C**) Data are presented as means ± S.D. n.s indicates a nonsignificant difference compared to the nontreated control (one-way ANOVA, Tukey's post hoc test) with argon (**B**) or helium (**C**). n = 3 independent samples per group.

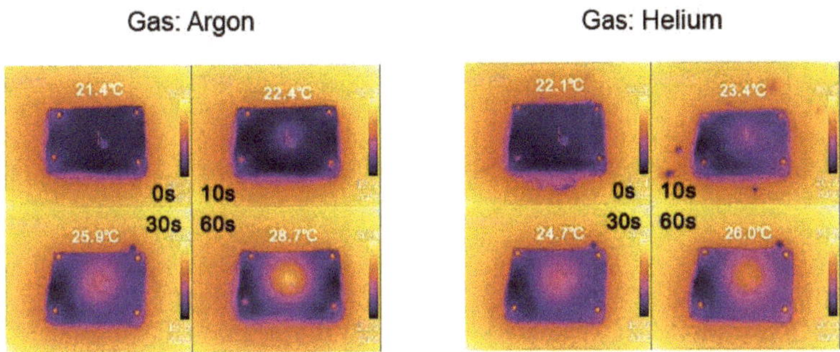

Figure 6. Temperature on porcine skin surface by argon and helium plasma irradiation for 0, 10, 30 or 60 s at a distance of 4 mm between the plasma jet and the surface of the skin.

3.3. Argon Plasma Irradiation Facilitates the Permeability of 70 kDa and 150 kDa Molecules in Keratinocytes

To investigate how to increase transdermal permeability by plasma, we further examined in vitro permeability in keratinocytes that were connected through cell–cell junctions in the epidermal layer. As high-molecular-weight molecules over 1 kDa have difficulty penetrating into the skin, we attempted to compare the penetration of 70 and 150 kDa molecules in keratinocytes with/without argon plasma irradiation. For this purpose, cultured HaCaT cells were first treated with our plasma device for 30 s or not. Subsequently, they were incubated in the medium containing 70 or 150 kDa dextran conjugated with fluorescein for 1 h. Thereafter, the cells were fixed with 4% paraformaldehyde, and the fluorescence intensity inside the cells was observed using a fluorescence microscope. As shown in Figure 7A, fluorescent signals of 70 and 150 kDa dextran were clearly observed in argon plasma-treated keratinocytes, whereas these fluorescent signals were not observed in nontreated cells. Indeed, statistical analysis revealed that argon plasma irradiation significantly increased the fluorescence intensity of 70 (22.0 ± 4.2) kDa (Figure 7B) and 150 (17.6 ± 3.6) kDa (Figure 7C) dextran in the cells compared to nontreated cells. Therefore, these results suggest that argon plasma irradiation facilitates the penetration of over 1 kDa molecules that generally have difficulty permeating the skin due to their high molecular weight.

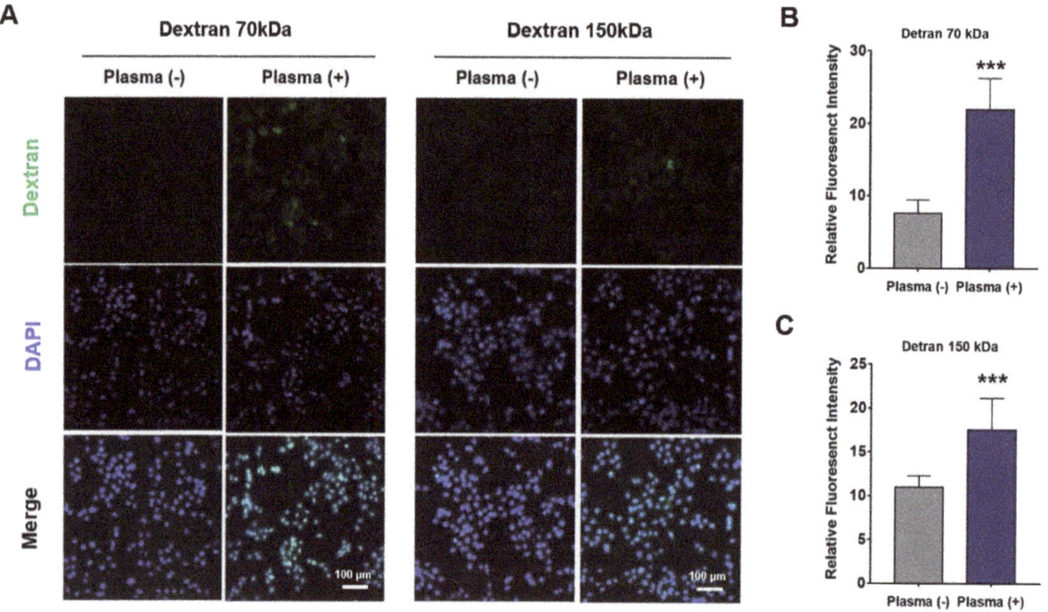

Figure 7. Cellular permeability of 70 kDa and 150 kDa dextran by argon plasma irradiation in keratinocytes. (**A**) Fluorescence images, HaCaT cells were treated with argon plasma for 30 s and then subjected to fluorescein-dextran 70 or 150 kDa. (**B,C**) Statistical results of relative fluorescence in the cells treated with 70 kDa (**B**) and 150 kDa (**C**) dextran in the presence and absence of argon plasma irradiation. Scale bar indicates 100 µm. The fluorescent intensity was averaged from 40 cells. Data are presented as means ± S.D. *** $p < 0.001$ compared to the nontreated control (unpaired two-tailed *t*-test).

3.4. Argon Plasma Irradiation Induces the Production of Nitric Oxide in Keratinocytes

Generally, nitric oxide is a free radical with an unpaired electron, and it is well known to regulate various epidermal functions, including epidermal proliferation, differentiation, wound healing, and barrier permeability in the skin [31–33]. In fact, nitric oxide regulates junctional, cytoskeletal, and extracellular matrix proteins. Hence, we believed that nitric oxide was involved in plasma-mediated permeability as a transdermal absorption donor. Therefore, we further investigated whether argon plasma irradiation induces the production of nitric oxide in keratinocytes. Argon plasma was applied to HaCaT cells for 10, 30, and 60 s, followed by incubation for 10, 30, 60, and 120 min. As shown in Figure 8, the exposure of plasma on the cells for 10, 30, and 60 s similarly caused the production of nitric oxide up to approximately 80 µM. Nitric oxide was significantly induced from 60 min after irradiation. These findings suggest that plasma-induced nitric oxides from keratinocytes are able to regulate the junctions between cells and the extracellular matrix in the epidermal layer, which would make the epidermal barrier permeable.

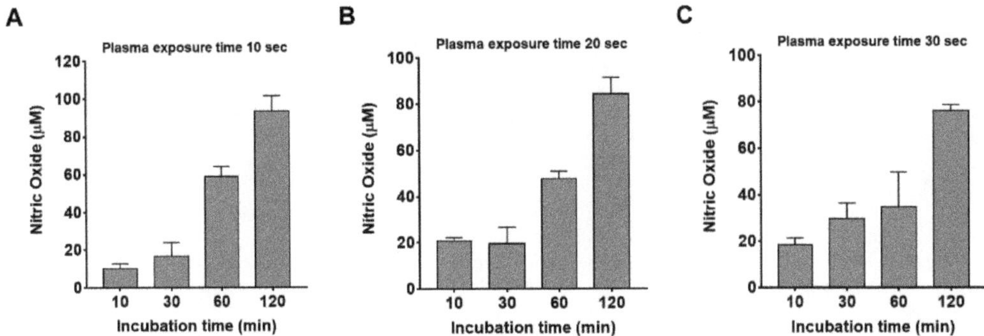

Figure 8. Plasma-induced nitric oxide in the keratinocytes. (**A–C**) Argon plasma was applied to keratinocytes for 10 (**A**), 30 (**B**), and 60 (**C**) seconds, followed by incubation for 10, 30, 60, and 120 min. Intracellular nitric oxide levels were measured by quantitative colorimetry at 540 nm from keratinocytes. Data are presented as means $\pm$ S.D. n = 3 independent samples per group.

4. Conclusions

The epidermal transmission of nutrients, cosmetics, and drugs is a painless and non-invasive delivery method that has the advantages of providing continuous and long-term release as well as improving patient adaptability. Among current percutaneous transmission methods and technologies, cold plasma can easily irradiate the desired portion of the skin without epidermal damage. Recently, several studies have provided the effect of cold plasma on transdermal permeability. In the present study, an atmospheric-pressure plasma jet device was fabricated to test the transdermal permeability, and its device effectively showed improved permeability of aniline blue dye on porcine skin. Furthermore, we showed a significant increase in the transmission of high-molecular-weight molecules (70 and 150 kDa) with the production of nitric oxide in plasma-treated keratinocytes. Further studies are warranted to clarify the molecular and cellular mechanism(s) underlying plasma-induced nitric oxide and transdermal absorption in keratinocytes.

Author Contributions: Conceptualization, S.L., J.C., Y.J. and T.H.L.; experiments and formal analysis, S.L., J.C. and J.K.; writing and visualization, S.L., J.C., Y.J. and T.H.L. All authors have read and agreed to the published version of the manuscript.

Funding: This research was funded by the MEB research fund of Hanyang University (HY-202000000002924).

Institutional Review Board Statement: Not applicable.

Informed Consent Statement: Not applicable.

Data Availability Statement: The data presented in this study are available on request from the corresponding author.

Conflicts of Interest: The authors declare that they have no conflict of interest.

References

1. Marwah, H.; Garg, T.; Goyal, A.K.; Rath, G. Permeation enhancer strategies in transdermal drug delivery. *Drug Deliv.* **2016**, *23*, 564–578. [CrossRef]
2. Akhtar, N.; Singh, V.; Yusuf, M.; Khan, R.A. Non-invasive drug delivery technology: Development and current status of transdermal drug delivery devices, techniques and biomedical applications. *Biomed. Eng. Biomed. Tech.* **2020**, *65*, 243–272. [CrossRef] [PubMed]
3. Al Hanbali, O.A.; Khan, H.M.S.; Sarfraz, M.; Arafat, M.; Ijaz, S.; Hameed, A. Transdermal patches: Design and current approaches to painless drug delivery. *Acta Pharmaceut.* **2019**, *69*, 197–215. [CrossRef]

4. Waghule, T.; Singhvi, G.; Dubey, S.K.; Pandey, M.M.; Gupta, G.; Singh, M.; Dua, K. Microneedles: A smart approach and increasing potential for transdermal drug delivery system. *Biomed. Pharmacother.* **2019**, *109*, 1249–1258. [CrossRef] [PubMed]

5. Toyoda, M.; Hama, S.; Ikeda, Y.; Nagasaki, Y.; Kogure, K. Anti-cancer vaccination by transdermal delivery of antigen peptide-loaded nanogels via iontophoresis. *Int. J. Pharmaceut.* **2015**, *483*, 110–114. [CrossRef] [PubMed]

6. Cordery, S.F.; Husbands, S.M.; Bailey, C.P.; Guy, R.H.; Delgado-Charro, M.B. Simultaneous Transdermal Delivery of Buprenorphine Hydrochloride and Naltrexone Hydrochloride by Iontophoresis. *Mol. Pharmaceut.* **2019**, *16*, 2808–2816. [CrossRef]

7. Azagury, A.; Khoury, L.; Enden, G.; Kost, J. Ultrasound mediated transdermal drug delivery. *Adv. Drug Deliver. Rev.* **2014**, *72*, 127–143. [CrossRef] [PubMed]

8. Daftardar, S.; Neupane, R.; Boddu, S.H.S.; Renukuntla, J.; Tiwari, A.K. Advances in Ultrasound Mediated Transdermal Drug Delivery. *Curr. Pharm. Des.* **2019**, *25*, 413–423. [CrossRef] [PubMed]

9. Wen, X.; Xin, Y.; Hamblin, M.R.; Jiang, X. Applications of cold atmospheric plasma for transdermal drug delivery: A review. *Drug Deliv. Transl. Res.* **2020**. [CrossRef] [PubMed]

10. Von Woedtke, T.; Schmidt, A.; Bekeschus, S.; Wende, K.; Weltmann, K.D. Plasma Medicine: A Field of Applied Redox Biology. *In Vivo* **2019**, *33*, 1011–1026. [CrossRef]

11. Semmler, M.L.; Bekeschus, S.; Schafer, M.; Bernhardt, T.; Fischer, T.; Witzke, K.; Seebauer, C.; Rebl, H.; Grambow, E.; Vollmar, B.; et al. Molecular Mechanisms of the Efficacy of Cold Atmospheric Pressure Plasma (CAP) in Cancer Treatment. *Cancers* **2020**, *12*, 269. [CrossRef]

12. Hwang, Y.; Jeon, H.; Wang, G.Y.; Kim, H.K.; Kim, J.-H.; Ahn, D.K.; Choi, J.S.; Jang, Y. Design and Medical Effects of a Vaginal Cleaning Device Generating Plasma-Activated Water with Antimicrobial Activity on Bacterial Vaginosis. *Plasma* **2020**, *3*, 204–213. [CrossRef]

13. Cordaro, L.; De Masi, G.; Fassina, A.; Gareri, C.; Pimazzoni, A.; Desideri, D.; Indolfi, C.; Martines, E. The Role of Thermal Effects in Plasma Medical Applications: Biological and Calorimetric Analysis. *Appl. Sci.* **2019**, *9*, 5560. [CrossRef]

14. Bernhardt, T.; Semmler, M.L.; Schafer, M.; Bekeschus, S.; Emmert, S.; Boeckmann, L. Plasma Medicine: Applications of Cold Atmospheric Pressure Plasma in Dermatology. *Oxid. Med. Cell Longev.* **2019**, *2019*, 3873928. [CrossRef] [PubMed]

15. Brany, D.; Dvorska, D.; Halasova, E.; Skovierova, H. Cold Atmospheric Plasma: A Powerful Tool for Modern Medicine. *Int. J. Mol. Sci.* **2020**, *21*, 2932. [CrossRef]

16. Lee, Y.; Ricky, S.; Lim, T.H.; Jang, K.S.; Kim, H.; Song, Y.; Kim, S.Y.; Chung, K.S. Wound Healing Effect of Nonthermal Atmospheric Pressure Plasma Jet on a Rat Burn Wound Model: A Preliminary Study. *J. Burn Care Res.* **2019**, *40*, 923–929. [CrossRef]

17. Boeckmann, L.; Schafer, M.; Bernhardt, T.; Semmler, M.L.; Jung, O.; Ojak, G.; Fischer, T.; Peters, K.; Nebe, B.; Muller-Hilke, B.; et al. Cold Atmospheric Pressure Plasma in Wound Healing and Cancer Treatment. *Appl. Sci.* **2020**, *10*, 6898. [CrossRef]

18. Shimizu, K.; Hayashida, K.; Blajan, M. Novel method to improve transdermal drug delivery by atmospheric microplasma irradiation. *Biointerphases* **2015**, *10*, 029517. [CrossRef] [PubMed]

19. Xin, Y.; Wen, X.; Hamblin, M.R.; Jiang, X. Transdermal delivery of topical lidocaine in a mouse model is enhanced by treatment with cold atmospheric plasma. *J. Cosmet. Dermatol.* **2020**. [CrossRef] [PubMed]

20. Shimizu, K.; Tran, A.N.; Kristof, J.; Blajan, M. Investigation of atmospheric microplasma for improving skin permeability. In Proceedings of the 2016 Electrostatics Joint Conference, West Lafayette, IN, USA, 13–16 June 2016; pp. 13–18.

21. Kristof, J.; Miyamoto, H.; Tran, A.N.; Blajan, M.; Shimizu, K. Feasibility of transdermal delivery of Cyclosporine A using plasma discharges. *Biointerphases* **2017**, *12*, 02B402. [CrossRef] [PubMed]

22. Alkilani, A.Z.; McCrudden, M.T.; Donnelly, R.F. Transdermal Drug Delivery: Innovative Pharmaceutical Developments Based on Disruption of the Barrier Properties of the stratum corneum. *Pharmaceutics* **2015**, *7*, 438–470. [CrossRef]

23. Parhi, R.; Suresh, P.; Patnaik, S. Physical means of stratum corneum barrier manipulation to enhance transdermal drug delivery. *Curr. Drug Deliv.* **2015**, *12*, 122–138. [CrossRef] [PubMed]

24. Marschewski, M.; Hirschberg, J.; Omairi, T.; Hofft, O.; Viol, W.; Emmert, S.; Maus-Friedrichs, W. Electron spectroscopic analysis of the human lipid skin barrier: Cold atmospheric plasma-induced changes in lipid composition. *Exp. Dermatol.* **2012**, *21*, 921–925. [CrossRef] [PubMed]

25. Schmidt, A.; Liebelt, G.; Striesow, J.; Freund, E.; von Woedtke, T.; Wende, K.; Bekeschus, S. The molecular and physiological consequences of cold plasma treatment in murine skin and its barrier function. *Free Radic. Biol Med.* **2020**, *161*, 32–49. [CrossRef] [PubMed]

26. Jang, Y.; Kim, E.K.; Shim, W.S.; Song, K.M.; Kim, S.M. Amniotic fluid exerts a neurotrophic influence on fetal neurodevelopment via the ERK/GSK-3 pathway. *Biol. Res.* **2015**, *48*, 44. [CrossRef] [PubMed]

27. Shao, T.; Zhang, C.; Wang, R.X.; Zhou, Y.X.; Xie, Q.; Fang, Z. Comparison of Atmospheric-Pressure He and Ar Plasma Jets Driven by Microsecond Pulses. *IEEE T. Plasma Sci.* **2015**, *43*, 726–732. [CrossRef]

28. Lee, H.Y.; Choi, J.H.; Hong, J.W.; Kim, G.C.; Lee, H.J. Comparative study of the Ar and He atmospheric pressure plasmas on E-cadherin protein regulation for plasma-mediated transdermal drug delivery. *J. Phys. D Appl. Phys.* **2018**, *51*, 215401. [CrossRef]

29. Bell, K.L.; Dalgarno, A.; Kingston, A.E. Penning Ionization by Metastable Helium Atoms. *J. Phys. Part B Atom. Mol. Phys.* **1968**, *1*, 18. [CrossRef]

30. Liu, Y.; Ni, H.Y.; Wargniez, W.; Gregoire, S.; Durand, I.; Roussel-Berlier, L.; Eilstein, J.; Jie, Q.; Ma, T.; Shen, T.; et al. Inter-laboratory study of the skin distribution of 4-n-butyl resorcinol in ex vivo pig and human skin. *J. Chromatogr. B* **2018**, *1093*, 77–79. [CrossRef] [PubMed]
31. Dang, E.L.; Man, G.; Zhang, J.C.; Lee, D.; Mauro, T.M.; Elias, P.M.; Man, M.Q. Inducible nitric oxide synthase is required for epidermal permeability barrier homeostasis in mice. *Exp. Dermatol.* **2020**, *29*, 1027–1032. [CrossRef] [PubMed]
32. Vaccaro, M.; Irrera, N.; Cutroneo, G.; Rizzo, G.; Vaccaro, F.; Anastasi, G.P.; Borgia, F.; Cannavo, S.P.; Altavilla, D.; Squadrito, F. Differential Expression of Nitric Oxide Synthase Isoforms nNOS and iNOS in Patients with Non-Segmental Generalized Vitiligo. *Int. J. Mol. Sci.* **2017**, *18*, 2533. [CrossRef] [PubMed]
33. Ikeyama, K.; Denda, M. Effect of endothelial nitric oxide synthase on epidermal permeability barrier recovery after disruption. *Br. J. Dermatol.* **2010**, *163*, 915–919. [CrossRef] [PubMed]

Article

Inhibitory Effect of Cold Atmospheric Plasma on Chronic Wound-Related Multispecies Biofilms

Maria Alcionéia Carvalho de Oliveira [1], Gabriela de Morais Gouvêa Lima [1], Thalita M. Castaldelli Nishime [2], Aline Vidal Lacerda Gontijo [1], Beatriz Rossi Canuto de Menezes [3], Marcelo Vidigal Caliari [4], Konstantin Georgiev Kostov [5] and Cristiane Yumi Koga-Ito [1,6,*]

[1] Oral Biopathology Graduate Program, Institute of Science and Technology, São Paulo State University (UNESP), São José dos Campos 12245-000, SP, Brazil; macoliveira12@gmail.com (M.A.C.d.O.); gabimoraisgl@gmail.com (G.d.M.G.L.); aline.gontijo@gmail.com (A.V.L.G.)

[2] Leibniz Institute for Plasma Science and Technology, 17489 Greifswald, Germany; thalita.nishime@inp-greifswald.de

[3] Plasmas and Process Laboratory (LPP), Technological Institute of Aeronautics (ITA), São José dos Campos 12228970, SP, Brazil; beatriz1menezes@gmail.com

[4] Department of General Pathology, University Federal de Minas (UFMG), Belo Horizonte 12516-41, MG, Brazil; caliari@ufmg.br

[5] Department of Physics, Guaratinguetá Faculty of Engineering, São Paulo State University (UNESP), Guaratinguetá 12516-410, SP, Brazil; konstantin.kostov@unesp.br

[6] Department of Environment Engineering, Institute of Science & Technology, São Paulo State University (UNESP), São José dos Campos 12247-016, SP, Brazil

* Correspondence: cristiane.koga-ito@unesp.br; Tel.: +55-1239479708

Citation: Oliveira, M.A.C.d.; Lima, G.d.M.G.; Nishime, T.M.C.; Gontijo, A.V.L.; Menezes, B.R.C.d.; Caliari, M.V.; Kostov, K.G.; Koga-Ito, C.Y. Inhibitory Effect of Cold Atmospheric Plasma on Chronic Wound-Related Multispecies Biofilms. *Appl. Sci.* **2021**, *11*, 5441. https://doi.org/10.3390/app11125441

Academic Editor: Andrei Vasile NASTUTA

Received: 18 April 2021
Accepted: 4 June 2021
Published: 11 June 2021

Publisher's Note: MDPI stays neutral with regard to jurisdictional claims in published maps and institutional affiliations.

Abstract: The presence of microbial biofilms in the wounds affects negatively the healing process and can contribute to therapeutic failures. This study aimed to establish the effective parameters of cold atmospheric plasma (CAP) against wound-related multispecies and monospecies biofilms, and to evaluate the cytotoxicity and genotoxicity of the protocol. Monospecies and multispecies biofilms were formed by methicillin-resistant *Staphylococcus aureus* (MRSA), *Pseudomonas aeruginosa* and *Enterococcus faecalis*. The monospecies biofilms were grown in 96 wells plates and multispecies biofilm were formed on collagen membranes. The biofilms were exposed to helium CAP for 1, 3, 5 and 7 min. In monospecies biofilms, the inhibitory effect was detected after 1 min of exposure for *E. faecalis* and after 3 min for MRSA. A reduction in *P. aeruginosa* biofilm's viability was detected after 7 min of exposure. For the multispecies biofilms, the reduction in the overall viability was detected after 5 min of exposure to CAP. Additionally, cytotoxicity and genotoxicity were evaluated by MTT assay and static cytometry, respectively. CAP showed low cytotoxicity and no genotoxicity to mouse fibroblastic cell line (3T3). It could be concluded that He-CAP showed inhibitory effect on wound-related multispecies biofilms, with low cytotoxicity and genotoxicity to mammalian cells. These findings point out the potential application of CAP in wound care.

Keywords: wounds; biofilm; plasma jet

1. Introduction

Chronic wounds, such as venous ulcers, diabetic foot ulcers, and pressure ulcers, are considered one of the most challenging problems in the medical area and affect millions of patients worldwide [1,2]. The prevalence of chronic wounds is estimated in 2.21 per 1000 in the population [3] and these rates tend to rise due to the ageing population and increasing prevalence of obesity and associated conditions, such as diabetes and cardiovascular diseases [4]. This condition has significant economic impact on health systems [2,5], with 2 to 3% of the budget in health applied in the treatment of chronic wounds in developed countries [1].

Microbial biofilms significantly affect wound healing and can contribute to therapeutic failures [6]. For these reasons, they have emerged as an important topic of discussion in the treatment of wounds [7]. The infections of wounds are usually polymicrobial, as their microenvironment is favorable for microbial growth [8,9]. Despite this, previous evidence showed that species diversity in infected wounds is relatively low. It has been reported that *Staphylococcus aureus*, *Enterococcus faecalis* and *Pseudomonas aeruginosa* are the most prevalent species [4,10], with occurrence of 93.5%, 71.7% and 52.2%, respectively [11]. Other bacterial species, such as *Proteus mirabilis*, *Klebsiella pneumoniae* and *Escherichia coli* were also isolated from wounds [12,13]. The increasing rates of antimicrobial resistance also contributes to the unfavorable scenario [14,15].

Thus, the control of the microbial biofilms is considered critical to the success of wound healing [7,16]. The conventional therapeutic alternatives are focused on the reduction or eradication of biofilms by using debridement techniques and treatments with negative pressure or ultrasound [17,18]. Other methods include the use of wound dressing impregnated with antimicrobial substances or compounds, such as gentian violet and methylene blue [15], honey [19], *Aloe vera* [20], nitric oxide releasing polymer [21], polyhexamethylenebiguanides [22] and ethylenediamine tetra acetic acid (EDTA) [23]. However, there is still need for alternatives to wound care, in particular, focusing on biofilm control [17,21,24].

Cold atmospheric plasma (CAP) has emerged as an attractive alternative approach to the biofilm control and is considered a rapid, environmental-friendly, energy saving, and versatile technology [25,26]. Plasma is described as the fourth state of matter and is usually generated from a gaseous precursor in which a high electric field is applied, producing an electric discharge [27]. Plasma is a complex mixture composed of ions, reactive radicals, neutral particles, molecules and electrons, as well as excited species and energetic photons [28,29]. These plasma components can damage cell structures, such as lipids, protein and DNA, and increase the levels of intracellular reactive species [30]. Thus, the biological activity of plasma is a product of synergetic action of several factors, such as neutral reactive species (oxygen reactive species—ROS, reactive nitrogen species—RNS) and UV radiation [31]. The anti-inflammatory and tissue repair inducing [32,33] effects have been considered highly promising for the treatment of several diseases.

The efficiency of CAP against the bacterial biofilms has been reported in the literature [34]. CAP was able to reduce the viability of monospecies biofilms of methicillin resistant *S. aureus* (MRSA) [35], *P. aeruginosa* [36] and *E. faecalis* [37]. Monospecies biofilms formed by *E. faecalis* [38,39] or *Escherichia coli* [40] were also inhibited by CAP. One of the main advantages of CAP when compared to conventional antibiotics is that the development of treatment resistance was not detected so far. Repeated exposure to CAP did not induce resistance in *S. aureus* [41]. The hypothesis is that as CAP does not have specific target structures [35], the chance of resistance development with plasma is lower since several targets are activated and not a specific one [42]. Besides the antimicrobial activity, anti-inflammatory and tissue repair inducing effects (stimulating cellular proliferation and migration and the pro-angiogenic effects) of CAP have been also reported [14,32,43,44].

The biological effects of CAP lead to the hypothesis that this technology can be useful for the treatment of infected wounds. However, to the best of our knowledge, little is known on the effects of CAP on wound-related multispecies biofilms. Thus, the present work aimed to study the effect of He-CAP on wound-related multispecies biofilms and confirm the safety of the protocol.

2. Materials and Methods

2.1. Plasma Jet Device

The plasma jet device used in this study consists of a plasma reactor with dielectric barrier discharge (DBD) configuration. The plasma generated in this primary DBD reactor is transported through a 1.0-m-long, flexible polyurethane tube (Kangaroo TM Nasogastric Feeding Tube, 10 Fr/Ch), connected to the reactor exit. A thin metal wire is installed inside the plastic tube acting as floating electrode. The wire penetrates few millimeters inside the primary discharge chamber, crosses the whole tube length and terminates about 2 mm before the plastic tube exit. When high voltage is applied to the primary DBD reactor, a remote plasma plume is ignited at the downstream end of the plastic tube. This device was already described in previous works [45–48]. In the present study, the reactor was connected to a Minipuls4 AC power supply (GBS Elektronik GmbH, Radeberg, Germany) and an amplitude modulated voltage signal (12 kV amplitude) with voltage duty cycle of 22% was employed. The system was fed with helium (99.5% purity) at a flow rate of 2.0 slm. The distance between the plastic tube tip and the surface to be treated was set at 1.5 cm.

The characteristic optical emission spectrum of this plasma jet was previously reported by Kostov et al. and Nishime et al., [45–48]. Briefly, weak emission lines of helium were detected with nitrogen emission bands being predominant in the spectrum, showing that excited nitrogen (N_2-second positive system) and ionized nitrogen species (N_2^+-first negative system) were produced in the plasma plume. Peaks associated to atomic oxygen (777 nm and 844 nm) and the OH band at 308 nm were also present indicating the production of ROS.

2.2. Bacterial Strains and Growth Conditions

The bacterial strains used in this study were: methicillin resistant *S. aureus* (MRSA) (ATCC 33591), *P. aeruginosa* (ATCC 27853) and *E. faecalis* (ATCC 29212). Strains were grown in tryptic soy agar for 24 h at 37 °C, under aerobiosis. After this period, standardized suspensions containing 10^8 cells/mL in sterile saline solution (NaCl 0.9%) were prepared with the aid of a spectrophotometer (*P. aeruginosa*: $\lambda = 600$, OD 0.115; *E. faecalis*: $\lambda = 760$, OD 1.278 and MRSA: $\lambda = 600$, OD 0.462). Then, these suspensions were diluted in TS broth obtaining the final concentration of 10^6 cells/mL.

2.3. Effect of CAP on Monospecies Biofilms

Monospecies bacterial biofilms of MRSA (ATCC 33591), *P. aeruginosa* (ATCC 27853) and *E. faecalis* (ATCC 29212) were grown in 96 wells plates, as these experiments aimed to standardize the ideal parameters of plasma device in a cost-effective manner. Standardized suspension (10^6 cells/mL) of each strain in Tryptic soy broth (200 µL) was added to each well. Plates were incubated for 90 min at 37 °C and 80 rpm for initial adherence. After this period, the culture medium was removed, and the wells were washed twice with physiologic solution (NaCl 0.9%). Then, 200 µL of fresh Tryptic soy broth medium were added to each well and the plates were incubated for 48 h. Subsequently, the culture medium was removed, and the biofilms were washed twice with physiologic solution (NaCl 0.9%). Biofilms were exposed to CAP for 1, 3, 5 and 7 min. A non-exposed treated group was included for comparative purposes. The number of viable cells was determined after serial dilution in physiologic solution and plating on brain heart infusion agar. The results were expressed as colony forming units per mL (CFU/mL). Three independent experiments were performed in triplicate.

2.4. Effect of CAP on Multispecies Biofilms

Multispecies biofilms were formed by MRSA (ATCC 33591), *P. aeruginosa* (ATCC 27853) and *E. faecalis* (ATCC 29212) on collagen membranes (Geistlich Bio-Gide®), according to the methodology described by [49,50]. First, standardized suspensions containing 10^6 cells/mL in physiologic solution (NaCl 0.9%) of each microorganism were prepared spectrophotometrically, using the same parameters described before. The collagen membranes were positioned on the surface of tryptic soy agar supplemented by fetal bovine serum solution (50% fetal bovine serum and 50% NaCl 0.9% solution). The smooth layer of the collagen membrane was considered for the growth of multispecies biofilm. Then, aliquots of 10 µL of each bacterial suspension were mixed up in a microtube and the total volume was transferred to the membrane's surface. Plates were incubated at 37 °C for 24 h. Then, the biofilms were exposed to CAP for 1, 3, 5 and 7 min. The biofilms were mechanically dispersed, and the number of viable cells was determined by plating method. The recovery of each species was carried out by plating the resuspended biofilm in selective culture media (mannitol salt agar, cetrimide agar and Enterococcosel agar, respectively). The experiments were performed in triplicate in three different experiments (n = 9). A negative control (exposed only to helium flow and without ignition) was included for comparative purposes.

2.5. Morphologic Analysis of the Biofilms by Scanning Electron Microscopy

Multispecies biofilms exposed to CAP for 1, 3, 5 and 7 min were analyzed morphologically by scanning electron microscopy. After exposure to CAP, the biofilms were prepared by standard fixation protocol 2.5% glutaraldehyde solution in 0.1 M phosphate buffer (pH 7.4) for 24 h and serial dehydration in ethanol (Synth, 99.8%), for 10 min in each concentration [51]. After 24 h under room temperature, the biofilms were metallized in gold. The specimens were analyzed by scanning electron microscopy with Field-emission gun (SEM-FEG). The analysis was performed using a Tescan Vega 3 microscope with accelerating voltage of 10 kV and magnification of 10 k×. Two fields per membrane were analyzed. The first was located at the central region of the membrane and the other at the peripheral region (approximately 2.5 mm from the center).

2.6. Cytotoxicity Evaluation

For evaluating the cytotoxicity of CAP, mouse fibroblastic cell line (3T3) were grown in DMEM (Dulbecco's Modified Eagle's medium) supplemented with 10% of inactivated fetal bovine serum (FBS), 100 IU mL^{-1} of penicillin, and 100 µg mL^{-1} of streptomycin. After, 200 µL of a suspension containing 3.6×10^4 cells/mL were transferred to the wells of 96-well plates. Plates were incubated for 24 h at 37 °C and in an atmosphere of 95% air and 5% CO_2. After, the culture medium was removed and 30 µL of Hanks' Balanced Salt Solution with 10 mM 4-(2-hydroxyethyl)-1-piperazineethanesulfonic acid were added. The cells were exposed to CAP for 1, 3, 5 and 7 min using the same parameters adopted for microbiological evaluations. The cell viability was determined immediately, 24 and 48 h after plasma exposure, using the MTT colorimetric method at 570 nm. For the analyses of later effects (24 and 48 h), the culture medium was immediately added after plasma jet exposure. The experiments were performed in triplicate in two independent experiments. The results were expressed as a percentage of viable cells (%), using the number of cells grown in wells without CAP exposure as the negative control.

The morphological grading of cytotoxicity, according to [52], was: 0 (not cytotoxic)—no reduction of cell growth; 1 (slight)—only slight growth inhibition observable; 2 (mild)—not more than 50% growth inhibition; 3 (moderate)–more than 50% growth inhibition observable; 4 (severe)—nearly complete or complete destruction of the cell layers.

2.7. Genotoxicity Evaluation

Genotoxicity was evaluated by static cytometry method for cell DNA ploidy analysis. NIH-3T3 cells were cultivated according to Felisbino et al. [53], with modifications. Cells (5×10^4 cells/mL) were seeded in 96-well plates in Dulbecco Modified Eagle Medium (DMEM) supplemented with 10% inactivated bovine fetal serum and penicillin/streptomycin 1% and maintained at 37 °C and 5% CO_2 for 24 h. Then, cells were washed once with Hanks' Balanced Salt Solution (HBSS) and 50 µL of HBSS were added per well for CAP treatment. Cells were exposed to CAP for 1, 3, 5, and 7 min. A non-treated group was included for comparative purposes. After treatment, residual HBSS was aspirated, culture medium was added, and plates were maintained at 37 °C and 5% CO_2 for 24 h. Cells were released from plate with trypsin, fixed in 4% formalin and transferred to glass slides. Next, slides were stained by Feulgen's method and covered with cover glass.

Nuclear IOD (integrated optical density) was performed through static cytometry. Images were digitalized with Olympus QColor3 camera in 400× magnification. Image analysis were performed in a Carl Zeiss image system and KS300 software (Carl Zeiss, Oberkochen, Germany) in black and white with a Feulgen filter for 570 nm wavelength.

DNA index (DI) was obtained dividing resulting IOD by the mean of IOD from 20 lymphocytes used as diploid control. DI was classified as diploid (DI = 0.9–1.1), slightly aneuploid (DI = 1.2–1.4), moderated aneuploid (DI = 1.5–1.7), and severely aneuploid (DI > 1.8), based on Lima et al. [54]. IOD is equivalent but not identical to DNA content. A total of 100 cells were analyzed.

2.8. Data Analysis

Data was analyzed for normality using Shapiro Wilk test. After, data of cell viability of biofilms (CFU/mL) were compared by Kruskal Wallis followed by Dunn's post hoc test. The values of cytotoxicity were obtained by percentage analysis. For the genotoxicity assays, from the DI values x the number of cells, histograms of the frequency of distribution of the DI values by group were obtained. The analyses were carried out using GraphPad Prism, version 7.0 (Graphpad Software Inc., La Jolla, CA, USA). The significance levels were set at 5%, 1% or 0.1% ($p < 0.05$, 0.01 and 0.001), according to the statistical analyses' outcomes.

3. Results

3.1. Effect of CAP on Wound-Related Monospecies Biofilms

Monospecies biofilms formed by MRSA, *P. aeruginosa* or *E. faecalis* were exposed to CAP. Figure 1 shows the results for antibiofilm effect. The median value (CFU/mL) of the non-treated control for *E. faecalis* was 1.0×10^9; for treatment groups the median values (CFU/mL) were 5.8×10^7 (reduction of 1.2 log, $p < 0.05$), 5.2×10^7 (reduction of 1.3 log, $p < 0.01$), 1.4×10^7 (reduction of 1.9 log, $p < 0.001$) and 3.3×10^7 (reduction of 1.5 log, $p < 0.01$) after 1, 3, 5 and 7 min of CAP exposure, respectively. The median value (CFU/mL) of the non-treated control for MRSA was 6.4×10^8; for treatment groups the median values (CFU/mL) were 8.0×10^6 (reduction of 1.9 log, $p > 0.05$), 3.0×10^6 (reduction of 2.3 log, $p < 0.05$), 8.0×10^5 (reduction of 2.9 log, $p < 0.001$) and 1.4×10^6 (reduction of 2.7 log, $p < 0.001$) after 1, 3, 5 and 7 min of CAP exposure, respectively. The median value (CFU/mL) of the non-treated control for *P. aeruginosa* was 5.0×10^7; for treatment group the median values (CFU/mL) were 4.0×10^6 (reduction of 1.1 log, $p > 0.05$), 4.0×10^6 (reduction of 1.1 log, $p > 0.05$), 4.0×10^6 (reduction of 1.1 log, $p > 0.05$) and 2.2×10^5 (reduction of 2.4 log, $p < 0.05$) after 1, 3, 5 and 7 min of CAP exposure, respectively.

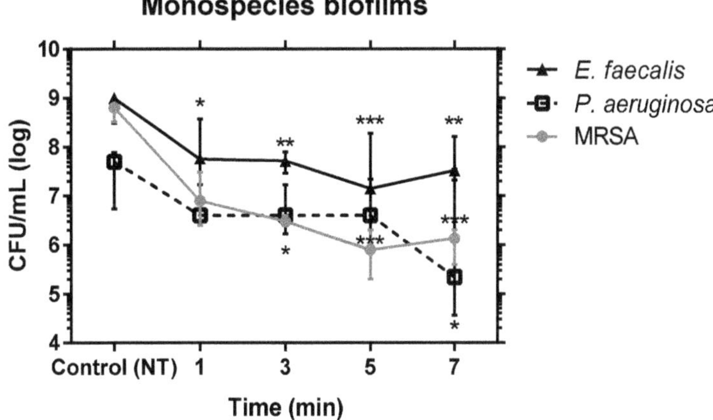

Figure 1. Median and interquartile range of MRSA, *P. aeruginosa* and *E. faecalis* viable cells (CFU/mL) recovered after exposure of monospecies biofilms to (CAP) for 1, 3, 5 and 7 min. Kruskal–Wallis test and Dunn's multiple comparison * $p < 0.05$; ** $p < 0.01$; *** $p < 0.001$.

3.2. Effect of CAP on Multispecies Biofilms

The effect of CAP on multispecies biofilms formed by MRSA, *P. aeruginosa* and *E. faecalis* can be observed in Figure 2. The median value (CFU/mL) of the non-treated control for *E. faecalis* was 2.3×10^9; for treatment groups the median values (CFU/mL) were 5.0×10^8 (reduction of 0.66 log, $p < 0.01$), 8.0×10^8 (reduction of 0.5 log, $p < 0.05$), 3.0×10^8 (reduction of 0.9 log, $p < 0.001$) and 7.0×10^8 (reduction of 0.5 log, $p < 0.05$) after 1, 3, 5 and 7 min of CAP exposure, respectively. The median value (CFU/mL) of the non-treated control for MRSA was 1.7×10^8; for treatment groups the median values (CFU/mL) were 1.1×10^8 (reduction of 0.2 log, $p > 0.05$), 3.7×10^7 (reduction of 0.66 log, $p > 0.05$), 2.0×10^7 (reduction of 0.9 log, $p < 0.05$) and 4.0×10^7 (reduction of 0.6 log, $p < 0.05$) after 1, 3, 5 and 7 min of CAP exposure, respectively. The median value (CFU/mL) of the non-treated control for *P. aeruginosa* was 3.0×10^7; for treatment group the median values (CFU/mL) were 6.0×10^6 (reduction of 0.7 log, $p > 0.05$), 5.0×10^6 (reduction of 0.8 log, $p > 0.05$), 1.0×10^6 (reduction of 1.5 log, $p < 0.001$) and 3.0×10^6 (reduction of 1.0 log, $p > 0.05$) after 1, 3, 5 and 7 min of CAP exposure, respectively.

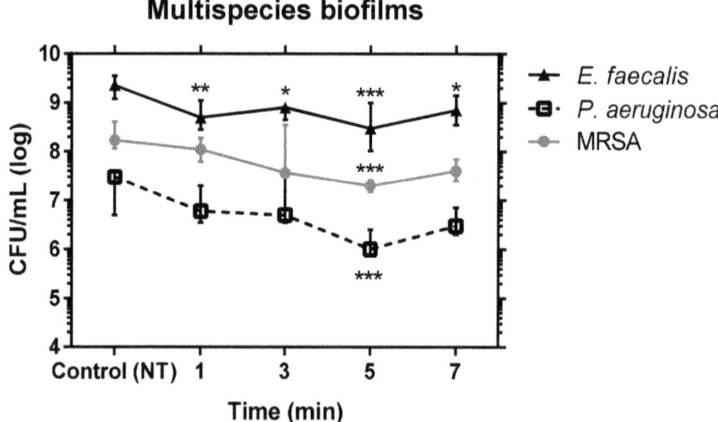

Figure 2. Median and interquartile range of MRSA, *P. aeruginosa* and *E. faecalis* viable cells (CFU/mL) recovered after exposure of multispecies biofilms to CAP for 1, 3, 5 and 7 min. Kruskal–Wallis test and Dunn's multiple comparison post hoc test * $p < 0.05$; ** $p < 0.01$; *** $p < 0.001$.

3.3. Morphologic Analysis of the Multispecies Biofilms by Scanning Electron Microscopy

The biofilms morphology was analyzed by scanning electron microscopy (SEM) in order to confirm presence of biofilm and corroborate the reduction in cell viability observed in microbiological analyses. To assess the effects of CAP in biofilm, a polymicrobial culture comprised of a 1:1:1 ratio of MRSA, *E. faecalis* and *P. aeruginosa* was grown.

Figure 3 shows the micrographs obtained without treatment and after exposure to CAP. Untreated biofilm (control) showed the presence of high cell density of the cocci and bacilli. No differences were observed between the images from central and lateral fields.

In the groups exposed to CAP, SEM images showed dispersed biofilm clusters with moderate differences in the amount of microorganisms according to the analyzed field. After 1 and 3 min of exposure to CAP (Figure 3C1,D1), there was a greater presence of diplococci in the central field, the cell density in the biofilm was lower than in the non-treated control. In the lateral field, the cell density was higher and the presence of bacilli could be noticed (Figure 3C2,D2). After 5 min of exposure to CAP, the images showed a reduction in the cell density in the biofilm mostly in the center field. On the other hand, 7 min exposure to CAP leads to a visible biofilm reduction over the entire analyzed area.

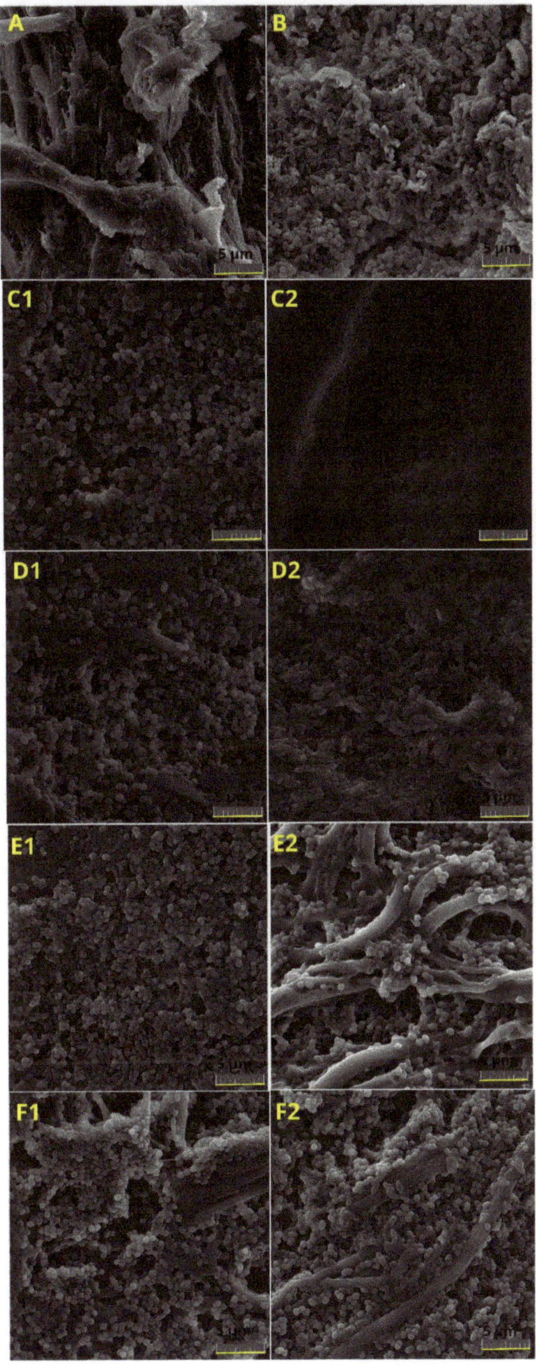

Figure 3. Scanning electron microscopy (SEM) of multispecies biofilms formed by MRSA, *P. aeruginosa* and *E. faecalis* non-treated or treated for 1–7 min: (**A**) Collagen membrane without biofilm, (**B**) negative

control (**C1**), biofilm after exposure to CAP for 1 min (central region), (**C2**) biofilm after exposure CAP for 1 min (lateral region), (**D1**) biofilm after exposure to CAP for 3 min (central region), (**D2**) biofilm after exposure to CAP for 3 min (lateral region), (**E1**) biofilm after exposure to CAP for 5 min (central region), (**E2**) biofilm after exposure to CAP for 5 min (lateral region), (**F1**) biofilm after exposure to CAP for 7 min (central region) and (**F2**) biofilm after treatment by CAP for 7 min (lateral region). Scale bare represents 5 μm.

3.4. Cytotoxicity

The 3T3 cells were exposed to the plasma jet for 1, 3, 5 and 7 min (Figure 4). The cytotoxicity immediately after exposure to plasma for all times was classified as Grade 0 (not cytotoxic), as the cell viability was 100% when compared to the unexposed control, except for 3 min, which was Grade 1 (slight), since the cell viability was 93.1% [52].

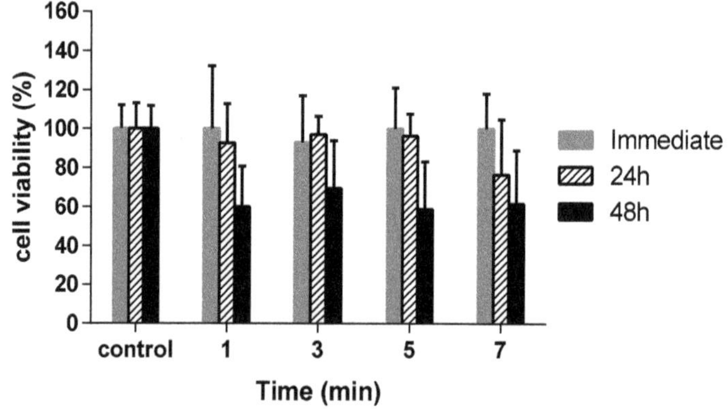

Figure 4. Cytotoxicity analysis of non-thermal plasma at atmospheric pressure expressed in cell viability. The 3T3 cells were exposed to CAP for 1, 3, 5 and 7 min. The cell viability was determined immediately (gray bar), 24 h (striped bar) and 48 h (black bar) after plasma exposure.

After 24 h of exposure to plasma, cytotoxicity was classified as Grade 1 (slight) for the exposure time for 1 to 5 min [52]. The exposure for 7 min was classified as Grade 2 (mild), since the cell viability was 76.6%. After 48 h of exposure to plasma, cytotoxicity was classified as Grade 2 (mild) for all times [52].

3.5. Genotoxicity

According to obtained results, all groups of 3T3 cells can be classified as slightly aneuploid (DI ≥ 1.2). Non-treated group (NT) presented DNA index (DI) mean of 1.27. These results suggest that CAP was not genotoxic. Table 1 shows DI median, mean, and standard deviation (SD) of studied groups. Figure 5 represents the distribution histogram of obtained DI × number of cells for all analyzed periods of time.

Table 1. Median, mean and standard deviation (SD) of DNA index (DI) obtained by static cytometry of 3T3 cells' nucleus when treated with cold atmospheric plasma for 1 (P1), 3 (P3), 5 (P5), and 7 (P7) minutes. NT = non-treated group.

	P1	**P3**	**P5**	**P7**	**NT**
Median	1.24	1.33	1.30	1.27	1.24
Mean	1.25	1.32	1.31	1.29	1.27
Standard deviation (SD)	0.12	0.14	0.11	0.11	0.09

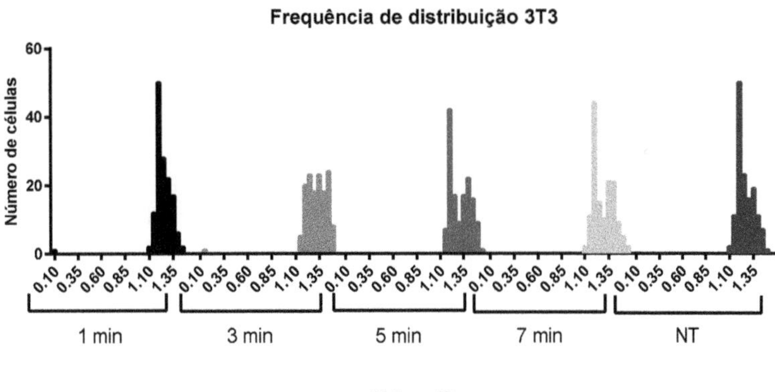

Figure 5. Histogram of distribution of DNA indices (ID) obtained × number of corresponding cells 4.

4. Discussion

The outcomes of this study showed that He-cold atmospheric plasma significantly reduced the viability of wound-related mono and multispecies biofilms. *S. aureus, P. aeruginosa* and *E. faecalis* are opportunistic pathogenic bacteria which cause chronic infections due a number of virulence factors and can interfere with wound healing process, in particular when strains resistant to antimicrobials are involved [55]. To the best of our knowledge, this is the first report on the activity of CAP on wound related multispecies biofilms, formed by MRSA, *P. aeruginosa and E. faecalis*, by using a model of wound-related biofilm.

Although the main goal of this study was to evaluate the effect of cold atmospheric plasma in multispecies biofilms, experiments with monospecies biofilms were carried out in order to establish the operation parameters of the plasma device that generated antimicrobial activity against the tested microbial species. For this purpose and taking in account the cost of membrane used in the multispecies experiments, monospecies biofilms were grown in 96-well plates. We observed that the exposure to CAP for 5 min reduced in 3 log the viable cells of MRSA monospecieis biofilms. This result corroborates the findings reported by Xu et al. [56], in which a plasma jet operating with 20 kV of voltage, frequency of 38 kHz, and He flow rate of 6.7 slm reduced in 3.6 log *S. aureus* biofilms after 10 min exposure. CAPs generated using other working gases also showed similar inhibitory effects. Matthes et al. [41] observed a reduction of 1.7 log in MRSA biofilms after repeated 20 s exposure to Ar plasma during 6 h. Besides, Wang et al. [57] observed a decrease of 4.67 ± 0.29 log in the viability of *S. aureus* biofilms after 10 min exposure to N_2 CAP.

We also detected a significant reduction in the viability of *E. faecalis* monospecies biofilm after 1 min exposure in the present study. The reduction varied from 1.2 to 1.9 log when compared to the untreated control group. This result is interesting, since *E. faecalis* monospecies biofilms appear to be resistant to other therapies. In the case of *E. faecalis* monospecies biofilm, it seems that other CAP generation setups lead to higher inhibitory effects when compared to He-CAP used in this study. Theinkon et al. [37] observed 3 log reduction in *E. faecalis* (ATCC 29212) biofilm viability after 5 min exposure to CAP generated by a surface micro-discharge reactor in open air operating with voltage of 3.5 kVp-p and a frequency of 4.0 kHz. It is important to highlight that in this study, differently than in ours, the treatment was applied indirectly (without direct contact between target and plasma) and the distribution of reactive species occurred by means of thermal convection and diffusion. Another study using CAP of Ar (98%) and O_2 (2%) observed 90% reduction in the viability of the 7-day biofilm of *E. faecalis* formed within the dental root canals of human teeth after 8 min of exposure [58].

For *P. aeruginosa* monospecies biofilms, we detected that He-CAP reduced in 2.4 log/CFU after 7 min of exposure. Alkawareed et al. [59] used a plasma jet with gas mixture (He/O_2) with voltage of 6 kV, frequency of 40 kHz and gas flow rate of 2 slm for treatment of *P. aerug-*

inosa and obtained a reduction of 4 log after treatment for 4 min. Besides, Gabriel et al. [36] reported a reduction of 5 log in *P. aeruginosa* counts after exposition for 90 s to microwave plasma jet (2.45 GHz) in air with flow rate of 5 slm, while we observed a reduction of 2.4 log after 7 min of exposure. However, it is important to mention that in both aforementioned studies the temperatures of CAP were above 40 °C, which could explain the higher reduction of the viable cells.

After determining the optimal parameters for monospecies biofilms, the effect of CAP on multispecies biofilms was studied. The study of mixed-species biofilms is relevant as, clinically, they can cause persistent infections and usually lead to worse infections [60–62]. The model of polymicrobial biofilm obtained on a porcine collagen membrane gave strong evidences that CAP can reduce the bacterial biofilm viability in complex models. The model described by Brackman e Coenye [8] for wound-related biofilms was selected because it mimics the conditions observed in wounds, by using a culture medium with a constitution close to the exudate and a nutrient flow diffusion from the bottom up. Besides, the model also mimics the air-wound-liquid interface that occurs in vivo [8].

The selection of the microbial species used to form the multispecies biofilms were based on their relevance in the context of chronic wounds infection. Previous studies reported that *S. aureus* and *P. aeruginosa* were the most prevalent species in wounds [4,11]. Besides, the presence of certain species, such as *E. faecalis*, seems to be related to worse clinical course of the wound [4,10,63].

In this study, *P. aeruginosa* was more susceptible to CAP in a multispecies biofilm, with reduction of 1.48 log. It is an unexpected result since previous report suggested that *P. aeruginosa* is one of the most benefited species in a mixed species culture, probably because it is present in a more protected niche of the biofilm [64].

It is known that CAP generates large variety of reactive species that can cause decontamination. Among them, ROS (e.g., O_3, O_2-, OH, O) are indicated as the major agent responsible for bactericidal effects of plasmas [65]. In plasma jets, the generated plasma plume excites and ionizes air molecules present in the surrounding ambient leading to formation of RONS [66]. The concentration of generated reactive species can be enhanced by adding molecular gases (e.g., air, O_2 or N_2) to the main carrier gas or applying a shielding gas curtain around the plasma plume [67]. In plasma jets, the produced reactive species are driven by the gas flow to the sample allowing short and long-living species to act on the treated surface, which can increase the treatment efficiency when compared to indirect methods. The plasma jet device used in the present study was operated with commercial low-grade purity helium that contains few impurities (mostly air) and ROS formation in the plasma plume was detected in previous studies [47,48]. Thus, the MRSA biofilm was significantly more susceptible than *P. aeruginosa* and *E. faecalis* under ROS-dominated conditions. Interestingly, the increased susceptibility of MRSA in a polymicrobial biofilm with *P. aeruginosa, E. faecalis and Klebsiella pneumoniae*, to RNS-dominated conditions were previously reported Modic et al. [31].

Our results showed that exposure to CAP for 1, 3, 5 and 7 min was not cytotoxic for fibroblasts 3T3 cells analyzed immediately after the exposure. Similarly, Borges et al. [45] found the cell viability to be higher than 80% immediately and 24 h after amplitude modulated cold atmospheric pressure plasma jet (AM-CAPPJ) exposure of Vero cells for 3 and 5 min. On the other hand, Borges et al. [45] showed that cell viability was maintained at around 80%, 48 h after AM-CAPPJ exposure, which was different from the present work where the cell viability has decreased to around 60% after 48 h. This could be explained by the internal cell signaling cascade that some reactive species may trigger as previously described [30,68,69]. Therefore, a potential cytotoxicity may be related to this lower cell viability after 48 h (late effects) in relation to the immediate time or 24 h, which could have side effects on cells. However, further chronic toxicity studies of CAP should be carried out. These late effects on fibroblasts could also exist for microorganisms and should be investigated in future studies.

Aiming the clinical application of CAP for wound care, it was extremally important to investigate whether plasma conditions would lead to DNA damage or not. In this study, all tested periods of treatments presented a DNA index between 1.2 and 1.33 (mean and median) which, according to the score, represents a slight ploidy, but taking into account that the CAP group was into the same score; median 1.24 and mean 1.27, we believe treatment was not genotoxic. Our results are coincident with others in the literature. Boxhammer et al. [70] observed no mutations, through EPRT assay, on V79 cells after 30 to 240 s of CAP treatment using a MiniFlatPlaSter device. Only a decrease in the proliferation rate was observed. Wende et al. [71] and Bekeschus et al. [72] tested the KINPen® trough different methods of evaluation, and found no signals of genotoxicity due to plasma treatment. Wende et al. [71] assays used up to 180 s of CAP in non-cancer cells (HaCat and MRC5 cells) and melanoma cells (SK-Mel-147 cells) and did not find mutagenic changes with CAP treatment as well. They used HPRT1 mutation assay, micronucleus formation assay, and clonogenic assay, to confirm their findings. Kludge et al. [73] used the KINPen MED™ to evaluate CAP genotoxicity using HET-MN model and also observed no genotoxic alterations on CAP treatment. In our study, static cytometry was chosen to calculate DNA index and, in that way, evaluate possible mutations caused by cold plasma treatment and found no genotoxic signals in cells expose up to 7 min (420 s). Different CAP sources, cell types, CAP treatment conditions, and assay models reported make it difficult to compare results, but it seems that CAP safety has been proven by literature.

The inhibitory effect of CAP on wound-related multispecies biofilm with the low cytotoxicity and no genotoxicity observed in this wound-related multispecies biofilm study encourage in vivo studies to evaluate the effects of CAP on infected wounds that can also investigate additional anti-inflammatory and tissue repair inducing effects.

5. Conclusions

In conclusion, exposure to He-CAP for 5 min showed inhibitory effect on wound-related multispecies biofilms formed by methicillin-resistant *S. aureus* (MRSA), *P. aeruginosa* and *E. faecalis*, with low cytotoxicity and genotoxicity to mammalian cells mainly immediately after exposure. These findings point out the application of this innovative technology in the control of infected chronic wounds.

Author Contributions: Conceptualization, C.Y.K.-I. and G.d.M.G.L.; methodology, M.A.C.d.O., G.d.M.G.L., T.M.C.N., A.V.L.G., B.R.C.d.M., M.V.C., K.G.K., C.Y.K.-I.; software, B.R.C.d.M., M.V.C., K.G.K.; validation, M.A.C.d.O., G.d.M.G.L., T.M.C.N., A.V.L.G., B.R.C.d.M., M.V.C., K.G.K. and C.Y.K.-I.; formal analysis, M.A.C.d.O., G.d.M.G.L., T.M.C.N., A.V.L.G., B.R.C.d.M., M.V.C., K.G.K., C.Y.K.-I.; investigation, M.A.C.d.O., G.d.M.G.L., T.M.C.N., A.V.L.G., B.R.C.d.M., M.V.C., K.G.K., C.Y.K.-I.; resources, C.Y.K.-I.; data curation, M.A.C.d.O., G.d.M.G.L.; writing—original draft preparation, M.A.C.d.O., G.d.M.G.L., T.M.C.N., A.V.L.G. and C.Y.K.-I.; writing—review and editing, M.A.C.d.O., G.d.M.G.L., T.M.C.N., A.V.L.G., B.R.C.d.M., M.V.C., K.G.K., C.Y.K.-I.; visualization, M.A.C.d.O., G.d.M.G.L. and C.Y.K.-I., supervision, G.d.M.G.L., C.Y.K.-I.; project administration, C.Y.K.-I.; funding acquisition, C.Y.K.-I., K.G.K. All authors have read and agreed to the published version of the manuscript.

Funding: This research was funded by the National Council for Scientific and Technological Development (CNPq), Grant numbers 405653/2016-6 and 308127/2018-8. This work received financial support from Coordenação de Aperfeiçoamento de Pessoal de Nível Superior—CAPES—Brazil.

Institutional Review Board Statement: Not applicable.

Informed Consent Statement: Not applicable.

Data Availability Statement: Data sharing is not applicable to this article.

Conflicts of Interest: The authors declare no conflict of interest.

References

1. Frykberg, R.G.; Banks, J. Challenges in the Treatment of Chronic Wounds. *Adv. Wound Care* **2015**, *4*, 560–582. [CrossRef] [PubMed]
2. Jones, R.E.; Foster, D.S.; Longaker, M.T. Management of Chronic Wounds—2018. *JAMA J. Am. Med. Assoc.* **2018**, *320*, 1481–1482. [CrossRef]
3. Martinengo, L.; Olsson, M.; Bajpai, R.; Soljak, M.; Upton, Z.; Schmidtchen, A.; Car, J.; Järbrink, K. Prevalence of chronic wounds in the general population: Systematic review and meta-analysis of observational studies. *Ann. Epidemiol.* **2019**, *29*, 8–15. [CrossRef] [PubMed]
4. Bjarnsholt, T. The role of bacterial biofilms in chronic infections. *APMIS. Suppl.* **2013**, 1–51. [CrossRef]
5. Olsson, M.; Järbrink, K.; Divakar, U.; Bajpai, R.; Upton, Z.; Schmidtchen, A.; Car, J. The humanistic and economic burden of chronic wounds: A systematic review. *Wound Repair Regen.* **2019**, *27*, 114–125. [CrossRef] [PubMed]
6. Wu, Y.K.; Cheng, N.C.; Cheng, C.M. Biofilms in Chronic Wounds: Pathogenesis and Diagnosis. *Trends Biotechnol.* **2019**, *37*, 505–517. [CrossRef]
7. Kim, P.J.; Steinberg, J.S. Wound care: Biofilm and its impact on the latest treatment modalities for ulcerations of the diabetic foot. *Semin. Vasc. Surg.* **2012**, *25*, 70–74. [CrossRef]
8. Brackman, G.; Coenye, T. In vitro and in vivo biofilm wound models and their application. *Adv. Exp. Med. Biol.* **2016**, *897*, 15–32. [CrossRef]
9. Tytgat, H.L.P.; Nobrega, F.L.; van der Oost, J.; de Vos, W.M. Bowel Biofilms: Tipping Points between a Healthy and Compromised Gut? *Trends Microbiol.* **2019**, *27*, 17–25. [CrossRef]
10. Rhoads, D.D.; Wolcott, R.D.; Sun, Y.; Dowd, S.E. Comparison of culture and molecular identification of bacteria in chronic wounds. *Int. J. Mol. Sci.* **2012**, *13*, 2535–2550. [CrossRef]
11. Gjødsbøl, K.; Christensen, J.J.; Karlsmark, T.; Jørgensen, B.; Klein, B.M.; Krogfelt, K.A. Multiple bacterial species reside in chronic wounds: A longitudinal study. *Int. Wound J.* **2006**, *3*, 225–231. [CrossRef]
12. Bessa, L.J.; Fazii, P.; Di Giulio, M.; Cellini, L. Bacterial isolates from infected wounds and their antibiotic susceptibility pattern: Some remarks about wound infection. *Int. Wound J.* **2015**, *12*, 47–52. [CrossRef]
13. Pallavali, R.R.; Degati, V.L.; Lomada, D.; Reddy, M.C.; Durbaka, V.R.P. Isolation and in vitro evaluation of bacteriophages against MDR-bacterial isolates from septic wound infections. *PLoS ONE* **2017**, *12*, e0179245. [CrossRef]
14. Haertel, B.; von Woedtke, T.; Weltmann, K.D.; Lindequist, U. Non-thermal atmospheric-pressure plasma possible application in wound healing. *Biomol. Ther.* **2014**, *22*, 477–490. [CrossRef]
15. Edwards, K. New Twist on an Old Favorite: Gentian Violet and Methylene Blue Antibacterial Foams. *Adv. Wound Care* **2016**, *5*, 11–18. [CrossRef]
16. Julák, J.; Scholtz, V.; Vaňková, E. Medically important biofilms and non-thermal plasma. *World J. Microbiol. Biotechnol.* **2018**, *34*, 1–15. [CrossRef]
17. Wolcott, R. Are chronic wounds, chronic infections? *J. Wound Care* **2016**, *25*, S33. [CrossRef] [PubMed]
18. Han, G.; Ceilley, R. Chronic Wound Healing: A Review of Current Management and Treatments. *Adv. Ther.* **2017**, *34*, 599–610. [CrossRef] [PubMed]
19. Oryan, A.; Alemzadeh, E.; Moshiri, A. Burn wound healing: Present concepts, treatment strategies and future directions. *J. Wound Care* **2017**, *26*, 5–19. [CrossRef]
20. Panahi, Y.; Izadi, M.; Sayyadi, N.; Rezaee, R.; Member, F.; Beiraghdar, F.; Member, F.; Zamani, A.; Sahebkar, A.; Member, F. Comparative trial of Aloe vera/olive oil combination cream versus phenytoin cream in the treatment of chronic wounds. *J. Wound Care* **2015**, *24*, 459–465. [CrossRef] [PubMed]
21. Craven, M.; Kasper, S.H.; Canfield, M.J.; Diaz-Morales, R.R.; Hrabie, J.A.; Cady, N.C.; Strickland, A.D. Nitric oxide-releasing polyacrylonitrile disperses biofilms formed by wound-relevant pathogenic bacteria. *J. Appl. Microbiol.* **2016**, *120*, 1085–1099. [CrossRef]
22. Napavichayanun, S.; Yamdech, R.; Aramwit, P. The safety and efficacy of bacterial nanocellulose wound dressing incorporating sericin and polyhexamethylene biguanide: In vitro, in vivo and clinical studies. *Arch. Dermatol. Res.* **2016**, *308*, 123–132. [CrossRef] [PubMed]
23. Finnegan, S.; Percival, S.L. EDTA: An Antimicrobial and Antibiofilm Agent for Use in Wound Care. *Adv. Wound Care* **2015**, *4*, 415–421. [CrossRef] [PubMed]
24. Thana, P.; Wijaikhum, A.; Poramapijitwat, P.; Kuensaen, C.; Meerak, J.; Ngamjarurojana, A.; Sarapirom, S.; Boonyawan, D. A compact pulse-modulation cold air plasma jet for the inactivation of chronic wound bacteria: Development and characterization. *Heliyon* **2019**, *5*, e02455. [CrossRef] [PubMed]
25. Bekeschus, S.; Favia, P.; Robert, E.; von Woedtke, T. White paper on plasma for medicine and hygiene: Future in plasma health sciences. *Plasma Process. Polym.* **2019**, *16*, 1–12. [CrossRef]
26. Šimončicová, J.; Kryštofová, S.; Medvecká, V.; Ďurišová, K.; Kaliňáková, B. Technical applications of plasma treatments: Current state and perspectives. *Appl. Microbiol. Biotechnol.* **2019**, *103*, 5117–5129. [CrossRef] [PubMed]
27. Laroussi, M.; Mendis, D.A.; Rosenberg, M. Plasma interaction with microbes. *New J. Phys.* **2003**, *5*. [CrossRef]
28. Fridman, G.; Friedman, G.; Gutsol, A.; Shekhter, A.B.; Vasilets, V.N.; Fridman, A. Applied plasma medicine. *Plasma Process. Polym.* **2008**, *5*, 503–533. [CrossRef]

29. Gay-Mimbrera, J.; García, M.C.; Isla-Tejera, B.; Rodero-Serrano, A.; García-Nieto, A.V.; Ruano, J. Clinical and Biological Principles of Cold Atmospheric Plasma Application in Skin Cancer. *Adv. Ther.* **2016**, *33*, 894–909. [CrossRef] [PubMed]

30. Laroussi, M. Plasma Medicine: A Brief Introduction. *Plasma* **2018**, *1*, 47–60. [CrossRef]

31. Modic, M.; McLeod, N.P.; Sutton, J.M.; Walsh, J.L. Cold atmospheric pressure plasma elimination of clinically important single- and mixed-species biofilms. *Int. J. Antimicrob. Agents* **2017**, *49*, 375–378. [CrossRef]

32. Chatraie, M.; Torkaman, G.; Khani, M.; Salehi, H.; Shokri, B. In vivo study of non-invasive effects of non-thermal plasma in pressure ulcer treatment. *Sci. Rep.* **2018**, *8*, 1–11. [CrossRef] [PubMed]

33. Gao, J.; Wang, L.; Xia, C.; Yang, X.; Cao, Z.; Zheng, L.; Ko, R.; Shen, C.; Yang, C.; Cheng, C. Cold atmospheric plasma promotes different types of superficial skin erosion wounds healing. *Int. Wound J.* **2019**, *16*, 1103–1111. [CrossRef]

34. Gupta, T.T.; Karki, S.B.; Matson, J.S.; Gehling, D.J.; Ayan, H. Sterilization of Biofilm on a Titanium Surface Using a Combination of Nonthermal Plasma and Chlorhexidine Digluconate. *Biomed. Res. Int.* **2017**, *2017*. [CrossRef]

35. Brun, P.; Bernabè, G.; Marchiori, C.; Scarpa, M.; Zuin, M.; Cavazzana, R.; Zaniol, B.; Martines, E. Antibacterial efficacy and mechanisms of action of low power atmospheric pressure cold plasma: Membrane permeability, biofilm penetration and antimicrobial sensitization. *J. Appl. Microbiol.* **2018**, *125*, 398–408. [CrossRef] [PubMed]

36. Gabriel, A.A.; Ugay, M.C.C.F.; Siringan, M.A.T.; Rosario, L.M.D.; Tumlos, R.B.; Ramos, H.J. Atmospheric pressure plasma jet inactivation of Pseudomonas aeruginosa biofilms on stainless steel surfaces. *Innov. Food Sci. Emerg. Technol.* **2016**, *36*, 311–319. [CrossRef]

37. Theinkom, F.; Singer, L.; Cieplik, F.; Cantzler, S.; Weilemann, H.; Cantzler, M.; Hiller, K.A.; Maisch, T.; Zimmermann, J.L. Antibacterial efficacy of cold atmospheric plasma against Enterococcus faecalis planktonic cultures and biofilms in vitro. *PLoS ONE* **2019**, *14*, e0223925. [CrossRef] [PubMed]

38. Üreyen Kaya, B.; Kececi, A.D.; Güldaş, H.E.; Çetin, E.S.; Öztürk, T.; Öksuz, L.; Bozduman, F. Efficacy of endodontic applications of ozone and low-temperature atmospheric pressure plasma on root canals infected with Enterococcus faecalis. *Lett. Appl. Microbiol.* **2014**, *58*, 8–15. [CrossRef] [PubMed]

39. Li, Y.; Sun, K.; Ye, G.; Liang, Y.; Pan, H.; Wang, G.; Zhao, Y.; Pan, J.; Zhang, J.; Fang, J. Evaluation of Cold Plasma Treatment and Safety in Disinfecting 3-week Root Canal Enterococcus faecalis Biofilm in Vitro. *J. Endod.* **2015**, *41*, 1325–1330. [CrossRef]

40. Han, L.; Patil, S.; Boehm, D.; Milosavljević, V.; Cullen, P.J.; Bourke, P. Mechanisms of inactivation by high-voltage atmospheric cold plasma differ for Escherichia coli and Staphylococcus aureus. *Appl. Environ. Microbiol.* **2016**, *82*, 450–458. [CrossRef]

41. Matthes, R.; Assadian, O.; Kramer, A. Repeated applications of cold atmospheric pressure plasma does not induce resistance in Staphylococcus aureus embedded in biofilms. *GMS Hyg. Infect. Control* **2014**, *9*. [CrossRef]

42. Jacofsky, M.C.; Lubahn, C.; McDonnell, C.; Seepersad, Y.; Fridman, G.; Fridman, A.; Dobrynin, D. Spatially resolved optical emission spectroscopy of a helium plasma jet and its effects on wound healing rate in a diabetic murine model. *Plasma Med.* **2014**, *4*, 177–191. [CrossRef]

43. Von Woedtke, T.; Schmidt, A.; Bekeschus, S.; Wende, K.; Weltmann, K.D. Plasma medicine: A field of applied redox biology. *In Vivo* **2019**, *33*, 1011–1026. [CrossRef]

44. Tort, S.; Demiröz, F.T.; Coşkun Cevher, Ş.; Sarıbaş, S.; Özoğul, C.; Acartürk, F. The effect of a new wound dressing on wound healing: Biochemical and histopathological evaluation. *Burns* **2020**, *46*, 143–155. [CrossRef] [PubMed]

45. Borges, A.C.; de Morais Gouvêa Lima, G.; Mayumi Castaldelli Nishime, T.; Vidal Lacerda Gontijo, A.; Kostov, K.G.; Koga-Ito, C.Y. Amplitude-modulated cold atmospheric pressure plasma jet for treatment of oral candidiasis: In vivo study. *PLoS ONE* **2018**, *13*, e0199832. [CrossRef] [PubMed]

46. Borges, A.C.; Nishime, T.M.C.; de Moura Rovetta, S.; Lima, G.d.M.G.; Kostov, K.G.; Thim, G.P.; de Menezes, B.R.C.; Machado, J.P.B.; Koga-Ito, C.Y. Cold Atmospheric Pressure Plasma Jet Reduces Trichophyton rubrum Adherence and Infection Capacity. *Mycopathologia* **2019**, *184*, 585–595. [CrossRef] [PubMed]

47. Kostov, K.G.; Nishime, T.M.C.; Machida, M.; Borges, A.C.; Prysiazhnyi, V.; Koga-Ito, C.Y. Study of Cold Atmospheric Plasma Jet at the End of Flexible Plastic Tube for Microbial Decontamination. *Plasma Process. Polym.* **2015**, *12*, 1383–1391. [CrossRef]

48. Nishime, T.M.C.; Wagner, R.; Kostov, K.G. Study of modified area of polymer samples exposed to a he atmospheric pressure plasma jet using different treatment conditions. *Polymers* **2020**, *12*. [CrossRef]

49. Werthén, M.; Henriksson, L.; Jensen, P.Ø.; Sternberg, C.; Givskov, M.; Bjarnsholt, T. An in vitro model of bacterial infections in wounds and other soft tissues. *APMIS. Suppl.* **2010**, *118*, 156–164. [CrossRef] [PubMed]

50. Hammond, A.A.; Miller, K.G.; Kruczek, C.J.; Dertien, J.; Colmer-Hamood, J.A.; Griswold, J.A.; Horswill, A.R.; Hamood, A.N. An in vitro biofilm model to examine the effect of antibiotic ointments on biofilms produced by burn wound bacterial isolates. *Burns* **2011**, *37*, 312–321. [CrossRef]

51. Gomes, L.C.; Mergulhão, F.J. SEM analysis of surface impact on biofilm antibiotic treatment. *Scanning* **2017**, *2017*. [CrossRef]

52. International Organization of Standardization. *ISO 10993-5:2009. Biological Evaluation of Medical Devices—Part 5: Tests for In Vitro Cytotoxicity*; ISO: Geneva, Switzerland, 2009; p. 13.

53. Felisbino, M.B.; Tamashiro, W.M.S.C.; Mello, M.L.S. Chromatin remodeling, cell proliferation and cell death in valproic acid-treated HeLa cells. *PLoS ONE* **2011**, *6*, e29144. [CrossRef]

54. Lima, C.F.; Alves, M.G.O.; Carvalho, B.F.d.C.; de Lima, T.A.; Coutinho-Camillo, C.M.; Soares, F.A.; Scholz, J.; Almeida, J.D. Is DNA ploidy related to smoking? *J. Oral Pathol. Med.* **2017**, *46*, 961–966. [CrossRef]

55. Kondeti, V.S.S.K.; Phan, C.Q.; Wende, K.; Jablonowski, H.; Gangal, U.; Granick, J.L.; Hunter, R.C.; Bruggeman, P.J. Long-lived and short-lived reactive species produced by a cold atmospheric pressure plasma jet for the inactivation of Pseudomonas aeruginosa and Staphylococcus aureus. *Free Radic. Biol. Med.* **2018**, *124*, 275–287. [CrossRef]

56. Xu, Z.; Shen, J.; Zhang, Z.; Ma, J.; Ma, R.; Zhao, Y.; Sun, Q.; Qian, S.; Zhang, H.; Ding, L.; et al. Inactivation effects of non-thermal atmospheric-pressure helium plasma jet on staphylococcus aureus biofilms. *Plasma Process. Polym.* **2015**, *12*, 827–835. [CrossRef]

57. Wang, J.; Yu, Z.; Xu, Z.; Hu, S.; Li, Y.; Xue, X.; Cai, Q.; Zhou, X.; Shen, J.; Lan, Y.; et al. Antimicrobial mechanism and the effect of atmospheric pressure N 2 plasma jet on the regeneration capacity of Staphylococcus aureus biofilm. *Biofouling* **2018**, *34*, 935–949. [CrossRef]

58. Pan, J.; Sun, K.; Liang, Y.; Sun, P.; Yang, X.; Wang, J.; Zhang, J.; Zhu, W.; Fang, J.; Becker, K.H. Cold plasma therapy of a tooth root canal infected with enterococcus faecalis biofilms in vitro. *J. Endod.* **2013**, *39*, 105–110. [CrossRef]

59. Alkawareek, M.Y.; Algwari, Q.T.; Laverty, G.; Gorman, S.P.; Graham, W.G.; O'Connell, D.; Gilmore, B.F. Eradication of Pseudomonas aeruginosa Biofilms by Atmospheric Pressure Non-Thermal Plasma. *PLoS ONE* **2012**, *7*, e44289. [CrossRef] [PubMed]

60. Sun, Y.; Dowd, S.E.; Smith, E.; Rhoads, D.D.; Wolcott, R.D. In vitro multispecies Lubbock chronic wound biofilm model. *Wound Repair Regen.* **2008**, *16*, 805–813. [CrossRef] [PubMed]

61. Trifilio, S.; Zhou, Z.; Fong, J.L.; Zomas, A.; Liu, D.; Zhao, C.; Zhang, J.; Mehta, J. Polymicrobial bacterial or fungal infections: Incidence, spectrum of infection, risk factors, and clinical outcomes from a large hematopoietic stem cell transplant center. *Transpl. Infect. Dis.* **2015**, *17*, 267–274. [CrossRef] [PubMed]

62. Rao, Y.; Shang, W.; Yang, Y.; Zhou, R.; Rao, X. Fighting Mixed-Species Microbial Biofilms with Cold Atmospheric Plasma. *Front. Microbiol.* **2020**, *11*, 1–11. [CrossRef] [PubMed]

63. Fazli, M.; Bjarnsholt, T.; Kirketerp-Møller, K.; Jørgensen, B.; Andersen, A.S.; Krogfelt, K.A.; Givskov, M.; Tolker-Nielsen, T. Nonrandom distribution of Pseudomonas aeruginosa and Staphylococcus aureus in chronic wounds. *J. Clin. Microbiol.* **2009**, *47*, 4084–4089. [CrossRef]

64. Touzel, R.E.; Sutton, J.M.; Wand, M.E. Establishment of a multi-species biofilm model to evaluate chlorhexidine efficacy. *J. Hosp. Infect.* **2016**, *92*, 154–160. [CrossRef]

65. Lu, X.; Ye, T.; Cao, Y.; Sun, Z.; Xiong, Q.; Tang, Z.; Xiong, Z.; Hu, J.; Jiang, Z.; Pan, Y. The roles of the various plasma agents in the inactivation of bacteria. *J. Appl. Phys.* **2008**, *104*. [CrossRef]

66. Laroussi, M. Low-Temperature Plasma Jet for Biomedical Applications: A Review. *IEEE Trans. Plasma Sci.* **2015**, *43*, 703–712. [CrossRef]

67. Schmidt-Bleker, A.; Winter, J.; Iseni, S.; Dünnbier, M.; Weltmann, K.D.; Reuter, S. Reactive species output of a plasma jet with a shielding gas device—Combination of FTIR absorption spectroscopy and gas phase modelling. *J. Phys. D Appl. Phys.* **2014**, *47*. [CrossRef]

68. Barekzi, N.; Laroussi, M. Fibroblast cell morphology altered by low-temperature atmospheric pressure plasma. *IEEE Trans. Plasma Sci.* **2014**, *42*, 2738–2739. [CrossRef]

69. Shahriar, M.A.; Morteza, A.; Hajar, R.; Reza, K.M.; Babak, S. Feasibility of Leukemia Cancer Treatment (K562) by Atmospheric Pressure Plasma Jet. *Int. J. Phys. Math. Sci.* **2014**, *8*, 719–722.

70. Boxhammer, V.; Li, Y.F.; Köritzer, J.; Shimizu, T.; Maisch, T.; Thomas, H.M.; Schlegel, J.; Morfill, G.E.; Zimmermann, J.L. Investigation of the mutagenic potential of cold atmospheric plasma at bactericidal dosages. *Mutat. Res. Genet. Toxicol. Environ. Mutagen.* **2013**, *753*, 23–28. [CrossRef]

71. Wende, K.; Bekeschus, S.; Schmidt, A.; Jatsch, L.; Hasse, S.; Weltmann, K.D.; Masur, K.; von Woedtke, T. Risk assessment of a cold argon plasma jet in respect to its mutagenicity. *Mutat. Res. Genet. Toxicol. Environ. Mutagen.* **2016**, *798–799*, 48–54. [CrossRef]

72. Bekeschus, S.; Schmidt, A.; Kramer, A.; Metelmann, H.R.; Adler, F.; von Woedtke, T.; Niessner, F.; Weltmann, K.D.; Wende, K. High throughput image cytometry micronucleus assay to investigate the presence or absence of mutagenic effects of cold physical plasma. *Environ. Mol. Mutagen.* **2018**, *59*, 268–277. [CrossRef] [PubMed]

73. Kluge, S.; Bekeschus, S.; Bender, C.; Benkhai, H.; Sckell, A.; Below, H.; Stope, M.B.; Kramer, A. Investigating the mutagenicity of a cold argon-plasma jet in an HET-MN model. *PLoS ONE* **2016**, *11*, e0160667. [CrossRef] [PubMed]

Article

Effects of O_2 Addition on the Discharge Parameters and Production of Reactive Species of a Transferred Atmospheric Pressure Plasma Jet

Fellype Nascimento *, **Kleber Petroski** and **Konstantin Kostov**

Faculty of Engineering in Guaratinguetá, São Paulo State University-UNESP, Guaratinguetá 12516-410, Brazil; kleber.a.petroski@gmail.com (K.P.); konstantin.kostov@unesp.br (K.K.)
* Correspondence: fellype@gmail.com

Abstract: The therapeutic effects of atmospheric pressure plasma jets (APPJs) have been associated with the presence of reactive species, mainly the reactive oxygen and nitrogen ones, generated in this kind of plasmas. Due to that, many studies attempting to enhance the production of reactive species in APPJs have been performed. The employment of gas admixtures, usually mixing a noble gas with oxygen (O_2) or water vapor, is one of the most common methods to achieve such goal. This work presents a study of how the addition of small amounts of O_2 affects the electrical parameters and the production of reactive species in a transferred APPJ produced at the tip of a long and flexible plastic tube. The study was carried out employing helium (He) as the working gas and applying a high voltage (HV) in the form of amplitude-modulated sine waveform (burst mode). With this configuration it was possible to verify that the O_2 addition reduces the discharge power and effective current, as a result of late ignition and shorter discharge duration. It was also found that the addition of O_2 to a certain content in the gas admixture makes the light emission from oxygen atoms increase, indicating an increment in oxygen related reactive species in the plasma jet. However, at the same time the light emitted from hydroxyl (OH) and nitric oxide (NO) exhibits the opposite behavior, i.e., decrease, indicating a reduction of such species in the APPJ. For these reasons, the addition of O_2 to the working gas seems to be useful for increasing the effectiveness of the plasma treatment only when the target modification effect is directly dependent on the content of atomic oxygen.

Keywords: DBD plasma; plasma jets; plasma properties; reactive species; RONS

Citation: Nascimento, F.; Petroski, K.; Kostov, K. Effects of O_2 Addition on the Discharge Parameters and Production of Reactive Species of a Transferred Atmospheric Pressure Plasma Jet. *Appl. Sci.* **2021**, *11*, 6311. https://doi.org/10.3390/app11146311

Academic Editor: Andrei Vasile Nastuta

Received: 14 June 2021
Accepted: 2 July 2021
Published: 8 July 2021

Publisher's Note: MDPI stays neutral with regard to jurisdictional claims in published maps and institutional affiliations.

1. Introduction

In recent years atmospheric pressure plasma jets (APPJs) produced in open environments have received a lot of attention, not only because of their versatility and ease of use, but also due to the encouraging results achieved in the most diverse applications [1–5]. Special attention has been given to medical and biological applications of APPJs, whose beneficial effects have been attributed to the reactive oxygen and nitrogen species (RONS) produced by the plasma jets [6–8]. The actions of the RONS are not only limited to biological materials, since they also contribute to the hydrophilization and surface activation processes of many materials treated with plasma [9,10].

Many studies indicated that the addition of small amounts of oxygen (O_2) to the gas employed to produce plasma, usually argon (Ar) or helium (He), can increase the amount of atomic oxygen (O) produced in the APPJ [11–17]. To verify this increase in the population of O most authors have measured the light emission from excited O atoms coming from the 777 nm triplet or, in some cases, from the 844 nm triplet [13]. In some works, it was found that there is an optimal O_2 percentage that maximizes the intensity of the O emission lines. There are also works in which the authors reported that some applications of the plasma jets, with both inorganic and biological materials as a target, showed better results when there was a small O_2 addition to the plasma. However, only a few works analyzed the production of other reactive species, such as OH and NO, together with the O production. In particular, Li et al [15] studied the production efficiency of RONS as a function of the O_2 content in the He-O_2 admixture, in the range from 0–2% of the total gas flow rate, and applying different voltage waveforms to generate APPJs. In that work, Li et al have found that the production efficiency of both OH and NO decreases with O_2 addition but the opposite occurs for the O_3 and NO_2 molecules. In the case of the production efficiency of O atoms, a peak value was observed for 0.5% of O_2 content. Those results were almost the same for any HV waveform applied to generate the APPJ.

Regarding the effects of additional O_2 on APPJ parameters, the main changes reported in the literature are decrease in jet length, reduction on discharge power and current, and variations in temperatures (rotational and vibrational) [18–24]. Some works also identified an increment in the applied voltage as a function of the O_2 content, even with the power source configured to apply a fixed voltage to the electrodes, a fact that suggests a change in the load impedance followed by an impedance matching [25,26]. Lazzaroni et al [27] performed an investigation on the stability of APPJs as a function of the O_2 content in the working gas which, based on a global model for APPJs produced with He-O_2 admixtures, has shown that there is no discharge equilibrium if the oxygen fraction exceeds 1% in a configuration that employs a power source with frequency of 13.56 MHz and a gap of 1 mm between the electrodes. The discharge dynamics in dielectric barrier discharge (DBD) plasmas with O_2 in the gas admixture was studied only using N_2-O_2 in a work aimed to investigate the transition from diffuse to filamentary regime [28].

Most studies found in the literature evaluate the production of RONS for only one fixed flow rate for the working gas, varying solely the amount of O_2 added to it. The present paper presents results using two different flow rates for the working gas together with the variation in the percentage of O_2 added to the plasma. The analysis of the RONS production as a function of the amount of O_2 present in the gas admixture is also performed for different distances from plasma outlet (in this case the end of a long and flexible plastic tube) to the target. This work also aims to evaluate the possible changes in the electrical parameters of plasma jets, mainly the discharge power and the effective current when small amounts of O_2 are added to the working gas. In addition, the unique combination of a plasma jet produced at the end tip of a long and flexible plastic tube with a high voltage (HV) presenting a sinusoidal-burst waveform can provide insights about the mechanisms involved in the interaction between the additional O_2 and the working gas in the generation of plasma jets.

2. Materials and Methods

Figure 1a shows the experimental setup used in this work, with detailed views of the light collection scheme shown in Figure 1b. Figure 1c presents an example of the waveform applied to the high voltage (HV) electrode without plasma ignition. The plasma jet device presented in Figure 1a consists mainly of a DBD type reactor, composed by a metal pin electrode placed inside a closed-end quartz tube, which in turn is placed inside a dielectric chamber. The working gas is fed into the chamber and flushed to the ambient air through a 1-meter long and flexible plastic tube connected to the reactor output. The plastic tube material is nylon-6, with outer and inner diameters equal to 4.0 mm and 2.0 mm, respectively. A conducting wire with 0.25 mm in diameter is put inside the plastic tube (see the floating electrode in Figure 1a). This floating electrode penetrates a few millimeters inside the reactor in a way that it does not touch the quartz tube. The other wire tip ends 2 mm before the plastic tube exit. When the primary discharge inside the DBD reactor is on it polarizes the floating electrode and a small plasma jet is ignited at the end of the plastic tube. This remote plasma jet was directed towards a copper plate, placed at a distance **d** from the plasma outlet, which is connected to an electric circuit whose impedance follows an international standard (IEC 60601-1). It is a simplified electrical circuit model aimed to mimic the electrical properties of the human body [29]. To obtain controllable gas admixtures of He and O_2 as a working gas, both gases have their flow rates controlled by mass flow controllers and are introduced into a gas mixer prior to entering the DBD reactor.

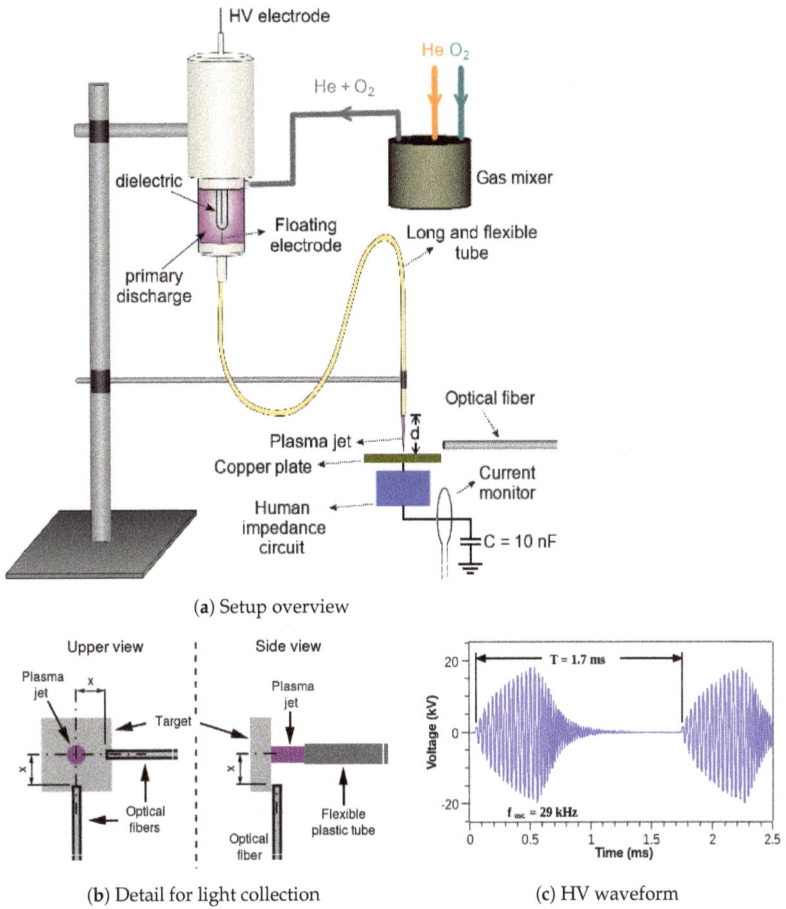

(**a**) Setup overview

(**b**) Detail for light collection   (**c**) HV waveform

Figure 1. (**a**) Schematic of the experimental setup. (**b**) Details of the arrangement for the optical fibers. (**c**) Example of a typical HV waveform applied to the pin electrode without producing plasma.

The power source employed to produce the plasma was a commercial AC generator (Minipuls4 GBS Elektronik GmbH, Germany). We choose to apply the specific HV waveform shown in Figure 1c instead of a pure sinusoidal signal because the former produces lower ohmic heating on the experimental setup components shown in Figure 1a. That HV waveform presents a sinusoidal HV "burst" with an oscillation frequency (f_{osc}) of 29 kHz followed by a voltage off period, which repeats at a repetition period (T) of 1.7 ms.

To obtain the mean discharge power (P_{dis}), simultaneous measurements of the voltage applied on the powered electrode and the voltage across a serial capacitor (C = 10 nF) were carried out. The calculation of P_{dis} values takes into account all voltage oscillations in each burst that produce a charge variation in C. To measure the applied voltage a 1000:1 voltage probe (Tektronix model P6015A) was used. The signals waveforms were recorded using a 200 MHz oscilloscope (Tektronix model 2024B). Then, the P_{dis} value is calculated by summing the area of the $q - V$ Lissajous figures formed between the voltage ($V(t)$) and charge ($q(t)$) signals, divided by the burst period T that is [30–32]:

$$P_{dis} = \frac{1}{T} \oint q(V)dV \tag{1}$$

By applying the Green's theorem to (1), P_{dis} can be calculated using:

$$P_{dis} = \frac{1}{2T} \int_{t_1}^{t_2} \left[V(t)q'(t) - V'(t)q(t) \right] dt \tag{2}$$

where $\{V'(t), q'(t)\} = d\{V(t), q(t)\}/dt$. Equation (2) has the advantage that it can be used to calculate the area of the Lissajous figure without the need to plot the $q - V$ curve.

We also measured the waveform for the current that passes through the system using a current monitor from Pearson^{TM} (model 4100). Then in each case the recorded signal was used to perform the calculation of effective discharge current (i_{RMS}).

To observe the light emission from multiple species at once and evaluate the production of the OH and NO species in the plasma a broad-band optical emission spectroscopy (OES) in the wavelength range from 200 nm to 750 nm was performed using a multi-channel spectrometer from Avantes (model AvaSpec-ULS2048X64T), with spectral resolution (FWHM) equal to 0.76 nm. More precise spectroscopic measurements in the 730–840 nm range were made with a multi-channel spectrometer from Horiba (model MicroHR), with FWHM equal to 0.42 nm, which is aimed to monitor mainly the line emissions from the O I ($^5P \to {}^5S$) triplet at $\lambda = 777$ nm and estimate the production of atomic oxygen in the plasma jet.

For both spectrometers, the light emitted by the plasma jet was collected using optical fibers placed parallel to the surface target. The collection scheme is depicted in Figure 1b. Both optical fibers have numerical apertures (NA) of 0.22, and were placed in the same plane at 90 degrees each other. The only relevant difference between the optical fibers is in their core diameters-1000 μm for that connected to the Avantes spectrometer and 100 μm for the other. The distance **x** between the center of the plasma column and the fiber optic light input are also the same and equal to 5 mm, so, by combining the **x** value, NA and core diameters of the optical fibers, the lengths of the plasma jet seen by each optical fiber are approximately 2.5 mm and 1.6 mm long for the Avantes and Horiba spectrometers, respectively.

Due to the absence of absolute calibration of the spectral intensity of the spectrometers employed to measure the intensities of emissions of lines and bands of the spectra, the analysis of the production of reactive species will be more qualitative than quantitative. However, we can use the relationship between intensity of emission and density of emitting species (I_{specie} and n_{specie}, respectively) to assess the number of species produced in the plasma, i.e.: $I_{specie} \propto n_{specie}$.

Spectroscopic measurements were also used to obtain rotational and vibrational temperature values (T_{rot} and T_{vib}, respectively) of N_2 molecules. To obtain those T_{rot} and T_{vib} values, we used spectroscopic emissions from the N_2 second positive system, $C^3\Pi_u, v' \to B^3\Pi_g, v''$ (referred as $N_2(C \to B)$ hereafter), with $\Delta v = v' - v'' = -1$, in the wavelength range from 345 to 360 nm [33–36]. Spectra simulations were performed using the massiveOES software [37,38]. Thus, comparisons between measured and simulated spectra are performed and temperature values are determined by those that generate simulated curves that best fit to the experimental spectra. The spectral resolution provided by the Avantes spectrometer is not sufficient to resolve the rotational levels of the N_2 molecules, which is a requirement to obtain accurate values for the T_{rot} parameter. However, there is a direct relationship between the shape and broadening of the N_2 vibrational bands and the variation of the T_{rot} values, being that the higher the T_{rot}, the larger the broadening and the higher the intensity of the rotational lines in the vibrational bands. Both effects cause a change in the shape of the vibrational bands in that part that degrades to violet, causing it to become higher and wider, allowing the estimation of the T_{rot} values using low-resolution spectrometers. Thus, even not very accurate, the obtained T_{rot} values can be good enough to show the trend of that parameter.

It is important to notice that the spectroscopic measurements used to obtain T_{rot} were performed in close proximity of the target surface. Therefore, in this case, the T_{rot} values may be different from the T_{gas} ones [34].

The results from both the electrical and the spectroscopic measurements were analyzed as a function of the O_2 percentage added to the He gas for different flow rates. To assess the jet electrical parameters the O_2 percentage ranged from 0 to 5% in steps of 1%, while for the OES the O_2 content ranged from 0 to 2.5% in steps lower than or equal to 0.5%. In addition, the discharge power was evaluated as a function of both He flow rate and distance **d** between the plasma outlet and the target surface, without additional O_2 in the two cases. OES measurements were also performed for different **d** values. In this work, we take the additional O_2 percentage as the ratio between the O_2 and He flow rates ($Q_{O_2}/Q_{He} \cdot 100\%$). The O_2 present in the atmosphere is not computed since its amount is supposed to remain unchanged regardless the amount of O_2 in the gas admixture.

3. Results and Discussion

3.1. Effects on APPJ Electrical Parameters

The behavior of P_{dis} as a function of different parameters varied in this work are shown in Figure 2, which also depicts the behavior of i_{RMS} as a function of the O_2 added to the plasma. In Figure 2a we can see that the P_{dis} values increase monotonically with the increment in He flow rate. In that case, the distance between plasma outlet and the target (**d**) was 7 mm. Regarding the influence of **d** in the P_{dis} values, it can be seen in Figure 2b that the higher the **d** the higher the P_{dis} until **d** is too large for a stable discharge to be formed. The data in Figure 2a,b were obtained using pure He as the working gas.

Concerning the influence of the additional O_2 content, the results obtained for P_{dis} and i_{RMS} shown in Figure 2c,d present the first attempt to understand how these quantities evolve as a function of the O_2 percentage added to the working gas. The measurements of P_{dis} and i_{RMS} for He flow rates of 1.5 and 2.0 L/min were performed changing the additional O_2 content from 0 to 5%, in steps of 1%. From the curves shown in Figure 2c it can be seen that in the range of 0–2% of additional O_2 only small variations in the P_{dis} values were detected. Above 2% of O_2 more pronounced decrease of mean discharge power was observed. Additionally, the values obtained for P_{dis} using a He flow rate of 2.0 L/min change considerably less than when the He flow rate is 1.5 L/min. On the other hand, it can be seen in Figure 2d that the i_{RMS} values decrease monotonically as more O_2 is added to the system, which is a behavior that can be considered an advantage for in vivo applications of the plasma jet.

To understand the reasons behind the behavior of the P_{dis} and i_{RMS} curves shown in Figure 2c,d, one can inspect the waveforms of the V, q and i signals that were acquired for different quantities of O_2 in the range of 0–5%. Those waveforms are presented in Figure 3, which shows an interesting time behavior of V, q and i depending on the additional concentration of O_2 present in the working gas. Notice that for better comparison of the evolution of these quantities as a function of O_2 content, all plots shown in Figure 3 are presented in the same scale (see the caption of Figure 3).

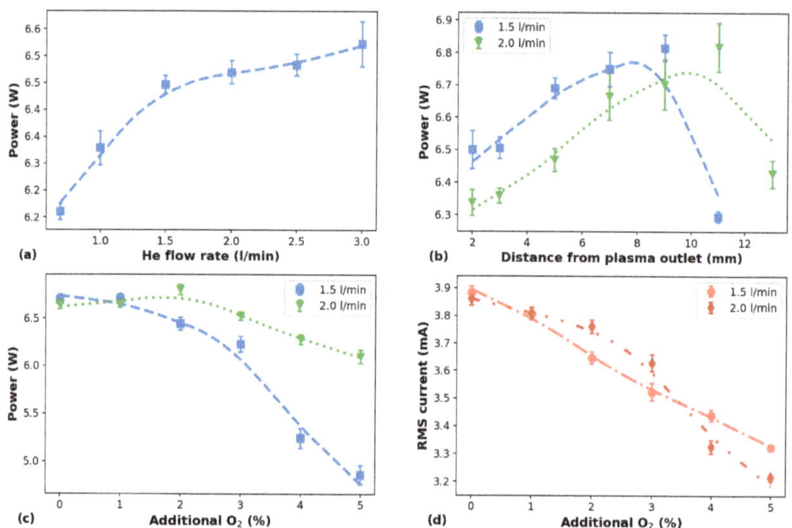

Figure 2. Curves of P_{dis} as a function of the He flow rate (**a**) and distance between plasma outlet and target (**b**), and curves of P_{dis} (**c**) and i_{RMS} (**d**) as a function of the O_2 percentage added to the working gas. **d** = 7 mm in (**a,c,d**). O_2 content is zero in (**a,b**).

From the waveforms presented in Figure 3 one can see that for an O_2 concentration higher than 1%, at least three main effects take place as more O_2 is added to the plasma: the peak values of the applied voltage (V_{peak}) slightly increase, the discharge ignition (indicated by sudden rises in the charge and current waveforms) occurs later and consequently the discharge duration tends to be shorter. It is very likely that the shorter duration of the discharge due to the delayed ignition is responsible for the decrease in the P_{dis} and i_{RMS} values presented in Figure 2c,d. The increase in the voltage values can explain the increment in the peak current measurements, which is another effect of the addition of O_2 observed in Figure 3. All waveforms in Figure 3 were obtained with a He flow rate of 1.5 L/min. However, very similar behaviors for the V, q and i waveforms were observed when different He flow rates, as well as different distances between the plasma outlet and the target were used.

The increase in the V_{peak} values with more O_2 present in the working gas, and also the reduction of P_{dis} and i_{RMS} values, are trends already observed in other works studying the addition of O_2 to the plasma [25,26]. However, since most of these experiments employed sinusoidal voltage waveforms to produce plasma jets, in none of those cases it was possible to observe a delay in the ignition of the discharge, even when a pulsed power source was used. Brandenburg et al [28] observed a delay in the active current when applying a sinusoidal HV to produce a N_2-O_2 DBD plasma in a closed environment. However, in that work the maximum O_2 concentration in the working gas was 1200 ppm (0.12%). Therefore, the results presented in Figure 3 provide more detailed information about the physical phenomena resulting from the addition of O_2 to the plasma.

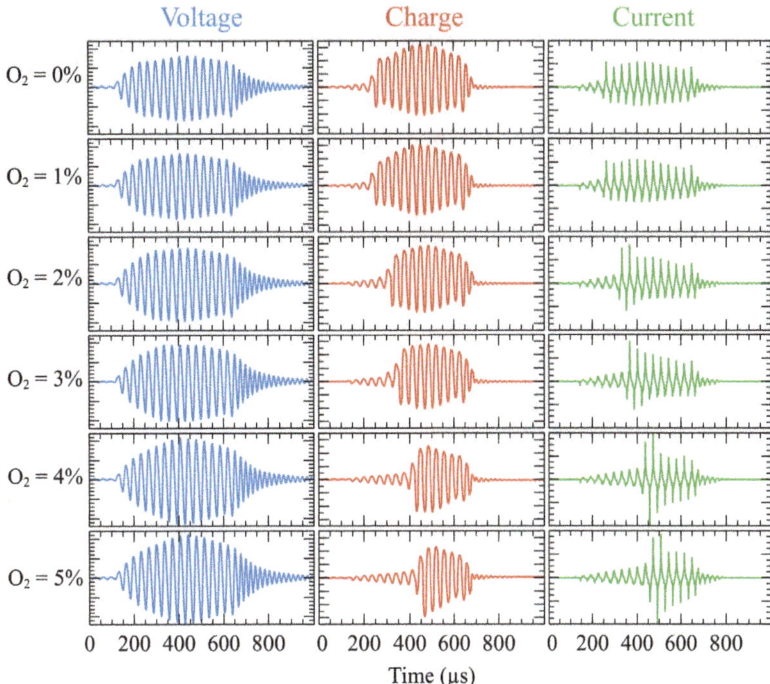

Figure 3. Waveforms of applied voltage, capacitor charge and current for various concentrations of additional O_2. The He flow rate was 1.5 L/min and the distance **d** was 7 mm. Vertical scales are constant for each parameter: Voltage: ±12 kV; Charge: ±70 nC; Current: ±40 mA.

Moreover, by inspecting the i waveforms measured for different O_2 concentrations, it is possible to observe the appearance of narrow peaks in the current signal when there is more O_2 in the plasma. This finding suggests that as more O_2 is added the plasma undergoes a transition from a diffuse to a filamentary regime, just as it happened in [28] using a N_2-O_2 gas admixture.

Part of the results presented in Figures 2c,d and 3, especially the reduction of the i_{RMS} values at higher in O_2 content, can be explained by the high electronegativity of the O atoms that tend to capture electrons produced in the plasma. This in turn probably reduces the number of free electrons that are responsible for establishing the electrical current in the plasma jet. However, following an argument given in [28], another possibility is that as more O_2 is introduced into the plasma, most He atoms in metastable states (He^M) first collide with the O_2 molecules, which reduces the population of secondary electrons that come into the plasma due to the collision of He^M atoms with the surfaces of the system.

Regarding the delay in the discharge ignition for higher O_2 concentrations, a possible explanation has also a relationship with the oxygen electronegativity, since the electrons accelerated by the variation of the electric field in the very first voltage oscillations of the burst signal may have been absorbed by the exceeding O_2 molecules and O atoms, making the discharge ignition more difficult until finally there are enough energetic electrons left to start the discharge.

3.2. Effects on the Production of Reactive Species

Figure 4 presents an overview of the emission spectra, without additional O_2 in the gas admixture, obtained using (a) the Avantes spectrometer, for the 200–750 nm wavelength range and (b) the Horiba one, for wavelengths between 730 nm and 840 nm. The employed He flow rate was 2.0 L/min and the distance between the plasma outlet and the target was

$\mathbf{d} = 3$ mm. The main emitting species observed in Figure 4a are NO, from 200 nm to 270 nm, OH at 288 nm, 296 nm and 308 nm, (the last two being jeopardized by N_2 emissions), N_2 from 298 nm to 450 nm and a He line emission at 587 nm. The emitting species observed in Figure 4b are the O I triplet at 777 nm and N_2 and N_2^+ in second order emissions.

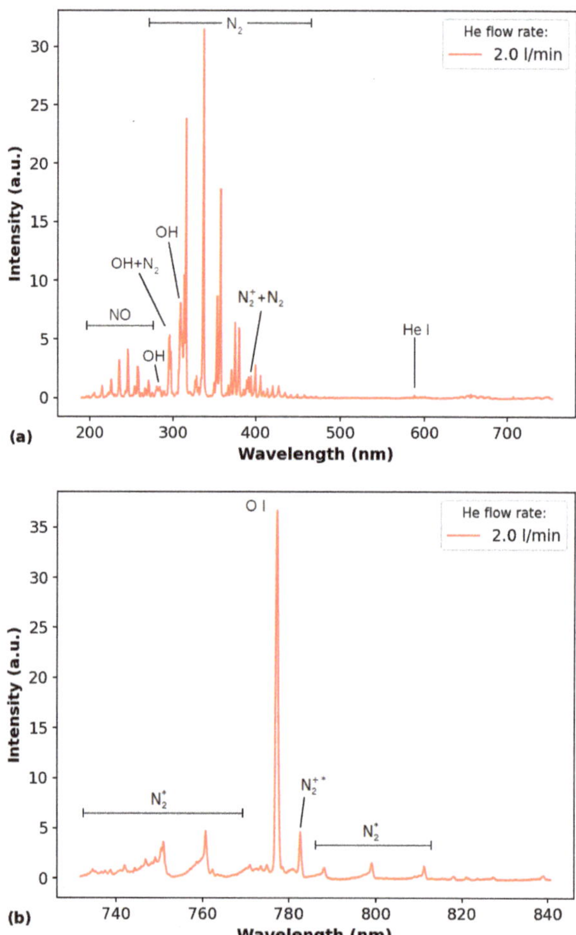

Figure 4. Overview of the emission spectrum of the plasma jet obtained using (**a**) the Avantes spectrometer and (**b**) the Horiba one with 2.0 L/min of He flow rate without additional O_2. The distance between the plasma outlet and the target was $\mathbf{d} = 3$ mm. The asterisks in the notations in (**b**) indicate second order emissions from N_2 and N_2^+.

To investigate how the addition of O_2 to the working gas affects the production of RONS in the plasma we performed measurements for different O_2 contents obtaining the intensity of the light emitted by the following species: $NO(A \rightarrow X)$ emitting at 247 nm (to be referred as NO_{247} hereafter), $OH(A \rightarrow X)$ emitting at 288 nm (OH_{288}) and O I at 777 nm (O_{777}). Although the emission at 288 nm is not the most intense coming from the OH molecule, it was chosen because it is isolated from emissions of other species, mainly those from N_2. Since the Avantes spectrometer, used for the measurements in the 200–750 nm wavelength range, has low resolution it would not be possible to separate emissions from different species emitting at close wavelengths.

In Section 3.1 we have found that although the discharge duration starts to decrease for more than 1% of O_2 in the working gas, the P_{dis} values do not change considerably for an O_2 percentage lower than 2%. Thus, since most applications require higher P_{dis} values and a longer discharge duration, we decided to analyze in more details the production of the NO, OH and O species only in the range from 0 to 2.5% of extra O_2 in the gas admixture. Figure 5 shows the emission intensities of the O_{777} triplet and the NO_{247} and OH_{288} bands as a function of the additional O_2 percentage for different He flow rates measured for a distance **d** = 3 mm in (a), and **d** = 5 mm in (b).

By comparing the curves obtained for the O_{777} triplet in Figure 5a, for a distance of 3 mm between the plasma outlet and the target at different He flow rates, it can be seen that neither the position of maximal emission intensity nor the profiles of the curves change significantly when He flow rate scales from 1.0 to 2.0 L/min. The intensity values of the O_{777} emissions increase significantly only when the He flow rate is 2.5 L/min. Regarding the gas flow rate effect on the NO emission intensities as functions of the oxygen content, they exhibit almost the same behavior as O_{777} emissions. On the other hand, the intensities of OH emissions seemingly do not change if the He flow rate is kept in the 1.5–2.5 L/min range, but at low O_2 concentration and 1.0 L/min gas flow rate the OH emission intensity decreases.

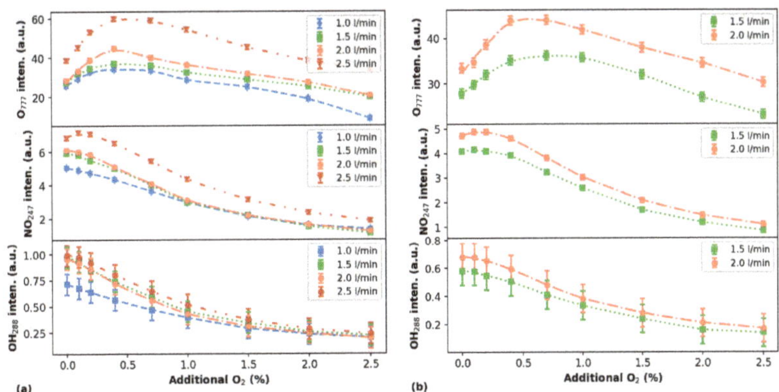

Figure 5. Intensities of selected lines/bands from O, NO, and OH radicals as a function of the additional O_2 percentage for different He flow rates (indicated in the curve legends). The data were acquired for distances **d** = 3 mm in (a) and **d** = 5 mm in (b). Symbols represent experimental data and lines are trend curves.

Concerning the shape of the curves shown in Figure 5a, in general both NO and OH emissions show similar trends that are different from the ones obtained for O_{777}. For instance, when the oxygen content is increased only the O_{777} emissions present a clear non-monotonical behavior, reaching a peak value at ~0.6%. By contrast, the NO and OH emissions do not present such well-defined maximum, with an exception for the NO emission when the He flow rate is 2.5 L/min. In this case, when the additional O_2 in the gas is close to 0.1–0.2% a small increase in NO emission is detected.

Regarding the curves in Figure 5b obtained for a distance of 5 mm between the plasma outlet and the target, it can be seen that the measured emission intensities employing He flow rates of 1.5 L/min and 2.0 L/min change significantly for all emitting species. Significant differences in the intensity values for the O_{777} emission are measured across the entire range, while the emissions differences for both NO and OH are notable only when the additional O_2 percentage is lower than 0.5% and, after that, the differences become smaller.

Comparing the NO and OH emissions in Figure 5a,b a point that is worth highlighting is that in the second case, for a higher distance between the plasma outlet and the target, the NO emissions present a maximum value in its intensity when a very small amount of

O_2 is added to the plasma. This behavior was predicted by some simulations performed using the global model for APPJs and verified in some experimental works [39]. On the other hand, the behavior of OH and NO emissions presented in Figure 5a, for the lower distance between the plasma outlet and the target, agrees with some results reported in the literature [15,40].

Figure 6 shows the profiles of RONS as a function of the additional O_2 for different distances between the plasma outlet and the target.

Regarding the intensity measurements of the light emitted by RONS for different **d** values an important remark must be made. It is that as **d** increases, different portions on the plasma are detected by both spectrometers, i.e., when **d** = 3 mm, the Avantes and Horiba spectrometers see ~83% and ~53% of the plasma jet, respectively, while when **d** = 7 mm, the first spectrometer sees ~36% and the second one only ~23%.

It is interesting to notice that using the particular setup presented in this work, the peak values of the O_{777} emissions obtained in all explored experimental conditions occurred for approximately 0.5–0.6% of O_2 content in the gas admixture. This finding coincides with the value where the highest efficiency in the production of O atoms was observed by Li et al. [15].

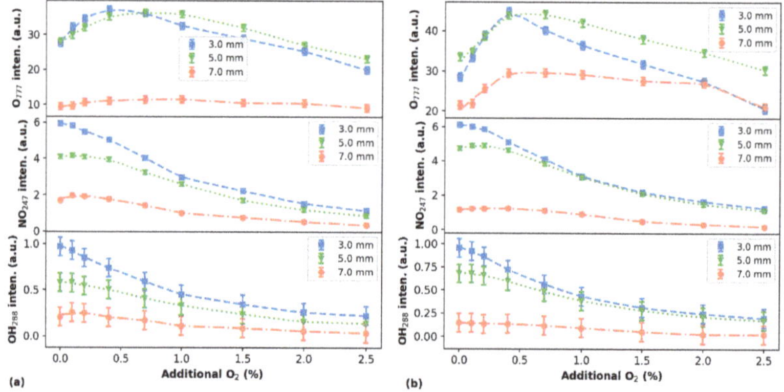

Figure 6. Emission intensities of selected lines/bands from O, NO, and OH radicals as a function of the additional O_2 percentage for different **d** values (indicated in the curve legends). The data were measured for He flow rates of 1.5 L/min in (**a**) and 2.0 L/min in (**b**). Symbols represent experimental data and lines are trend curves.

3.3. Effects on APPJ Thermal Parameters

The rotational and vibrational temperatures (T_{rot} and T_{vib}, respectively) can be considered the main thermal parameters of APPJs, since T_{rot} has a close relationship with the gas temperature (T_{gas}) and in most cases it is being assumed that $T_{rot} \approx T_{gas}$. In APPJs the T_{vib} is related to the rate of chemical reactions, i.e., higher T_{vib} values increase the likelihood of chemical reactions between the plasma and the target under treatment [41,42]. Of course, the electron temperature (T_e) is a thermal parameter that also plays an important role in the plasma dynamics, and it is probably the one most affected by the additional O_2 injected into the plasma. However, the variations on the electron temperature could not be analyzed in this work because we do not have the appropriated tools to perform T_e measurements. It can be seen in Figure 7 that the T_{rot} values are not significantly affected by the additional O_2 in the plasma for all distances between the plasma outlet and the target and for the two He flow rates used, presenting variations around 550 K in almost all measurements. On the other hand, the T_{vib} curves present clear downward trends as more O_2 is added to the plasma. Additionally, the T_{vib} tends to decrease as **d** increases. In addition, no significant variation in T_{vib} values are observed when the He flow rate was changed.

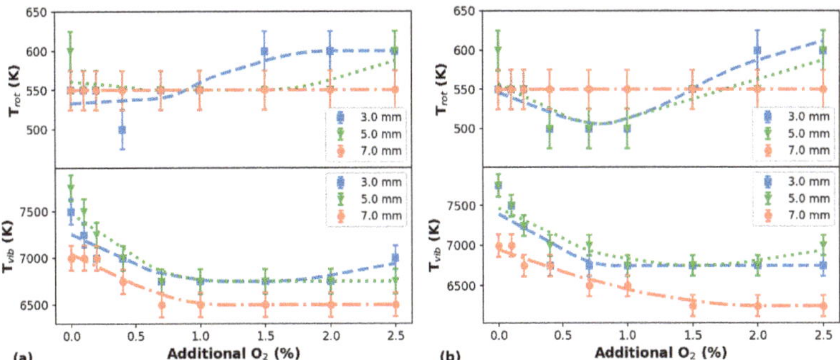

Figure 7. Variation of T_{rot} and T_{vib} values as a function of the additional O_2 percentage for different **d** values. The temperatures were measured for He flow rates of 1.5 L/min in (**a**) and 2.0 L/min in (**b**). Symbols represent experimental data and lines are trend curves.

By comparing the profiles of T_{rot} and T_{vib} values measured as a function of the additional O_2 in the plasma, shown in Figure 7a,b, it can be seen that at different He flow rates the behavior of the temperatures as a function of the O_2 content do not present significant changes. Regarding the magnitude of the temperature values, those obtained for T_{rot} are within the expected range for APPJs produced in He with the plasma jet impinging a metallic target [43]. However, those T_{rot} values are higher than the room temperature ($\sim$300 K) and are likely being overestimated due to the low-resolution spectrometer employed to perform the measurements. Additionally, one should consider that performing OES measurements of plasma portions that are close to the target's surface, can artificially increase the T_{rot} values. The fact that the plasma jet is directly impinging on the metallic target results in a strong plasma-material interaction, causing the release of a lot of secondary electrons to the plasma, which influences the population of higher energy rotational levels leading to higher T_{rot} values. Furthermore, since the floating wire placed inside the plastic tube is a conducting material, the plasma jet may present a corona discharge component, which can also be one of the causes of the higher T_{rot} values obtained.

Concerning the T_{vib} values obtained in this work, they are much higher than those obtained for APPJs in similar conditions, being more than ten times higher than the corresponding T_{rot} values. A possible explanation for this observation is that these T_{vib} values have a close relationship with the small diameter of the plastic tube, which causes an increase in the gas pressure inside the reactor leading to an increment in T_{vib} [44]. At the same time, the plasma-material interaction may be influencing the T_{vib} values as well.

In our preliminary experiments we used an infrared thermometer to monitor the target temperature during the plasma jet application. It was found that the target temperature increased by $\sim$3 K after the first 15 s of plasma exposure and with another $\sim$7 K after two minutes of operation. This total increase of 10 K in the target temperature due to the plasma impinging did not change significantly during 10 min of operation. Therefore, the variation in the target temperature is not high enough to cause any thermal damage to the target (even if we consider thermo-sensitive materials). Furthermore, the increase in the target temperature will not affect the T_{rot} and T_{vib} measurements since the target temperature is far from the values obtained for those parameters and its variation is within the uncertainties in the T_{rot} and T_{vib} values.

4. Conclusions

In this work we have demonstrated that the addition of O_2 to the working gas for generation of APPJs affects most of the plasma parameters, the production of RONS and the discharge duration and its time of ignition. Regarding the electrical parameters, the discharge power starts to decrease significantly if more than 2% of O_2 is added to the gas admixture while the effective current is always reduced when any amount of O_2 is added.

Concerning the plasma thermal parameters, we have seen that the T_{rot} values remains almost unchanged towards the addition of O_2, while the T_{vib} ones presented downward tendency when small amounts of O_2 are added (less than 1%) followed by stable T_{vib} values for higher O_2 percentages. An interesting observation is that the trends exhibited by those parameters did not change if the operating conditions (mainly distance between the plasma outlet and the target and the He flow rate) are modified. On the other hand, the production of RONS presented different behaviors in their variations depending on the specie under study and on the operating conditions. As an example, the trend in the production of NO as a function of the O_2 content seems to be affected by both He flow rate and distance between the plasma outlet and the target.

The aim of most works reported in the literature as well as the current one, has been to enhance the production of RONS in the plasma by addition of O_2 to the working gas. However, using the configuration presented in this work it was verified that the O_2 fulfills this goal only considering the production of atomic oxygen, which showed significant increases in almost all studied conditions when $\sim$0.5–0.6% of O_2 is added to the gas admixture. One can speculate that the production of other RONS, such as O_3 or NO_2, is probably increasing with the increment of O atoms in the plasma jet, although we have not carried out the measurements that would prove that.

Then, as a general conclusion, we can say that the addition of O_2 to the working gas used to generate APPJs certainly provides an increase in the amount of O atoms in the plasma jet. However, using the configuration described in this work, the simple act of adding O_2 to the plasma does not necessarily imply in an increment of other RONS species, such as OH and NO, which also play important roles when the interaction between plasma and materials occurs. However, we have found that there are also ways to improve the amount of RONS other than O atoms when O_2 is added to the plasma by setting appropriately and carefully select some operation parameters such as the distance between the plasma outlet and the target. Nevertheless, it is important to take into account what kind of reactive specie is required by the application in question and the target under treatment to choose the best setup for the occasion.

Author Contributions: Conceptualization, F.N., K.P. and K.K.; methodology, F.N.; formal analysis, F.N.; investigation, F.N., K.P. and K.K.; resources, K.K.; data curation, F.N.; writing—original draft preparation, F.N. and K.K.; writing—review and editing, F.N. and K.K.; supervision, K.K.; project administration, K.K. All authors have read and agreed to the published version of the manuscript.

Funding: This research was supported by CAPES (Programs DS and PrInt) and FAPESP (grants #2019/05856-7 and #2020/09481-5). Financial support was also provided by São Paulo State University–UNESP (Edital Propg 5/2021).

Data Availability Statement: Data are contained within the article. Raw data are available from the authors under reasonable request.

Acknowledgments: The authors thank to the students Ananias Barbosa and Ana Almeida for helping to acquire some of the data used in this work.

Conflicts of Interest: The authors declare no conflict of interest.

References

1. Brandenburg, R. Dielectric barrier discharges: Progress on plasma sources and on the understanding of regimes and single filaments. *Plasma Sources Sci. Technol.* **2017**, *26*, 53001. [CrossRef]
2. Bekeschus, S.; Favia, P.; Robert, E.; Woedtke, T.V. White paper on plasma for medicine and hygiene: Future in plasma health sciences. *Plasma Process. Polym.* **2019**, *16*, 1800033. [CrossRef]
3. Busco, G.; Robert, E.; Chettouh-Hammas, N.; Pouvesle, J.M.; Grillon, C. The emerging potential of cold atmospheric plasma in skin biology. *Free. Radic. Biol. Med.* **2020**, *161*, 290–304. [CrossRef] [PubMed]
4. Laroussi, M. Cold Plasma in Medicine and Healthcare: The New Frontier in Low Temperature Plasma Applications. *Front. Phys.* **2020**, *8*. [CrossRef]
5. Liu, D.; Zhang, Y.; Xu, M.; Chen, H.; Lu, X.; Ostrikov, K.K. Cold atmospheric pressure plasmas in dermatology: Sources, reactive agents, and therapeutic effects. *Plasma Process. Polym.* **2020**, e1900218. [CrossRef]

6. Lu, X.; Naidis, G.V.; Laroussi, M.; Reuter, S.; Graves, D.B.; Ostrikov, K. Reactive species in non-equilibrium atmospheric-pressure plasmas: Generation, transport, and biological effects. *Phys. Rep.* **2016**, *630*, 1–84. [CrossRef]

7. Cheng, X.; Sherman, J.; Murphy, W.; Ratovitski, E.; Canady, J.; Keidar, M. The Effect of Tuning Cold Plasma Composition on Glioblastoma Cell Viability. *PLoS ONE* **2014**, *9*, e98652. Publisher: Public Library of Science. [CrossRef] [PubMed]

8. Chen, T.W.; Liu, C.T.; Chen, C.Y.; Wu, M.C.; Chien, P.C.; Cheng, Y.C.; Wu, J.S. Analysis of Hydroxyl Radical and Hydrogen Peroxide Generated in Helium-Based Atmospheric-Pressure Plasma Jet and in Different Solutions Treated by Plasma for Bioapplications. *ECS J. Solid State Sci. Technol.* **2020**, *9*, 115002. [CrossRef]

9. Kim, D.H.; Park, C.S.; Shin, B.J.; Seo, J.H.; Tae, H.S. Uniform Area Treatment for Surface Modification by Simple Atmospheric Pressure Plasma Treatment Technique. *IEEE Access* **2019**, *7*, 103727–103737. [CrossRef]

10. Nishime, T.M.C.; Wagner, R.; Kostov, K.G. Study of Modified Area of Polymer Samples Exposed to a He Atmospheric Pressure Plasma Jet Using Different Treatment Conditions. *Polymers* **2020**, *12*, 1028. [CrossRef]

11. Wang, X.; Tan, Z.; Pan, J.; Chen, X. Effects of Oxygen Concentration on Pulsed Dielectric Barrier Discharge in Helium-Oxygen Mixture at Atmospheric Pressure. *Plasma Sci. Technol.* **2016**, *18*, 837–843. [CrossRef]

12. Dai, Y.; Zhang, M.; Li, Q.; Wen, L.; Wang, H.; Chu, J. Separated Type Atmospheric Pressure Plasma Microjets Array for Maskless Microscale Etching. *Micromachines* **2017**, *8*, 173. [CrossRef]

13. Fang, Z.; Wang, X.; Shao, T.; Zhang, C. Influence of Oxygen Content on Argon/Oxygen Dielectric Barrier Discharge Plasma Treatment of Polyethylene Terephthalate Film. *IEEE Trans. Plasma Sci.* **2017**, *45*, 310–317. [CrossRef]

14. Yue, Y.; Pei, X.; Gidon, D.; Wu, F.; Wu, S.; Lu, X. Investigation of plasma dynamics and spatially varying O and OH concentrations in atmospheric pressure plasma jets impinging on glass, water and metal substrates. *Plasma Sources Sci. Technol.* **2018**, *27*, 064001. [CrossRef]

15. Li, J.; Wu, F.; Nie, L.; Lu, X.; Ostrikov, K. The Production Efficiency of Reactive Oxygen and Nitrogen Species (RONS) of AC and Pulse-DC Plasma Jet. *IEEE Trans. Plasma Sci.* **2020**, *48*, 4204–4214. [CrossRef]

16. Korolov, I.; Steuer, D.; Bischoff, L.; Hübner, G.; Liu, Y.; Gathen, V.S.v.d.; Böke, M.; Mussenbrock, T.; Schulze, J. Atomic oxygen generation in atmospheric pressure RF plasma jets driven by tailored voltage waveforms in mixtures of He and O2. *J. Phys. D Appl. Phys.* **2021**, *54*, 125203. [CrossRef]

17. Liu, F.; Chu, H.; Zhuang, Y.; Fang, Z.; Zhou, R.; Cullen, P.J.; Ostrikov, K.K. Uniform and stable plasma reactivity: Effects of nanosecond pulses and oxygen addition in atmospheric-pressure dielectric barrier discharges. *J. Appl. Phys.* **2021**, *129*, 033302. [CrossRef]

18. Huang, C.; Yu, Q.; Hsieh, F.H.; Duan, Y. Bacterial Deactivation Using a Low Temperature Argon Atmospheric Plasma Brush with Oxygen Addition. *Plasma Process. Polym.* **2007**, *4*, 77–87. [CrossRef]

19. Fang, Z.; Zhou, Y.; Yao, Z. The Influences of Water and Oxygen Contents on Length of Atmospheric Pressure Plasma Jets in Ar/H2O and Ar/O2 Mixtures. *IEEE Trans. Plasma Sci.* **2014**, *42*, 2618–2619. [CrossRef]

20. Sarani, A.; Nicula, C.; Gonzales, X.F.; Thiyagarajan, M. Characterization of Kilohertz-Ignited Nonthermal He and He/O2 Plasma Pencil for Biomedical Applications. *IEEE Trans. Plasma Sci.* **2014**, *42*, 3148–3160. [CrossRef]

21. Niu, J.; Liu, D.; Ji, L.; Xia, Y.; Bi, Z.; Song, Y.; Ma, Y.; Huang, Z.; Wang, W.; Yang, W. Propagation of Brush-Shaped He/O2 Plasma Plumes in Ambient Air. *IEEE Trans. Plasma Sci.* **2015**, *43*, 1993–1998. [CrossRef]

22. Fatima, S.S.; Rehman, N.U.; Younus, M.; Ahmad, I. Optical characterization of atmospheric-pressure plasma needle. *Contrib. Plasma Phys.* **2017**, *57*, 387–394. [CrossRef]

23. Barkhordari, A.; Ganjovi, A.; Mirzaei, I.; Falahat, A. Study of the physical discharge properties of a Ar/O2 DC plasma jet. *Indian J. Phys.* **2018**, *92*, 1177–1186. [CrossRef]

24. Wu, S.; Wu, F.; Liu, X.; Mao, W.; Zhang, C. A Bipolar DC-Driven Touchable Helium Plasma Jet Operated in Self-Pulsed Mode. *IEEE Trans. Plasma Sci.* **2018**, *46*, 4091–4098. [CrossRef]

25. Wang, S.; Wan, J. Oxygen Effects on a He/O2 Plasma Jet at Atmospheric Pressure. *IEEE Trans. Plasma Sci.* **2009**, *37*, 551–554. [CrossRef]

26. Pang, H.; Chen, Q.; Li, B.; Fei, F.; Yang, S. The Role of Oxygen in a Large Area of RF-Powered Atmospheric Pressure Dielectric Barrier Glow Discharge Plasma in Sterilization. *IEEE Trans. Plasma Sci.* **2011**, *39*, 1689–1694. [CrossRef]

27. Lazzaroni, C.; Chabert, P. A comparison between micro hollow cathode discharges and atmospheric pressure plasma jets in Ar/O2gas mixtures. *Plasma Sources Sci. Technol.* **2016**, *25*, 65015. [CrossRef]

28. Brandenburg, R.; Maiorov, V.A.; Golubovskii, Y.B.; Wagner, H.E.; Behnke, J.; Behnke, J.F. Diffuse barrier discharges in nitrogen with small admixtures of oxygen: Discharge mechanism and transition to the filamentary regime. *J. Phys. D Appl. Phys.* **2005**, *38*, 2187–2197. [CrossRef]

29. Stancampiano, A.; Chung, T.; Dozias, S.; Pouvesle, J.; Mir, L.M.; Robert, E. Mimicking of Human Body Electrical Characteristic for Easier Translation of Plasma Biomedical Studies to Clinical Applications. *IEEE Trans. Radiat. Plasma Med. Sci.* **2020**, *4*, 335–342. [CrossRef]

30. Hołub, M. On the measurement of plasma power in atmospheric pressure DBD plasma reactors. *Int. J. Appl. Electromagn. Mech.* **2012**, *39*, 81–87. [CrossRef]

31. Ashpis, D.E.; Laun, M.C.; Griebeler, E.L. Progress Toward Accurate Measurement of Dielectric Barrier Discharge Plasma Actuator Power. *AIAA J.* **2017**, *55*, 2254–2268. [CrossRef] [PubMed]

32. Pipa, A.V.; Brandenburg, R. The Equivalent Circuit Approach for the Electrical Diagnostics of Dielectric Barrier Discharges: The Classical Theory and Recent Developments. *Atoms* **2019**, *7*, 14. [CrossRef]
33. Moon, S.Y.; Choe, W. A comparative study of rotational temperatures using diatomic OH, O2 and N2+ molecular spectra emitted from atmospheric plasmas. *Spectrochim. Acta Part B At. Spectrosc.* **2003**, *58*, 249–257. [CrossRef]
34. Bruggeman, P.J.; Sadeghi, N.; Schram, D.C.; Linss, V. Gas temperature determination from rotational lines in non-equilibrium plasmas: A review. *Plasma Sources Sci. Technol.* **2014**, *23*, 023001. [CrossRef]
35. Zhang, Q.Y.; Shi, D.Q.; Xu, W.; Miao, C.Y.; Ma, C.Y.; Ren, C.S.; Zhang, C.; Yi, Z. Determination of vibrational and rotational temperatures in highly constricted nitrogen plasmas by fitting the second positive system of N2 molecules. *AIP Adv.* **2015**, *5*, 057158. [CrossRef]
36. Ono, R. Optical diagnostics of reactive species in atmospheric-pressure nonthermal plasma. *J. Phys. D Appl. Phys.* **2016**, *49*, 083001. [CrossRef]
37. Voráč, J.; Synek, P.; Potočňáková, L.; Hnilica, J.; Kudrle, V. Batch processing of overlapping molecular spectra as a tool for spatio-temporal diagnostics of power modulated microwave plasma jet. *Plasma Sources Sci. Technol.* **2017**, *26*, 025010. [CrossRef]
38. Voráč, J.; Synek, P.; Procházka, V.; Hoder, T. State-by-state emission spectra fitting for non-equilibrium plasmas: OH spectra of surface barrier discharge at argon/water interface. *J. Phys. D Appl. Phys.* **2017**, *50*, 294002. [CrossRef]
39. Verreycken, T.; Bruggeman, P.J. OH Dynamics in a Nanosecond Pulsed Plasma Filament in Atmospheric Pressure He–H2O upon the Addition of O2. *Plasma Chem Plasma Process* **2014**, *34*, 605–619. [CrossRef]
40. Deng, G.; Jin, Q.; Yin, S.; Zheng, C.; Liu, Z.; Yan, K. Experimental study on bacteria disinfection using a pulsed cold plasma jet with helium/oxygen mixed gas. *Plasma Sci. Technol.* **2018**, *20*, 115503. [CrossRef]
41. Lambert, J.D. Vibration–vibration energy transfer in gaseous collisions. *Q. Rev. Chem. Soc.* **1967**, *21*, 67–78. [CrossRef]
42. Smith, R.R.; Killelea, D.R.; DelSesto, D.F.; Utz, A.L. Preference for Vibrational over Translational Energy in a Gas-Surface Reaction. *Science* **2004**, *304*, 992–995. [CrossRef]
43. Wang, R.; Xu, H.; Zhao, Y.; Zhu, W.; Ostrikov, K.K.; Shao, T. Effect of dielectric and conductive targets on plasma jet behaviour and thin film properties. *J. Phys. D Appl. Phys.* **2018**, *52*, 074002. [CrossRef]
44. Nascimento, F.d.; Moshkalev, S.; Machida, M. The role of vibrational temperature variations in a pulsed dielectric barrier discharge plasma device. *Contrib. Plasma Phys.* **2020**, *60*, e202000046. [CrossRef]

Article

The Role of HNO$_2$ in the Generation of Plasma-Activated Water by Air Transient Spark Discharge

Mário Janda *, Karol Hensel, Peter Tóth, Mostafa E. Hassan and Zdenko Machala

Faculty of Mathematics, Physics and Informatics, Comenius University Mlynská Dolina, 84248 Bratislava, Slovakia; karol.hensel@fmph.uniba.sk (K.H.); toth169@uniba.sk (P.T.); mostafa.hassan@fmph.uniba.sk (M.E.H.); zdenko.machala@fmph.uniba.sk (Z.M.)
* Correspondence: mario.janda@fmph.uniba.sk

Abstract: Transient spark (TS), a DC-driven self-pulsing discharge generating a highly reactive atmospheric pressure air plasma, was employed as a rich source of NOx. In dry air, TS generates high concentrations of NO and NO$_2$, increasing approximately linearly with increasing input energy density (E_d), reaching 1200 and 180 ppm of NO and NO$_2$, at E_d = 400 J/L, respectively. In humid air, the concentration of NO$_2$ decreased down to 120 ppm in favor of HNO$_2$ that reached approximately 100 ppm at E_d = 400 J/L. The advantage of TS is its capability of simultaneous generation of the plasma and the formation of microdroplets by the electrospray (ES) of water directly inside the discharge zone. The TS discharge can thus efficiently generate plasma-activated water (PAW) with high concentration of H$_2$O$_2{}^-$$_{(aq)}$, NO$_2{}^-$$_{(aq)}$ and NO$_3{}^-$$_{(aq)}$, because water microdroplets significantly increase the plasma-liquid interaction interface. This enables a fast transfer of species such as NO, NO$_2$, HNO$_2$ from the gas into water. In this study, we compare TS with water ES in a one stage system and TS operated in dry or humid air followed by water ES in a two-stage system, and show that gaseous HNO$_2$, rather than NO or NO$_2$, plays a major role in the formation of NO$_2{}^-$$_{(aq)}$ in PAW that reached the concentration up to 2.7 mM.

Keywords: non-thermal plasma; transient spark; electrospray; plasma-activated water; nitrous acid; nitrites

Citation: Janda, M.; Hensel, K.; Tóth, P.; Hassan, M.E.; Machala, Z. The Role of HNO$_2$ in the Generation of Plasma-Activated Water by Air Transient Spark Discharge. *Appl. Sci.* **2021**, *11*, 7053. https://doi.org/10.3390/app11157053

Academic Editor: Andrei Vasile Nastuta

Received: 8 July 2021
Accepted: 28 July 2021
Published: 30 July 2021

Publisher's Note: MDPI stays neutral with regard to jurisdictional claims in published maps and institutional affiliations.

1. Introduction

Highly reactive non-thermal (cold) plasmas (NTP) in atmospheric air can be generated by various electrical discharges. Their high chemical activity is due to the presence of high energy electrons. The NTP are therefore very commonly used in various environmental, surface processing, or biomedical applications [1–3]. The research interest focused on the cold plasma applications in biology and medicine has dramatically grown during the last decades, because NTP can efficiently inactivate bacteria and other dangerous microbes and induce interesting therapeutic effects, e.g., against cancer cells [4,5].

There are several biocidal agents provided by non-thermal plasma treatments: UV radiation, ions, and reactive neutral species [6–8]. The role of individual agent can vary depending on the used plasma source, input energy density, or gaseous environment. The reactive oxygen and nitrogen species (RONS) were demonstrated as the key biocidal agents produced in the transient spark (TS) air discharge extensively studied in our group [9]. We previously reported that the dominant gaseous products in the air treated by TS were nitric oxide (NO) and nitrogen dioxide (NO$_2$) [10].

Biomedical applications of NTP often lead to the interaction of plasmas with liquid media (e.g., water). The generation of NTP in contact with water has therefore become a hot topic over the last few years [11]. Transport of active species from the plasma (gas phase) into the liquid phase leads to the production of so-called plasma-activated water (PAW), with various potential applications in medicine (e.g., wound healing or inactivation of cancer cells) or in agriculture (e.g., seed germination and plant growth promotion, pest control) [12–16]. With PAW, the biocidal effects of plasma can be indirectly applied in cases where a direct plasma treatment is not possible.

The composition of PAW and thus also its potential application varies with the used plasma source and working gas [15,17]. The liquid phase RONS in PAW include long-lived species: ozone ($O_{3(aq)}$), hydrogen peroxide ($H_2O_{2(aq)}$), nitrites ($NO_2^-{}_{(aq)}$), or nitrates ($NO_3^-{}_{(aq)}$), besides short-lived species that are challenging for diagnostics: hydroxyl OH, peroxyl HO_2 radicals, atomic O, N and H, singlet molecular O_2 ($^1\Delta$), superoxide anion O_2^- and other ions. Transport of reactive species from gas to water is the primary source of the liquid RONS, but the composition of liquid RONS does not perfectly mirror the concentrations of gaseous RONS. First, the concentration of liquid phase RONS depends on the solubility of their gas phase counterparts. Second, the composition of PAW is influenced by a rich set of the subsequent chemical reactions in the liquid phase [18,19].

The highest achievable (saturated) concentration of species in liquid phase c_i^{satur} is determined by Henry's law:

$$c_i^{satur} = k_H^i p_i \tag{1}$$

Here p_i is partial pressure and k_H^i is the Henry's law coefficient of i-species. For instance, H_2O_2 (with $k_H = 108{,}000$ mol.kg^{-1} atm^{-1} at 298.15 K [20]) is much more soluble than many other gas phase RONS, having k_H by seven orders of magnitude higher than O_3 and NO_2, and by eight orders of magnitude higher than NO. The high solubility of H_2O_2 leads to the depletion of its gas phase concentration near the liquid interface [21,22]. The highest achievable concentration of $H_2O_{2(aq)}$ in the liquid phase is thus limited by the number of available H_2O_2 molecules in the gas phase. On the other hand, for weakly soluble molecules, such as O_3, the assumption of the constant concentration and partial pressure in the gas phase is valid. Here, the saturated concentration in the liquid phase could be achieved without a noticeable decrease of O_3 concentration in the gas phase. In plasma-liquid interactions, the situation is much more complicated, because solubility determined by Henry's law coefficient is valid for systems in thermodynamic equilibrium. This is certainly not the case of NTP interacting with water. Electric fields, temperature gradients, electrohydrodynamic spray of liquids, charging of water droplets, ionic wind, chemical reactions among various species, and other phenomena make solvation of reactive species from NTP to water very complicated process that is still not completely understood [11,21].

The plasma-liquid interface surface area is a key parameter maximizing the contact between the plasma and the treated water solution, thus determining the obtained plasma chemical effects [23]. The transformation of bulk water into fine droplets results in an increase of this interfacial surface area and thus accelerates the transport of reactive plasma species into the water. The electrohydrodynamic spray of liquids, here simply called electrospray (ES), is a simple process to produce microdroplets from bulk liquid by a strong electric field. Despite the pioneering experimental studies describing several ES modes were conducted more than a century ago [24,25], due to a high application potential of ES in many areas [26–31], the research of ES continues [32–34], and several reviews were published recently focused on different aspects of ES [35–37].

The idea of using ES microdroplets to increase plasma-water interface area has been adopted by several research groups [21,38–40], as well as in our previous works [15,41]. For instance, TS was also used to prepare PAW by the electrospray of fine aerosol droplets directly through the active plasma zone, which resulted in a very efficient transfer of gaseous RONS into water. The PAW generated by TS contains besides $H_2O_{2(aq)}$ high concentration of $NO_2^-{}_{(aq)}$ and $NO_3^-{}_{(aq)}$ [15]. This was attributed to the fact that the dominant gas phase species in air TS are NO and NO_2.

In this paper we focus on the possible role of gas phase nitrous acid (HNO_2) on the PAW generation by TS. Even with much lower gas phase concentration than NO_2 and NO, HNO_2 may influence PAW generation significantly since the k_H of HNO_2 (50 mol kg^{-1} atm^{-1} at 298.15 K [42]) is four and five orders of magnitude higher then k_H of NO_2 (0.007 mol kg^{-1} atm^{-1} at 295 K [43]) and NO (0.0018 mol kg^{-1} atm^{-1} at 298.15 K [44]), respectively. In our previous work, we were able to detect small amounts of HNO_2 in the gas phase using the Fourier transform infrared (FT-IR) spectroscopic technique [45]. A new diagnostic technique (direct UV-Vis absorption spectroscopy) allowed us to perform a more precise HNO_2 diagnosis in the TS-treated air when installed much closer to the plasma reactor. Moreover, the experiments where PAW is prepared by the direct ES of microdroplets directly through the discharge zone in a common one-stage system are compared to the experiment with a two-stage system, with one reactor for the TS air treatment followed by the ES of water microdroplets in the second reactor.

2. Materials and Methods

Several types of experiments were performed in a one-stage system (1SS) with a single reactor (Figure 1), or in two-stage system (2SS) with two reactors in series (Figure 2). In 1SS, the single reactor enables a simultaneous generation of TS with a direct ES of water. This reactor can be also used for generation of TS without ES, or vice-versa, for generation of ES without TS. In 2SS, the gas is treated by TS inside the first reactor, and subsequently passed into the second reactor with water electrosprayed on the microdroplets. In 2SS, the second reactor is the same as the single reactor used in 1SS (Figure 2).

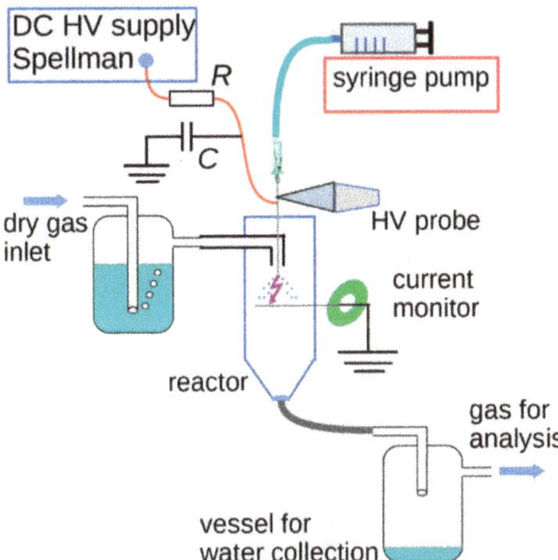

Figure 1. Simplified schematic of the experimental setup of one-stage system (1SS), with a single reactor enabling simultaneous TS and ES generation; *R*—resistor, *C*—capacitor, HV—high voltage.

The treated air was either dry synthetic air (99.999% purity, Messer Tatragas, Bratislava, Slovakia), or humidified synthetic air. In addition, dry and humidified O_2 (99.95% purity, Linde Gas, Bratislava, Slovakia) were used in two comparative experiments. In order to moisturize the gas, the dry air was bubbled through a bubbler filled with deionised (DI) water (with conductivity < 3 µS/cm). Typically, the relative humidity around 94–96% was achieved, verified by humidity sensor. The gas flow rate 1.1 L/min was constant in all experiments, controlled by rotameters Aalborg. We also performed additional comparative experiments in 1SS focused on NO and NO_2 solvation to ES water microdroplets without discharge. For this purpose, as well as for calibration of spectrometers, two special pressure tanks with calibration gases were used: one with 2000 ppm of NO in N_2, the second one with 1000 ppm of NO_2 in synthetic air. Lower concentration of NO and NO_2 was achieved by mixing of gases from these bottles with N_2 or synthetic air, respectively. More specific description of the used discharge and individual components of our experimental setups follow.

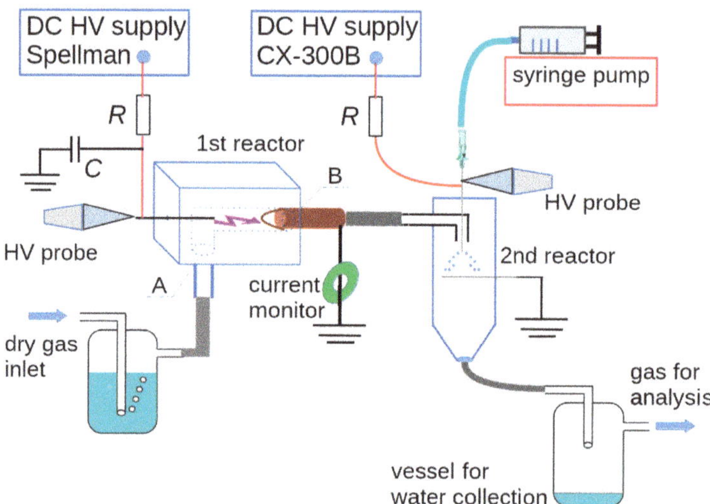

Figure 2. Simplified schematic of the experimental setup of two-stage system (2SS), with the first reactor for TS generation and the second reactor for water electrospray (ES); *R*—resistor, *C*—capacitor, HV—high voltage, A—opening for the gas inlet, B—opening for the gas outlet and insertion of the grounded electrode.

2.1. Reactor Details

The first reactor in 2SS (Figure 2) used for TS generation, has shape of a block with dimensions $16 \times 11 \times 43$ mm made of polytetrafluoroethylene (PTFE). Two intersecting perpendicular holes A and B (diameter 6 mm with M6 thread) are drilled in this block, so that the gas can pass through it, making 90 degree turn. One of these openings (B) serves for insertion of grounded electrode into the reactor. The grounded electrode is a copper wire with diameter 1 mm connected to the end of copper tube used for gas outlet, with the inner diameter 5 mm and outer diameter 6 mm and with outer M6 thread. There is another hole, with diameter 2 mm and M2 thread, opposite to the opening B. The M2 hole serves for insertion of stressed (anode) electrode, namely an iron M2 screw with a sharpened tip. The gap between the electrodes is approximately 10 mm.

The second reactor, used for ES in 2SS experiments (Figure 2) and for generation of both TS and ES in 1SS experiments (Figure 1), has a cylindrical shape with an inner diameter 2.9 cm and its length is 13 cm. The high voltage is applied to the top blunt hollow needle electrode (anode), having outer diameter 0.7 mm and inner diameter 0.5 mm. When this reactor is used as a second stage for generation of ES without TS (2SS experiments), a CX-300B DC HV power supply (unbranded, China) is used, connected to the anode via series resistor $R = 13.5$ MΩ. The applied voltage is 6 kV (below TS onset voltage) and it is monitored by a N2771A HV probe (Agilent, Santa Clara, CA, USA). The current was not measured in the second reactor when used only for ES generation without TS, because it was below the measurable limits of our current 2877 monitor (Pearson Electronics, Palo Alto, CA, USA).

The grounded electrode is also a medical needle with the outer diameter 0.7 mm going through the body of the reactor, with the gap between the electrodes being 10 mm. The top needle electrode (anode) is also used for the ES water input, continuously supplied with the deionized water (pH ~5.4, conductivity ~3 µS/cm) by a NE-300 syringe pump (New Era Pump Systems, Farmingdale, NY, USA) through silicon tubing under controlled flow rate ($Q_w = 100$–500 µL/min). The treated gas enters the reactor in the upper part in direction parallel to the HV needle electrode, where it comes to contact with electrosprayed microdroplets. The water and the gas leave the reactor together at the bottom of the cylindrical reactor. The water is collected in a vessel behind, while the gas continues towards the analytical part of the system. The openings in the reactor where needle electrodes are inserted are sealed by a vacuum grease to avoid leaks of a treated gas.

2.2. TS Generation and Diagnostics

A high voltage (HV) DC power supply (SL30P300 Spellman, Hauppauge, NY, USA) connected to the reactor via series resistor $R = 3.2$ MΩ is used to generate positive polarity TS between two metallic electrodes in a pin-to-wire configuration. The distance between electrodes in both reactors is approximately 10 mm. The TS discharge characteristics were measured by a P6015A HV probe (Tektronix, Berkshire, UK) and a Pearson Electronics 2877 current monitor, and then processed by a TBS2104 digital oscilloscope (Tektronix).

Despite using DC power supply, TS is a self-pulsing discharge characterized by short high current pulses (Figure 3), with a typical repetition frequency $f = 1$–10 kHz [46,47]. Energy delivered to the plasma per pulse can be calculated from the measured current I and voltage V waveforms by the following formula:

$$E_p = \int_T V \times I dt \tag{2}$$

with the integration period T covering all the spark current pulse produced by discharging of the electric circuit capacity C. In our previous works we relied on internal capacity of the circuit, provided mostly by the HV cable connecting the anode with the ballast resistor R, with a length 1–2 m. Here we used a much shorter cable and thus we used an external capacitor $C = 50$ pF to provide a sufficient discharging capacity. Still, 50 pF is low enough to avoid classical spark discharge that generates a thermal plasma, because TS current pulses have too short duration and E_p is only around 2 mJ/pulse. Knowing E_p, the discharge power (p) can be obtained as a product of the repetition frequency times the energy delivered per pulse: $P = f \times E_p$. Finally, the input energy density E_d in [J/L] can be calculated as follows:

$$E_d = 60 \times P/q \tag{3}$$

where q is a gas flow rate in [L/min].

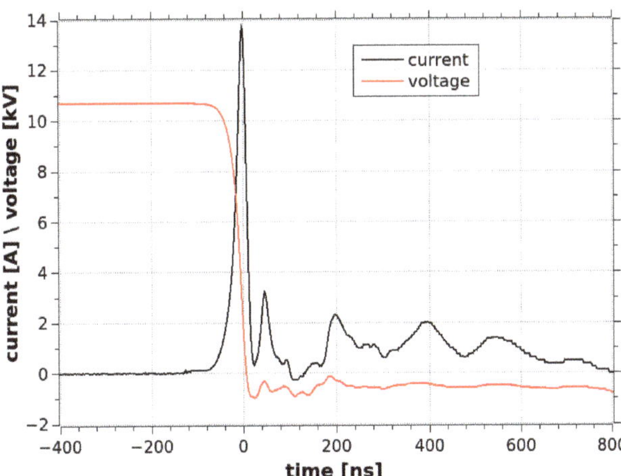

Figure 3. Typical current and voltage waveforms of the TS, short time scale showing the voltage drop and the current pulse associated with the gas breakdown.

2.3. Gas Phase Diagnostics

The analytical system for gas composition analysis consists of three stages. First, the gas leaving the plasma reactor entered a 10 cm long cell with inner diameter 2.5 mm, used for UV-Vis absorption spectroscopy. As a light source, an AvaLight-D-S deuterium lamp (Avantes, Louisville, CO, USA) is used, along with an Avantes AvaSpec-Mini4096CL spectrometer for spectrally resolved light detection. Spectral resolution is 0.4–0.5 nm and spectra in the range 190–650 nm were recorded. Next, the analysed gas passes to the second, 75 cm long cell for UV-Vis absorption spectroscopy, with the same inner diameter 2.5 mm. As a light source, an Insight PX-2 pulsed Xe lamp (Ocean Insight, Orlando, FL, USA) is used here, with an Ocean Insight STS-UV spectrometer for spectrally resolved light detection. Spectral resolution is 3 nm and spectra in the range 185–665 nm are recorded. The UV-Vis absorption technique enables quantitative simultaneous detection of NO, NO_2, HNO_2 and O_3. Measurement in two absorption cells with different lengths serves to expand the dynamic range of the studied RONS from a few ppm up to more than 1000 ppm. For example, the 75 cm cell enables quantitative detection of NO approximately in the range 5–200 ppm, while the 10 cm cell expands the upper detection limit of NO up to ~1400 ppm.

The third technique used for analysis of treated gas was Fourier transform infrared (FT-IR) spectrophotometry (IRAffinity-1S with wavenumber range 7800–350 cm^{-1} and a best spectral resolution of 0.5 cm^{-1}, Shimadzu, Kyoto, Japan). The absorption spectroscopy in IR region is more versatile, i.e., more compounds can be found in the spectra than in the UV-Vis region. Besides NO, NO_2, HNO_2 and O_3, detectable also in UV-Vis region, we can detect HNO_3, N_2O, CO, CO_2 and almost all volatile hydrocarbons. In practice, the variations of water vapor concentration during experiments with humid air and with ES makes analysis of FT-IR spectra more complicated. The obtained spectra are noisy, with more complicated baseline shifts. Moreover, the used spectrometer is very sensitive to the electromagnetic noise from TS.

Both FT-IR and UV-Vis absorption techniques are absolute. The concentration of studied RONS in the gas were obtained by fitting of measured spectra with synthetic spectra. In order to obtain synthetic FT-IR spectra, we downloaded set of absorption lines for NO, NO_2, H_2O, HNO_3, O_3 and H_2O_2 molecules from HITRAN database [48]. Next, we convoluted these absorption lines to match the calculated spectra with the spectra measured by our spectrometer for spectral resolution 1 cm^{-1}. We verified this approach by measurement of NO and NO_2 with known concentration from calibration pressure tanks. As there is no suitable set of absorption lines for HNO_2, we used absorption cross sections downloaded from supplemental HITRAN database [49].

The UV-Vis absorption cross sections of NO, NO_2, HNO_2, HNO_3, N_2O_4, O_3 and H_2O_2 were downloaded from the MPI-Mainz UV/VIS Spectral Atlas [50]. Next, we convoluted all absorption cross sections measured with better spectral resolution to fit spectral resolution of our spectrometers, keeping the area under the curve constant. Our approach was verified by using calibration pressure tanks with the known concentration of NO and NO_2.

Figure 4 shows an example of the experimentally obtained spectrum, fitted by synthetic spectra of NO_2, HNO_2, H_2O_2 and NO. We achieved very good agreement between experimental and synthetic spectra by combined absorption of NO_2, HNO_2, NO and H_2O_2 molecules. The H_2O_2 absorption cross sections helped us to fit the spectrum in the range 200–280 nm, although the obtained concentrations of H_2O_2 are not presented in this paper because the H_2O_2 absorption cross section has no specific pattern and N_2O, N_2O_4, or HNO_3 have very similar cross sections. Absorbance in this part of the spectrum is also influenced by humidity inside the absorption cell.

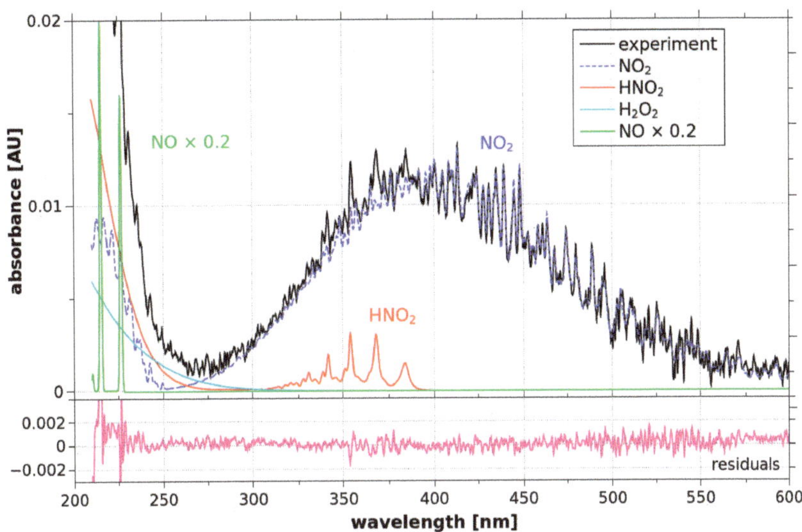

Figure 4. Typical UV-Vis absorption spectrum measured in 10 cm long cell, humid air treated by TS, mean current 2 mA, compared with synthetic spectra of NO_2, HNO_2, H_2O_2 and NO. Bottom chart shows residuals in experimental spectrum after subtraction of all synthetic spectra.

2.4. Analysis of Water

The pH of collected plasma-activated water (PAW) was measured by a calibrated pH-EC-TDS-TEMP pH multimeter (EZ-9908, unbranded, China), and the concentrations of $NO_2^-{}_{(aq)}$ and $H_2O_{2(aq)}$ formed in water were detected by the colorimetric methods using the Shimadzu UV-1800 UV/VIS absorption spectrophotometer. A Nitrate/Nitrite Colorimetric Assay Kit (# 780001, Cayman Chemicals, Ann Arbor, MI, USA) with ready-to-use Griess reagents was used for a quantitative analysis of $NO_2^-{}_{(aq)}$. The Griess assay method is the most commonly used for $NO_2^-{}_{(aq)}$ detection in plasma-activated solutions [51–57]. In the Griess diazotation reaction, $NO_2^-{}_{(aq)}$ reacts with sulfanilic acid under acidic conditions to form a diazonium ion, which couples to α-naphthylamine to form a readily water-soluble, deep purple colored azo dye, with the absorption maximum at 540 nm. The specificity and accuracy of the Griess assay for the detection of $NO_2^-{}_{(aq)}$ in PAW was confirmed by ion chromatography [58].

Measurement of $H_2O_{2(aq)}$ was performed by the titanium oxysulfate assay based on the reaction of $H_2O_{2(aq)}$ with the titanium (IV) ions under acidic conditions. The yellow-colored product of pertitanic acid H_2TiO_4 is formed with the absorption maximum at 407 nm [59]. The concentration of $H_2O_{2(aq)}$ is proportional to the absorbance according to Lambert–Beer's law (molar extinction coefficient $\varepsilon = 6.89 \times 10^2$ L mol^{-1} cm^{-1}). Because of the possible $H_2O_{2(aq)}$ decomposition by $NO_2^-{}_{(aq)}$ under acidic conditions, sodium azide (NaN_3, 60 mM) was added to the sample prior to mixing with the titanium oxysulfate reagent [19]. Sodium azide immediately reduces $NO_2^-{}_{(aq)}$ into molecular nitrogen and preserves the $H_2O_{2(aq)}$ concentration intact.

3. Results and Discussion

TS generates a highly reactive non-equilibrium plasma (despite a very short temporary gas temperature increase during the spark phase [47]), with chemical effects comparable to plasmas generated by short HV pulses [60]. One of the advantages of the TS in comparison with discharges generated by short HV pulses is its capability of simultaneous generation of the plasma and the ES of water through the discharge zone. ES cannot be induced by short voltage pulses applied to the needle. In the TS, the voltage on the hollow needle electrode is not constant, but it is high enough to induce formation of microdroplets most of the time, except during the short period after voltage drops associated with gas breakdown (Figure 5a).

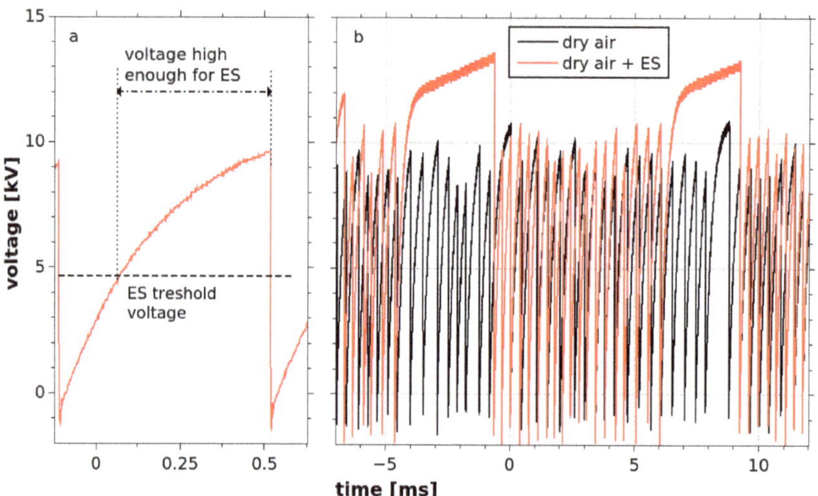

Figure 5. Typical long timescale TS voltage waveforms; (**a**) voltage increase regrowth between two voltage drops (gas breakdowns, i.e., TS pulses), showing that voltage is high enough for formation of microdroplets by ES most of the time; (**b**) voltage waveforms showing several tens of breakdowns, comparison of voltage waveforms without (black) and with ES (red; water flow rate Q_w = 300 µL/min) where periods without breakdowns and TS current pulses appear (e.g., time period 6–9 ms).

However, we must consider that TS and ES influence each other. This mutual influence causes instabilities of both ES and TS. Stable steady-state ES can be achieved only without a discharge or with a pulseless discharge, such as glow corona [61]. Vice-versa, the water flow needed for the ES causes instabilities of the TS, as shown in Figure 5b with periods where voltage increases well above the TS onset voltage (~10 kV), but no voltage drops associated with TS current pulses (e.g., time period 6–9 ms on Figure 5b). During these periods, only corona discharge may be generated, with much lower power and almost no generation of nitrogen oxides. Moreover, due to these corona discharge periods, the uncertainty of f in TS combined with ES is higher than in TS without ES. As a result, the uncertainty of the input energy density E_d is higher, causing a worse repeatability of experimental results. Nevertheless, it is still possible to compare generation of RONS by TS with and without ES, as shown in the Section 3.1.

The formation of RONS in the gas phase as described in Section 3.1 is just the first step in PAW generation. The second step is the solvation of gaseous RONS into water microdroplets. The changes of gaseous RONS concentrations caused by ES of water are described in Section 3.2. Finally, the Section 3.3 shows results from the analysis of generated PAW, explaining importance and contribution of individual RONS generated by TS (NO, NO_2 and HNO_2) on the achieved concentration of $NO_2^-{}_{(aq)}$.

3.1. Generation of RONS in the Gas Phase

The analytical method we used enables the detection of various RONS. In the dry air treated by the TS we detected only three long-lived species: NO, NO_2 and traces of HNO_2. Other possible RONS, such as O_3 or HNO_3 were not detected. NO is a dominant product of the air TS plasma with the highest concentration achieved. The concentration of NO grows almost linearly with increasing input energy density, and the concentration of NO using either dry or humid air are almost identical (Figure 6).

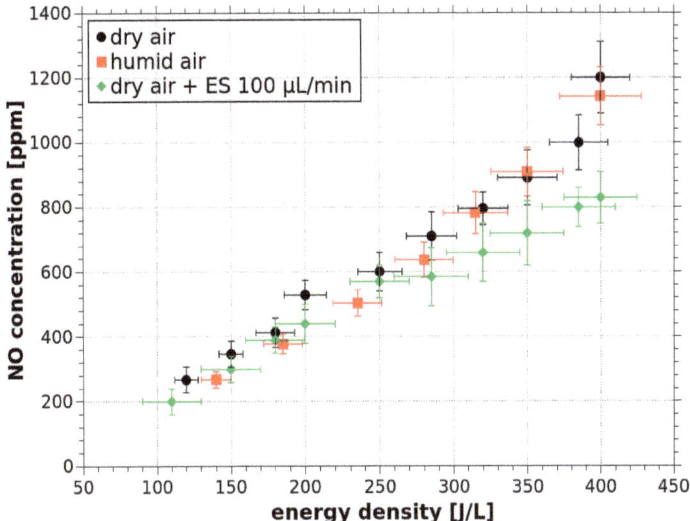

Figure 6. NO concentration as a function of the TS input energy density, measured by UV-Vis absorption technique, short absorption cell, 1SS.

Based on preliminary results from chemical kinetic model of TS [62], we assume that NO is generated by the modified Zeldovich mechanism. The original Zeldovich mechanism is initiated by the thermal decomposition of N_2 and O_2 into their atomic states at high temperature (above ~1500 K). Despite the fact that TS generates 'cold' non-equilibrium plasma, the temperature during the short spark phase of the TS can be as high as ~3000 K [47]. Next, both N and O atomic radicals are able to produce NO:

$$O + N_2 \rightarrow NO + N \tag{4}$$

$$N + O_2 \rightarrow NO + O \tag{5}$$

The rate coefficient of Reaction (5) with N atoms is four orders of magnitude higher at 3000 K than at room temperature, but the production of N atomic radicals via the thermal decomposition of N_2 is very slow, even at 3000 K (rate coefficient $\sim 1.5 \times 10^{-24} \text{cm}^3\,\text{s}^{-1}$) [63]. As a result, the thermal decomposition of N_2 is a rate-limiting step of the thermal NO formation. In the TS discharge, the thermal mechanism of NO formation is bypassed by alternative reaction pathways for the generation of N and O atomic radicals. The influence of spark on plasma induced chemistry is not based on gas heating only, more important is an achievement of a high degree of ionization, with the electron density $n_e > 10^{17} \text{ cm}^{-3}$ [64].

The high degree of ionization results in a high degree of atomization thanks to the dissociative electron-ion recombination reactions:

$$e + O_2^+ \rightarrow O + O \qquad (O(^1D) \sim 40\%, O(^1S) \sim 5\%) \tag{6}$$

$$e + N_2^+ \rightarrow N + N \qquad (N(^2D) \sim 45\%, N(^2P) \sim 5\%) \tag{7}$$

In the next step, the products of reactions (6) and (7) enhance the NO synthesis, especially by N production bypassing the rate limiting step in the Zeldovich mechanism by Reaction (4). Moreover, reactions (4) and (5) can be much faster if one of the reactants is in an excited state [63,65]. There are many other reactions that can contribute to N or O atoms production. For example, additional O atoms can also be created from O_2 molecules by dissociative electron attachment reaction.

A few microseconds after the spark current pulse, air cools down and the N concentration decreases. Consequently, the NO production stops. Remaining O atoms start to generate ozone O_3:

$$O + O_2 + M \rightarrow O_3 + M \tag{8}$$

Alternatively, the atomic oxygen can oxidize NO to form NO_2 in a three-body Reaction (9):

$$NO + O + M \rightarrow NO_2 + M \tag{9}$$

However, Reaction (8) producing O_3 is faster. Consequently, O_3 also oxidizes NO to NO_2:

$$NO + O_3 \rightarrow NO_2 + O_2 \tag{10}$$

As a result, NO_2 is a product with the second highest measured concentration (Figure 7). In dry air, there are no other significant products of TS.

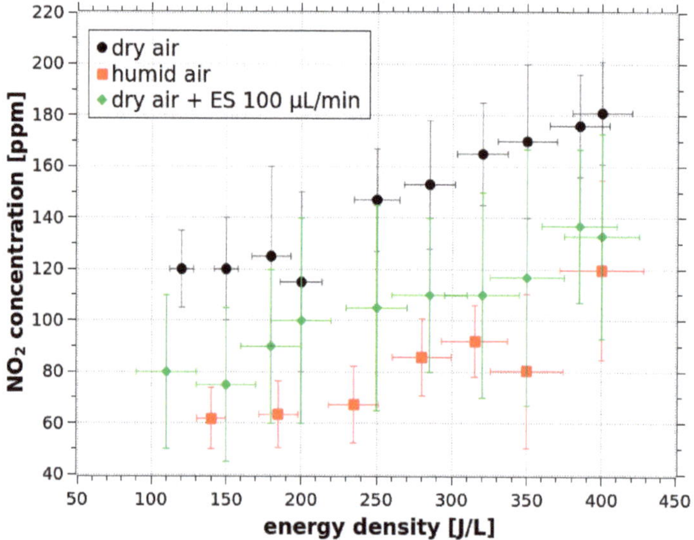

Figure 7. NO_2 concentration as function of the TS input energy density, measured by UV-Vis absorption technique, short absorption cell, 1SS.

We observed no O_3 at the outlet, probably because Reaction (10) is fast enough and there is a sufficient amount of NO so that O_3 is fully consumed by the oxidation of NO to NO_2 before the gas enters the closest UV absorption cell (the estimated delay between the moment when air is treated by the discharge and when it is analyzed in the short UV cell is 17–20 s). If we assume that Reaction (10) is a dominant source of NO_2, the achieved NO_2 concentration (~100–200 ppm, Figure 7) should be approximately equal to concentration of formed and immediately consumed O_3. To understand O_3 formation without its depletion by NO, we performed experiments in 1SS with TS operating in dry O_2. These experiments proved that TS can really generate high concentration of O_3 (Figure 8), when it is not consumed by reaction with NO.

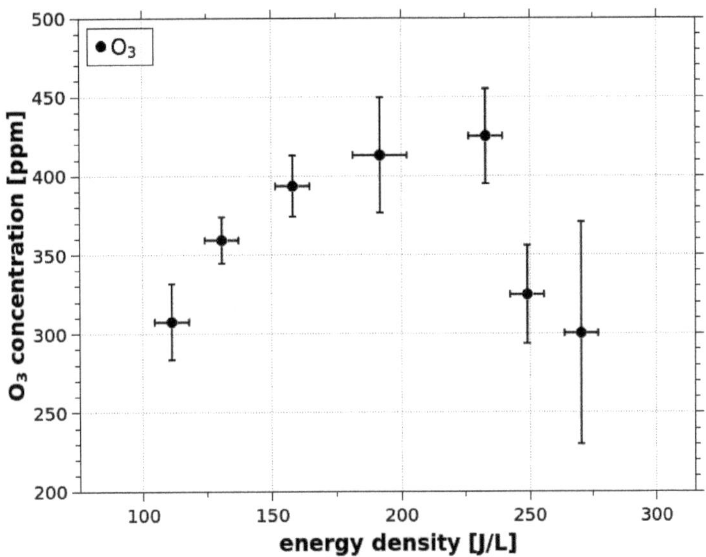

Figure 8. O_3 concentration as a function of the TS input energy density, dry O_2 treated by TS in 1SS without ES, measured by UV-Vis absorption technique, short absorption cell.

We also performed experiments in humid O_2, where the electron impact reaction with H_2O can result into several radicals (OH, H) or ions (O^-, OH^-, H_2O^+) [66,67]. The OH radicals can be also created by collisions of H_2O with $O(^1D)$ species. Next, H and OH radicals can decompose O_3 in the reactions:

$$H + O_3 \rightarrow OH + O_2 \tag{11}$$

$$OH + O_3 \rightarrow HO_2 + O_2 \tag{12}$$

For this reason, the concentration of O_3 generated in humid O_2 was approximately 15 times lower than in dry O_2. We can also expect lower production of O_3 in the humid air compared to dry air. In turn, lower production of O_3 could explain why the presence of humidity in air decreased the NO_2 concentration (Figure 7). However, the concentration of NO_2 in humid air was not 15 times lower than in dry air. It is probably because NO_2 may be also generated by different reaction channels, for example by the reaction with HO_2 radical. Sufficient amount of HO_2 radicals may be provided by reaction of OH radicals with O_3 (Equation (12)), or with H_2O_2.

The presence of OH radicals influences the reaction mechanism even more significantly by the reaction:

$$NO + OH + M \rightarrow HNO_2 + M \tag{13}$$

leading to the formation of HNO_2. In the humid air treated by TS, the HNO_2 concentration becomes comparable to NO_2 concentration (Figure 9). Some HNO_2 was observed in dry air as well, showing a significant influence of even minor H_2O impurities (from the carrier gas and/or due to desorption from surfaces) on NO, NO_2 and HNO_2 generation chemistry.

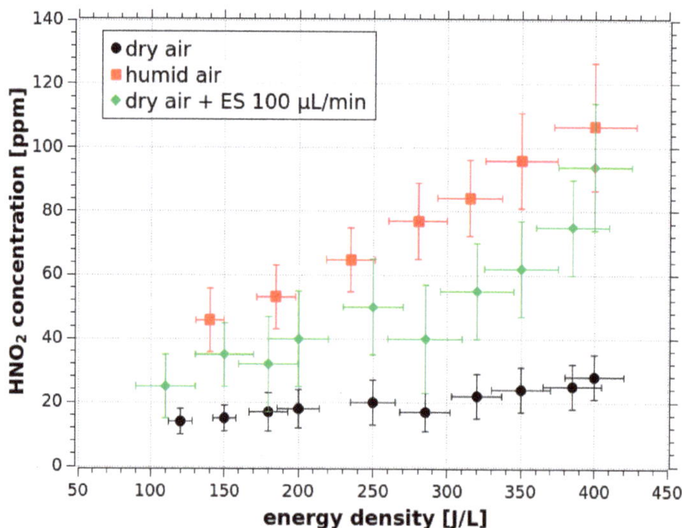

Figure 9. HNO_2 concentration as function of the TS input energy density, measured by UV-Vis absorption technique, short cell, 1SS.

Overall, the sum of NO_2 and HNO_2 concentrations (H_yNO_2) is almost equal in humid and dry air, only below E_d ~250 J/L it is slightly lower in humid air compared to dry air. The total amount of H_yNO_x ($NO + NO_2 + HNO_2$) is within the experimental uncertainty the same in the entire range of E_d (Figure 10).

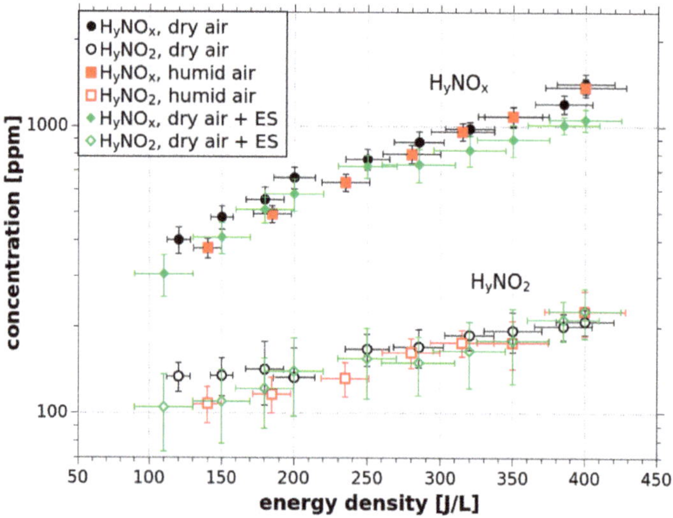

Figure 10. H_yNO_2 ($HNO_2 + NO_2$) and H_yNO_x ($HNO_2 + NO_2 + NO$) concentrations as functions of the TS input energy density, comparison of dry air, humid air, and dry air + ES, measured by UV-Vis absorption technique, short cell, 1SS.

The total amount of H_yNO_x is constant in time as well, based on comparison of concentrations obtained in short UV-Vis absorption cell and concentrations measured by FT-IR spectrometry. However, the ratio of individual components slightly changes. The

NO_2 concentration increases, while the concentrations of NO and HNO_2 slightly decrease. The decomposition of HNO_2 may proceed via the following reaction:

$$HNO_2 + HNO_2 \rightarrow NO_2 + NO + H_2O. \tag{14}$$

Taking into account the instability of HNO_2 molecules, we can deduce that their concentration is higher inside the reactor than when measured in the short UV absorption cell. Nonetheless, the measured concentration of HNO_2 is so high that HNO_2 probably plays a major role in PAW generation by TS with ES, because it has a much higher solubility than NO_2 and NO. Thus, the concentration of RONS in TS with ES is not influenced only by chemical reactions (11)–(14), but also by the solvation processes discussed in Section 3.2.

3.2. Solvation of Gaseous RONS to ES Microdroplets

The solubility of gases depends on their Henry's law coefficient. NO, NO_2 and HNO_2 generated by TS differ significantly from the solubility point of view. The solubility of NO is the lowest (with Henry's law coefficient $k_H = 0.0018$ mol kg^{-1} atm^{-1} at 298.15 K [44]). It is easy to show that in an experiment with gas flow 1.1 L/min and water flow rate 100 µL/min, even if we achieve a saturated NO solution, the decrease of the gas phase concentration of NO is not detectable as it is much smaller than the accuracy of our analytic techniques. Thus, the decrease of NO concentration generated by TS in dry air with ES compared to the results without ES (Figure 6), cannot be explained by the solvation of NO into water.

We assume that there are other processes responsible for this NO decrease. We must consider instabilities of TS caused by the water flow supplied for the ES, as shown in Figure 5b with long time scale voltage waveforms. Taking into account the appearance of short periods without spark pulses and energy consumed on evaporation of the sprayed microdroplets, we assume that water sprayed into the discharge has a cooling effect. This could explain a lower efficiency of TS to generate NO at higher input energy densities (Figure 6).

The solubility of NO_2 is slightly higher than the solubility of NO, almost by about one order of magnitude. However, the decrease of NO_2 in the experiment with ES compared to the dry air (Figure 7) also cannot be dominantly due to the NO_2 solvation to liquid water. Moreover, there was even less NO_2 in the experiment in humid air than in dry air or air with ES. We thus suppose that the observed changes of NO_2 concentration are rather related to the shift in NO_2/HNO_2 formation chemistry due to the presence of H_2O (reactions (11)–(14)).

The solubility of HNO_2 is much higher (by three orders of magnitude) than the solubility of NO_2. Thus, it is possible that the measured gaseous HNO_2 concentration is influenced by the solvation effect. However, it is questionable since the sum of NO_2 and HNO_2 (H_yNO_2) is the same in all experiments (Figure 10). It is therefore possible that the decrease of HNO_2 in the experiment in dry air with ES compared to the experiment in humid air is caused by the change in NO_2/HNO_2 chemistry due to differences in the humidity.

In the single reactor experiments in 1SS, when comparing the results with and without ES, it is impossible to separate changes of NO_2 and HNO_2 concentrations due to solvation to water, changes in chemical pathways of their formation and decay, cooling effect of water, or discharge instabilities. In order to assess the role of individual RONS in PAW generation, additional experiments were performed in 2SS with two reactors. The synthetic air (dry or humidified) was treated by TS in the first reactor. This treated air containing HyNOx was lead as the inlet gas into the second reactor with ES but without TS. We measured changes of the gas composition and aqueous $NO_2^-{}_{(aq)}$ and $H_2O_{2(aq)}$ in the accumulated water (see Section 3.3).

Figure 11 shows the comparison of NO concentration generated by TS in dry and humid air in 2SS with and without ES treatment in the second reactor (water flow rate 500 µL/min).

These results confirm a low solubility of NO, because its concentration in gas is within the experimental uncertainty not influenced by ES, both for dry and humid air.

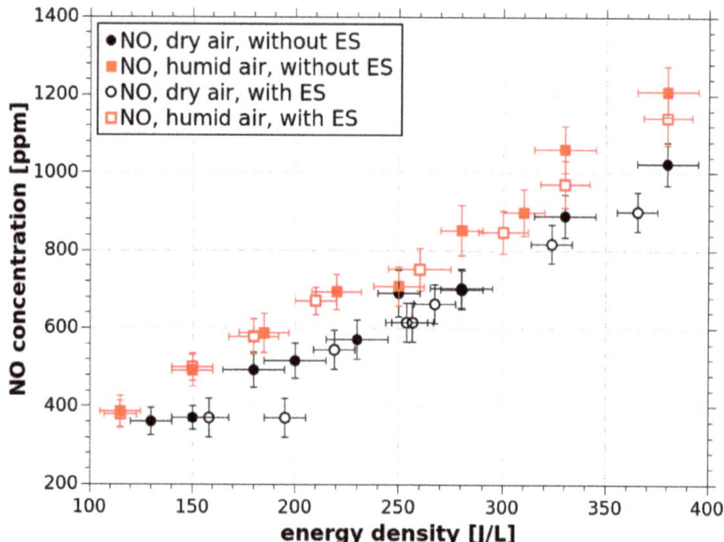

Figure 11. Comparison of NO concentration produced by TS in the first reactor, with and without ES (Q_w = 500 µL/min) in the second reactor (2SS), measured by UV-Vis absorption technique, short cell.

The solubility of NO_2 is almost by one order of magnitude higher than the solubility of NO. There is probably some decrease of NO_2 due to post-discharge ES treatment, especially at higher energy densities, i.e., when the NO_2 concentration is higher (Figure 12).

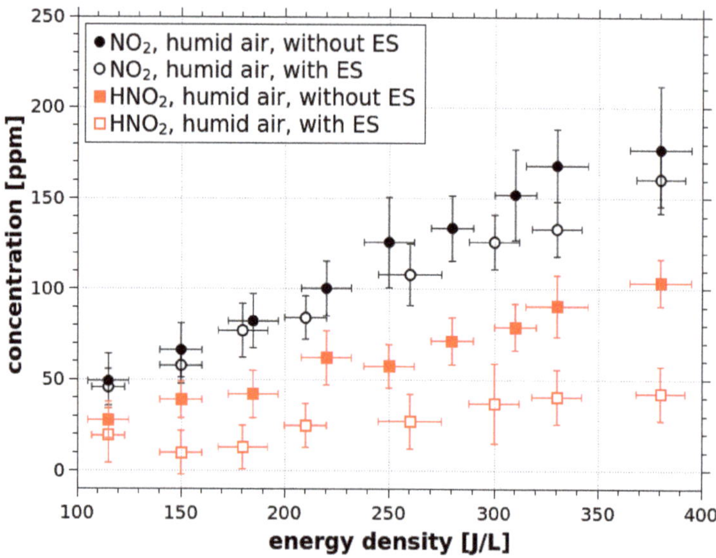

Figure 12. Comparison of NO_2 and HNO_2 concentration with and without post discharge ES treatment, measured by UV-Vis absorption technique, short cell, 2SS.

However, the decrease is still not significant with respect to the experimental uncertainty. The decrease of HNO_2 due to ES is more remarkable, despite that after the ES treatment in the second reactor, the HNO_2 concentration is already quite low and the relative experimental uncertainty is higher. These results indicate the key importance of HNO_2 for PAW generation, which is further confirmed by the PAW analysis.

3.3. Analysis of Plasma-Activated Water

We measured $NO_2^-{}_{(aq)}$ and $H_2O_{2(aq)}$ in plasma-activated water collected during 1SS experiments with TS combined with ES, as well as in water collected during 2SS experiments, with TS and ES separated into two reactors. As for control, we also performed experiments in 1SS with ES, but without TS. The used gas was either synthetic air (dry or humidified), or NO and NO_2 from calibration pressure tanks, mixed with N_2 or synthetic air, respectively.

In the control experiments with ES, the applied voltage (6 kV) was not sufficient for TS generation, but it was above the onset voltage of the corona discharge. Due to this, there was probably some formation of reactive species inside the ES reactor, and we detected a low concentration of $H_2O_{2(aq)}$, ~20 μM. This weak corona discharge in air, however, produced no detectable amount of nitrogen oxides and no $NO_2^-{}_{(aq)}$ was detected in water.

Relatively small amount of $NO_2^-{}_{(aq)}$ (~80 μM, Figure 13, red point) was detected in control experiment in 1SS with ES and without TS, using gas from the pressure tank containing ~1400 ppm NO in N_2. This proves a low solubility of NO. The formation of $NO_2^-{}_{(aq)}$ from the dissolved $NO_{(aq)}$ proceeds via the following reaction:

$$NO_{(aq)} + NO_{2(aq)} + H_2O \rightarrow 2NO_2^-{}_{(aq)} + 2H^+ \tag{15}$$

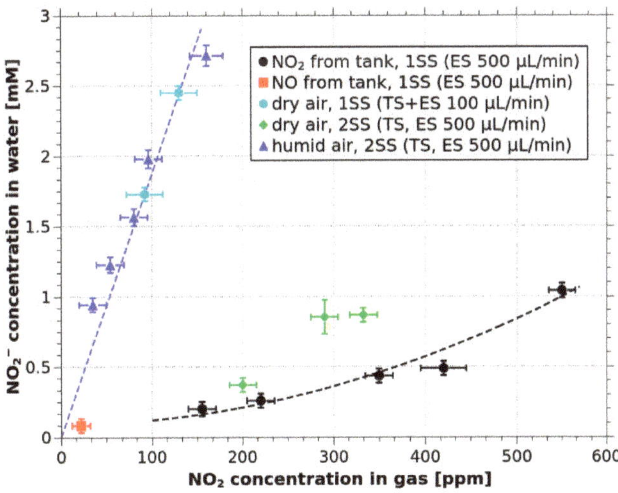

Figure 13. Concentration of $NO_2^-{}_{(aq)}$ in collected water as function of gas phase NO_2 concentration, ES with Q_w = 500 uL/min, different gas mixtures, and combined TS with ES (Q_w = 100 uL/min).

Thus, the dissolved $NO_{2(aq)}$ is also needed to produce $NO_2^-{}_{(aq)}$. In the gas phase only around 25 ppm NO_2 was detected and this limited the $NO_2^-{}_{(aq)}$ production. We can conclude that without NO_2, even very high concentration of NO has almost a negligible effect on the aqueous RONS formation. On the other hand, the solvation of NO_2 from the gas phase into water can lead to $NO_2^-{}_{(aq)}$ formation even without a presence of $NO_{(aq)}$, via the following reaction:

$$NO_{2(aq)} + NO_{2(aq)} + H_2O \rightarrow NO_2^-{}_{(aq)} + NO_3^-{}_{(aq)} + 2H^+ \tag{16}$$

One $NO_2^-{}_{(aq)}$ molecule is produced from two dissolved $NO_2^-{}_{(aq)}$ molecules. If we assume that $NO_2^-{}_{(aq)}$ is proportional to the gas phase NO_2 concentration, Reaction (16) could explain why the measured concentration of $NO_2^-{}_{(aq)}$ in water can be well fitted as a quadratic function of the gas phase NO_2 concentration (Figure 13, black points), when NO_2 from the pressure tank mixed with dry synthetic air was used.

In the dry synthetic air treated by TS in the first reactor and ES in the second reactor (2SS), the measured concentration of $NO_2^-{}_{(aq)}$ in water was higher, above the fitting curve valid for NO_2 from the pressure tank (Figure 13, green points). We assume that it can be explained by the presence of both NO and NO_2 in the sprayed gas, i.e., both NO and NO_2 can also contribute to the $NO_2^-{}_{(aq)}$ formation via Reactions (15) and (16).

If NO_2 in the gas phase plays a dominant role on $NO_2^-{}_{(aq)}$ formation in water, one could expect less $NO_2^-{}_{(aq)}$ in the water collected during 2SS experiments in humid synthetic air, where the gas phase NO_2 concentration produced by TS in the first reactor was lower compared to that generated by TS in the dry air. On the contrary, significantly higher $NO_2^-{}_{(aq)}$ concentration was observed (Figure 13, blue points). Here, the TS discharge in the first reactor generated not only NO and NO_2, but also directly by gaseous HNO_2, which can then form $NO_2^-{}_{(aq)}$ directly by its rapid solvation into water microdroplets generated by ES in the second reactor. Figure 13 shows an approximately linear dependence of $NO_2^-{}_{(aq)}$ on gaseous NO_2 when applying ES in humid air treated by TS. This most likely indicates a linear dependence of $NO_2^-{}_{(aq)}$ on the gaseous HNO_2 concentration, since HNO_2 concentration is proportional to the NO_2 concentration (Figure 12).

If we assume that the amount of $NO_2^-{}_{(aq)}$ in the liquid phase is proportional to the depletion of HNO_2 from the gas phase, we can estimate the expected $NO_2^-{}_{(aq)}$ concentration in the water. The highest HNO_2 decrease observed was approximately 50 ± 10 ppm. Taking into account gas and water flow rates, the corresponding estimated $NO_2^-{}_{(aq)}$ concentration in the liquid phase should be 4.4 ± 0.9 mM. The measured $NO_2^-{}_{(aq)}$ concentration was somewhat lower: 2.7 ± 0.2 mM. This difference could be explained by the instability of $NO_2^-{}_{(aq)}$ at acidic conditions, because the measured pH = 2.5–2.8 in the collected water. The disproportionation reaction:

$$3NO_2^-{}_{(aq)} + 3H^+ \rightarrow 2NO_{(aq)} + NO_3^-{}_{(aq)} + H_3O^+ \tag{17}$$

leads to exponential-like NO_2^- decrease in time. In PAW generated by TS, the observed characteristic decay time was approximately 30–40 min [58]. In our experiments in 2SS, the collected water sample was mixed with Griess reagent within 5 min after the end of the experiment. During this time, the NO_2^- concentration could decrease approximately by about 20% from the concentration reached at the moment when the experiment was stopped. Moreover, conversion of $NO_2^-{}_{(aq)}$ to $NO_3^-{}_{(aq)}$ also occurs during the time of experiment (4 min) inside the vessel where the water accumulates. Taking into account the $NO_2^-{}_{(aq)}$ concentration decrease due to Reaction (17) and the experimental uncertainty, the agreement between the measured $NO_2^-{}_{(aq)}$ concentration and the estimated one from the HNO_2 depletion from the air is satisfactory.

In the previous paragraphs we have discussed and compared differences of NO, NO_2 and HNO_2 solubility, and we based our reasoning mainly on differences of Henry's law coefficients of these three species. As mentioned in the Introduction, the Henry's law coefficient is valid for systems in a thermodynamic equilibrium and in plasma-water interactions, the solvation process may be influenced by many other phenomena. Here we must emphasize that in 2SS, the plasma and water did not interact directly. Direct interaction of plasma with water microdroplets occurred only in 1SS. However, the concentration of $NO_2^-{}_{(aq)}$ in PAW generated in 1SS with a direct contact of TS with microdroplets is not higher than $NO_2^-{}_{(aq)}$ concentration achieved in 2SS experiments in humid air with TS and ES separated into two reactors. It shows the same trend and similar values (Figure 13, cyan vs. blue points). This result justifies the reasons to compare solubility of species with long lifetime based on their Henry's law coefficient.

The major difference between PAW generated by TS with simultaneous ES (1SS), and water prepared in 2SS by spraying humid air treated by TS in the first reactor, is the concentration of produced $H_2O_{2(aq)}$. In the latter case the $H_2O_{2(aq)}$ concentration was only around 20 μM, while the concentration of $H_2O_{2(aq)}$ in PAW generated by TS with ES was higher, 60–80 μM. This result can be interpreted so that $H_2O_{2(aq)}$ is created mainly by solvation of OH radicals with short lifetime and therefore their concentration in the second reactor of 2SS is negligible. However, we must admit that this is just an assumption that must be verified by additional measurements.

Another interesting fact concerning $H_2O_{2(aq)}$ is its very low concentration compared to $NO_2^-{}_{(aq)}$. In our previous research we showed that TS in an open-air reactor can generate PAW with significantly higher $H_2O_{2(aq)}$ concentration, exceeding the concentration of $NO_2^-{}_{(aq)}$ [15]. The $NO_2^-{}_{(aq)}/H_2O_{2(aq)}$ ratio was approximately 3:7. In a closed reactor, with higher concentration of NOx in the gas phase, this ratio was much different, approximately 15:1 in favor of $NO_2^-{}_{(aq)}$. Here, with even higher concentrations of NOx in the gas phase, the $NO_2^-{}_{(aq)}/H_2O_{2(aq)}$ ratio is even higher, approximately 30–40:1. We therefore suppose that new results presented in this paper are consistent with the previous observations. However, we must admit that the low concentration of $H_2O_{2(aq)}$ in PAW with very high concentration of $NO_2^-{}_{(aq)}$ and acidic pH can be also explained by a partial depletion of $H_2O_{2(aq)}$ by its reaction with $NO_2^-{}_{(aq)}$ during the liquid sample accumulation and processing.

4. Conclusions

Transient spark (TS) discharge in air turns out to be a very efficient source for generation of nitrogen oxides NO and NO_2. More than 1000 ppm of NO was generated with an input energy density of 400 J/L. It was shown that in humid air, the TS discharge also generates a significant amount of HNO_2, providing 10–100 ppm of HNO_2 depending on the input energy density.

The air TS discharge in contact with water is very suitable for the generation of plasma-activated water (PAW), because TS can operate in a direct contact with water electrosprayed (ES) into the discharge zone. The formation of microdroplets with high interface surface area facilitates the transfer of reactive species from plasma into water. HNO_2 was also detected in air treated by TS in combination with water ES in one-stage system (1SS). This indicates an important role of HNO_2 in PAW generation, because HNO_2, although directly produced at lower concentrations, is much more soluble than two other major gas phase species generated by TS, NO and NO_2.

To assess the role of HNO_2, measurements were performed in two-stage system (2SS), where dry or humid air was treated by TS in the first reactor and water ES was applied in the second reactor. The dry air treated by TS in the first reactor contained mainly NO and NO_2 with only traces of HNO_2, while in the humid air, NO_2 decreased and HNO_2 reached almost the same concentration of as that of NO_2. For comparison, we also applied ES on NO and NO_2 from pressure tanks diluted in N_2 or air, respectively. These experiments confirmed that solvation to ES microdroplets caused a significant depletion of HNO_2 from the gas phase, while the depletion of NO was not observed, and the decrease of the NO_2 concentration was questionable.

The importance of HNO_2 for PAW generation by TS in air was proved by the measured concentrations of $NO_2^-{}_{(aq)}$ in water. The concentration of $NO_2^-{}_{(aq)}$ was much higher in humid air containing HNO_2 than in dry air with higher concentration of NO_2, but almost no HNO_2 in 2SS experiments with TS in the first reactor and ES in the second reactor.

The $NO_2^-{}_{(aq)}$ concentration in 2SS experiments in humid air was similar (or even higher) than $NO_2^-{}_{(aq)}$ concentration in PAW generated by a direct contact of the plasma with microdroplets in 1SS. The advantage of using 2SS is the stability and repeatability of the process. The TS in humid air without ES microdroplets inside the first reactor is relatively stable and the stable ES of water with higher water flow rate can be also achieved without TS in the second reactor. On the contrary, in 1SS with a direct contact of the discharge with water, the ES and TS negatively influence each other and both ES and TS are less stable. However, the advantage of using 1SS with generation of TS and ES in the same reactor is reaching a higher concentration of H_2O_2 in the water.

The optimization of the HNO_2 generation by TS and of the 2SS with a shorter distance between TS and ES reactors are planned in future. Further research of undergoing chemistry is also desirable, both experimentally and by chemical kinetic modeling. Comparison of biocidal effects of PAW generated in 1SS and 2SS should be also performed to investigate the role of low vs. high $H_2O_{2(aq)}$ concentration in addition to $NO_2^-{}_{(aq)}$. Nevertheless, this experimental study showed that gaseous HNO_2 produced in plasma discharges operated in humid air is a dominant contributor of nitrites in the plasma-activated water.

Author Contributions: Conceptualization, M.J., Z.M. and K.H.; methodology, M.J., K.H., P.T. and Z.M.; software, M.J.; validation, M.J. and P.T.; formal analysis, M.J. and P.T.; investigation, M.J., M.E.H. and P.T.; resources, M.J., K.H. and Z.M.; data curation, M.J., M.E.H. and P.T.; writing—original draft preparation, M.J.; writing—review and editing, Z.M., K.H., M.J. and M.E.H.; visualization, M.J.; supervision, Z.M., K.H. and M.J.; project administration, K.H. and M.J.; funding acquisition, Z.M. and M.J. All authors have read and agreed to the published version of the manuscript.

Funding: This research was funded by Slovak Research and Development Agency, grant number APVV-17-0382, and Slovak Grant Agency VEGA, grant number 1/0419/18.

Institutional Review Board Statement: Not applicable.

Informed Consent Statement: Not applicable.

Data Availability Statement: The data that supports the findings of this study are available within the article. Additional data that support the findings of this study are available from the corresponding author upon reasonable request.

Conflicts of Interest: The authors declare no conflict of interest.

References

1. Brandenburg, R.; Bogaerts, A.; Bongers, W.; Fridman, A.; Fridman, G.; Locke, B.R.; Miller, V.; Reuter, S.; Schiorlin, M.; Verreycken, T.; et al. White paper on the future of plasma science in environment, for gas conversion and agriculture. *Plasma Process. Polym.* **2019**, *16*, 1700250. [CrossRef]
2. Cvelbar, U.; Walsh, J.L.; Cernak, M.; De Vries, H.W.; Reuter, S.; Belmonte, T.; Corbella, C.; Miron, C.; Hojnik, N.; Jurov, A.; et al. White paper on the future of plasma science and technology in plastics and textiles. *Plasma Process. Polym.* **2019**, *16*, 1700228. [CrossRef]
3. Bekeschus, S.; Favia, P.; Robert, E.; Von Woedtke, T. White paper on plasma for medicine and hygiene: Future in plasma health sciences. *Plasma Process. Polym.* **2019**, *16*, 1800033. [CrossRef]
4. Von Woedtke, T.; Reuter, S.; Masur, K.; Weltmann, K.-D. Plasmas for Medicine. *Phys. Rep.* **2013**, *530*, 291–320. [CrossRef]
5. Keidar, M. Plasma for cancer treatment. *Plasma Sources Sci. Technol.* **2015**, *24*, 033001. [CrossRef]
6. Lu, X.; Ye, T.; Cao, Y.; Sun, Z.; Xiong, Q.; Tang, Z.; Xiong, Z.; Hu, J.; Jiang, Z.; Pan, Y. The roles of the various plasma agents in the inactivation of bacteria. *J. Appl. Phys.* **2008**, *104*, 053309. [CrossRef]
7. Lukes, P.; Clupek, M.; Babicky, V.; Sunka, P. Ultraviolet Radiation from the Pulsed Corona Discharge in Water. *Plasma Sources Sci. Technol.* **2008**, *17*, 024012. [CrossRef]
8. Dobrynin, D.; Fridman, G.; Friedman, G.; Fridman, A. Physical and Biological Mechanisms of Direct Plasma Interaction with Living Tissue. *New J. Phys.* **2009**, *11*, 115020. [CrossRef]
9. Machala, Z.; Chládeková, L.; Pelach, M. Plasma Agents in Bio-Decontamination by Dc Discharges in Atmospheric Air. *J. Phys. D Appl. Phys.* **2010**, *43*, 222001. [CrossRef]
10. Janda, M.; Martišovitš, V.; Hensel, K.; Machala, Z. Generation of Antimicrobial Nox by Atmospheric Air Transient Spark Discharge. *Plasma Chem. Plasma Process.* **2016**, *36*, 767–781. [CrossRef]
11. Bruggeman, P.J.; Kushner, M.J.; Locke, B.R.; Gardeniers, J.G.E.; Graham, W.G.; Graves, D.B.; Hofman-Caris, R.C.H.M.; Maric, D.; Reid, J.P.; Ceriani, E.; et al. Plasmaliquid Interactions: A Review and Roadmap. *Plasma Sources Sci. Technol.* **2016**, *25*, 053002. [CrossRef]
12. Puač, N.; Gherardi, M.; Shiratani, M. Plasma Agriculture: A Rapidly Emerging Field. *Plasma Process. Polym.* **2018**, *15*, 1700174. [CrossRef]
13. Thirumdas, R.; Kothakota, A.; Annapure, U.; Siliveru, K.; Blundell, R.; Gatt, R.; Valdramidis, V.P. Plasma Activated Water (PAW): Chemistry, Physico-Chemical Properties, Applications in Food and Agriculture. *Trends Food Sci. Technol.* **2018**, *77*, 21–31. [CrossRef]

14. Brisset, J.-L.; Pawlat, J. Chemical Effects of Air Plasma Species on Aqueous Solutes in Direct and Delayed Exposure Modes: Discharge, Post-Discharge and Plasma Activated Water. *Plasma Chem. Plasma Process.* **2016**, *36*, 355–381. [CrossRef]
15. Machala, Z.; Tarabová, B.; Sersenová, D.; Janda, M.; Hensel, K. Chemical and Antibacterial Effects of Plasma Activated Water: Correlation with Gaseous and Aqueous Reactive Oxygen and Nitrogen Species, Plasma Sources and Air Flow Conditions. *J. Phys. D Appl. Phys.* **2018**, *52*, 034002. [CrossRef]
16. Kaushik, N.; Ghimire, B.; Li, Y.; Adhikari, M.; Veerana, M.; Kaushik, N.; Jha, N.; Adhikari, B.; Lee, S.-J.; Masur, K.; et al. Biological and Medical Applications of Plasma-Activated Media, Water and Solutions. *Biol. Chem.* **2018**, *400*, 39–62. [CrossRef] [PubMed]
17. Lu, P.; Boehm, D.; Bourke, P.; Cullen, P.J. Achieving Reactive Species Specificity Within Plasma-Activated Water Through Selective Generation Using Air Spark and Glow Discharges. *Plasma Process. Polym.* **2017**, *14*, 1600207. [CrossRef]
18. Lukes, P.; Locke, B.R.; Brisset, J.-L. Aqueous-phase chemistry of electrical discharge plasma in water and in gas–liquid environments. In *Plasma Chemistry and Catalysis in Gases and Liquids*; Wiley-VCH Verlag & Co. KGaA: Weinheim, Germany, 2012; pp. 243–308. ISBN 978-3-527-64952-5.
19. Lukes, P.; Dolezalova, E.; Sisrova, I.; Clupek, M. Aqueous-Phase Chemistry and Bactericidal Effects from an Air Discharge Plasma in Contact with Water: Evidence for the Formation of Peroxynitrite Through a Pseudo-Second-Order Post-Discharge Reaction of H_2O_2 and HNO_2. *Plasma Sources Sci. Technol.* **2014**, *23*, 015019. [CrossRef]
20. Huang, D.M.; Chen, Z.M. Reinvestigation of the Henry's Law Constant for Hydrogen Peroxide with Temperature and Acidity Variation. *J. Environ. Sci.* **2010**, *22*, 570–574. [CrossRef]
21. Kruszelnicki, J.; Lietz, A.M.; Kushner, M.J. Atmospheric Pressure Plasma Activation of Water Droplets. *J. Phys. D Appl. Phys.* **2019**, *52*, 355207. [CrossRef]
22. Hassan, M.E.; Janda, M.; Machala, Z. Transport of Gaseous Hydrogen Peroxide and Ozone into Bulk Water vs. Electrosprayed Aerosol. *Water* **2021**, *13*, 182. [CrossRef]
23. Stratton, G.R.; Bellona, C.L.; Dai, F.; Holsen, T.M.; Thagard, S.M. Plasma-Based Water Treatment: Conception and Application of a New General Principle for Reactor Design. *Chem. Eng. J.* **2015**, *273*, 543–550. [CrossRef]
24. Zeleny, J. The Electrical Discharge from Liquid Points, and a Hydrostatic Method of Measuring the Electric Intensity at Their Surfaces. *Phys. Rev.* **1914**, *3*, 69–91. [CrossRef]
25. Zeleny, J. Instability of Electrified Liquid Surfaces. *Phys. Rev.* **1917**, *10*, 1–6. [CrossRef]
26. Dwivedi, P.; Matz, L.M.; Atkinson, D.A.; Hill, H.H., Jr. Electrospray ionization-ion mobility spectrometry: A rapid analytical method for aqueous nitrate and nitrite analysis. *Analyst* **2004**, *129*, 139–144. [CrossRef]
27. Jaworek, A. Electrospray droplet sources for thin film deposition. *J. Mater. Sci.* **2007**, *42*, 266–297. [CrossRef]
28. Carotenuto, C.; Di Natale, F.; Lancia, A. Wet Electrostatic Scrubbers for the Abatement of Submicronic Particulate. *Chem. Eng. J.* **2010**, *165*, 35–45. [CrossRef]
29. Cui, H.; Li, N.; Peng, J.; Cheng, J.; Zhang, N.; Wu, Z. Modeling the Particle Scavenging and Thermal Efficiencies of a Heat Absorbing Scrubber. *Build. Environ.* **2017**, *111*, 218–227. [CrossRef]
30. Jaworek, A.; Sobczyk, A.T.; Krupa, A. Electrospray Application to Powder Production and Surface Coating. *J. Aerosol Sci.* **2018**, *125*, 57–92. [CrossRef]
31. Boda, S.K.; Li, X.; Xie, J. Electrospraying an Enabling Technology for Pharmaceutical and Biomedical Applications: A Review. *J. Aerosol Sci.* **2018**, *125*, 164–181. [CrossRef]
32. Ball, A.K.; Roy, S.S.; Kisku, D.R.; Murmu, N.C.; Dos Santos Coelho, L. Optimization of Drop Ejection Frequency in EHD Inkjet Printing System Using an Improved Firefly Algorithm. *Appl. Soft Comput.* **2020**, *94*, 106438. [CrossRef]
33. Kim, S.; Jung, M.; Choi, S.; Lee, J.; Lim, J.; Kim, M. Discharge Current of Water Electrospray with Electrical Conductivity under High-Voltage and High-Flow-Rate Conditions. *Exp. Therm. Fluid Sci.* **2020**, *118*, 110151. [CrossRef]
34. Kim, Y.; Jung, S.; Kim, S.; Choi, S.T.; Kim, M.; Lee, H. Heat Transfer Performance of Water-Based Electrospray Cooling. *Int. Commun. Heat Mass Transf.* **2020**, *118*, 104861. [CrossRef]
35. Borra, J.-P. Review on Water Electro-Sprays and Applications of Charged Drops with Focus on the Corona-Assisted Cone-Jet Mode for High Efficiency Air Filtration by Wet Electro-Scrubbing of Aerosols. *J. Aerosol Sci.* **2018**, *125*, 208–236. [CrossRef]
36. Rosell-Llompart, J.; Grifoll, J.; Loscertales, I.G. Electrosprays in the Cone-Jet Mode: From Taylor Cone Formation to Spray Development. *J. Aerosol Sci.* **2018**, *125*, 2–31. [CrossRef]
37. Jaworek, A.; Gañán-Calvo, A.M.; Machala, Z. Low Temperature Plasmas and Electrosprays. *J. Phys. D Appl. Phys.* **2019**, *52*, 233001. [CrossRef]
38. Kanev, I.L.; Mikheev, A.Y.; Shlyapnikov, Y.M.; Shlyapnikova, E.A.; Morozova, T.Y.; Morozov, V.N. Are Reactive Oxygen Species Generated in Electrospray at Low Currents? *Anal. Chem.* **2014**, *86*, 1511–1517. [CrossRef] [PubMed]
39. Burlica, R.; Grim, R.G.; Shih, K.-Y.; Balkwill, D.; Locke, B.R. Bacteria Inactivation Using Low Power Pulsed Gliding Arc Discharges with Water Spray. *Plasma Process. Polym.* **2010**, *7*, 640–649. [CrossRef]
40. Oinuma, G.; Nayak, G.; Du, Y.; Bruggeman, P.J. Controlled Plasmadroplet Interactions: A Quantitative Study of OH Transfer in Plasmaliquid Interaction. *Plasma Sources Sci. Technol.* **2020**, *29*, 095002. [CrossRef]
41. Kovalova, Z.; Leroy, M.; Kirkpatrick, M.J.; Odic, E.; Machala, Z. Corona Discharges with Water Electrospray for *Escherichia coli* Biofilm Eradication on a Surface. *Bioelectrochemistry* **2016**, *112*, 91–99. [CrossRef]
42. Becker, K.; Kleffmann, J.; Kurtenbach, A.; Wiesen, P. Solubility of Nitrous Acid (HONO) in Sulfuric Acid Solutions. *J. Phys. Chem.* **1996**, *100*, 14984–14990. [CrossRef]

43. Lee, Y.N.; Schwartz, S.E. Reaction Kinetics of Nitrogen Dioxide with Liquid Water at Low Partial Pressure. *J. Phys. Chem.* **1981**, *85*, 840–848. [CrossRef]

44. Armor, J.N. Influence of pH and Ionic Strength Upon Solubility of Nitric Oxide in Aqueous Solution. *J. Chem. Eng. Data* **1974**, *19*, 82–84. [CrossRef]

45. Kučerová, K.; Machala, Z.; Hensel, K. Transient Spark Discharge Generated in Various N_2/O_2 Gas Mixtures: Reactive Species in the Gas and Water and Their Antibacterial Effects. *Plasma Chem. Plasma Process.* **2020**, *40*, 749–773. [CrossRef]

46. Janda, M.; Martišovitš, V.; Machala, Z. Transient Spark: A Dc-Driven Repetitively Pulsed Discharge and Its Control by Electric Circuit Parameters. *Plasma Sources Sci. Technol.* **2011**, *20*, 035015. [CrossRef]

47. Janda, M.; Machala, Z.; Niklová, A.; Martišovitš, V. The Streamer-to-Spark Transition in a Transient Spark: A Dc-Driven Nanosecond-Pulsed Discharge in Atmospheric Air. *Plasma Sources Sci. Technol.* **2012**, *21*, 045006. [CrossRef]

48. Gordon, I.E.; Rothman, L.S.; Hill, C.; Kochanov, R.V.; Tan, Y.; Bernath, P.F.; Birk, M.; Boudon, V.; Campargue, A.; Chance, K.V.; et al. The HITRAN2016 Molecular Spectroscopic Database. *J. Quant. Spectrosc. Radiat. Transf.* **2017**, *203*, 3–69. [CrossRef]

49. Kochanov, R.V.; Gordon, I.E.; Rothman, L.S.; Shine, K.P.; Sharpe, S.W.; Johnson, T.J.; Wallington, T.J.; Harrison, J.J.; Bernath, P.F.; Birk, M.; et al. Infrared Absorption Cross-Sections in HITRAN2016 and Beyond: Expansion for Climate, Environment, and Atmospheric Applications. *J. Quant. Spectrosc. Radiat. Transf.* **2019**, *230*, 172–221. [CrossRef]

50. Keller-Rudek, H.; Moortgat, G.K.; Sander, R.; Sörensen, R. The MPI-Mainz UV/VIS Spectral Atlas of Gaseous Molecules of Atmospheric Interest. *Earth Syst. Sci. Data* **2013**, *5*, 365–373. [CrossRef]

51. Oehmigen, K.; Hähnel, M.; Brandenburg, R.; Wilke, C.; Weltmann, K.-D.; Von Woedtke, T. The Role of Acidification for Antimicrobial Activity of Atmospheric Pressure Plasma in Liquids. *Plasma Process. Polym.* **2010**, *7*, 250–257. [CrossRef]

52. Chauvin, J.; Judée, F.; Yousfi, M.; Vicendo, P.; Merbahi, N. Analysis of Reactive Oxygen and Nitrogen Species Generated in Three Liquid Media by Low Temperature Helium Plasma Jet. *Sci. Rep.* **2017**, *7*, 4562. [CrossRef] [PubMed]

53. Anderson, C.E.; Cha, N.R.; Lindsay, A.D.; Clark, D.S.; Graves, D.B. The Role of Interfacial Reactions in Determining Plasma–Liquid Chemistry. *Plasma Chem. Plasma Process.* **2016**, *36*, 1393–1415. [CrossRef]

54. Wende, K.; Williams, P.; Dalluge, J.; Gaens, W.V.; Aboubakr, H.; Bischof, J.; Von Woedtke, T.; Goyal, S.M.; Weltmann, K.D.; Bogaerts, A.; et al. Identification of the biologically active liquid chemistry induced by a nonthermal atmospheric pressure plasma jet. *Biointerphases* **2015**, *10*, 029518. [CrossRef]

55. Tian, Y.; Ma, R.; Zhang, Q.; Feng, H.; Liang, Y.; Zhang, J.; Fang, J. Assessment of the Physicochemical Properties and Biological Effects of Water Activated by Non-Thermal Plasma above and Beneath the Water Surface. *Plasma Process. Polym.* **2015**, *12*, 439–449. [CrossRef]

56. Lu, P.; Boehm, D.; Cullen, P.; Bourke, P. Controlled Cytotoxicity of Plasma Treated Water Formulated by Open-Air Hybrid Mode Discharge. *Appl. Phys. Lett.* **2017**, *110*, 264102. [CrossRef]

57. Kovačević, V.V.; Dojčinović, B.P.; Jović, M.; Roglić, G.M.; Obradović, B.M.; Kuraica, M.M. Measurement of Reactive Species Generated by Dielectric Barrier Discharge in Direct Contact with Water in Different Atmospheres. *J. Phys. D Appl. Phys.* **2017**, *50*, 155205. [CrossRef]

58. Tarabová, B.; Lukeš, P.; Janda, M.; Hensel, K.; Šikurová, L.; Machala, Z. Specificity of Detection Methods of Nitrites and Ozone in Aqueous Solutions Activated by Air Plasma. *Plasma Process. Polym.* **2018**, *15*, 1800030. [CrossRef]

59. Eisenberg, G. Colorimetric Determination of Hydrogen Peroxide. *Ind. Eng. Chem. Anal. Ed.* **1943**, *15*, 327–328. [CrossRef]

60. Janda, M.; Machala, Z.; Lacoste, D.; Stancu, G.D.; Laux, C.O. Stabilization of A Lean Methane-Air Flame Using Transient Spark Discharge. In Proceedings of the 13th International Symposium on High Pressure Low Temperature Plasma Chemistry (Hakone XIII), Kazimierz Dolny, Poland, 9–14 September 2012; pp. 185–189.

61. Borra, J.-P.; Ehouarn, P.; Boulaud, D. Electrohydrodynamic Atomisation of Water Stabilised by Glow Discharge—Operating Range and Droplet Properties. *J. Aerosol Sci.* **2004**, *35*, 1313–1332. [CrossRef]

62. Janda, M.; Hensel, K.; Machala, Z. Chemical Kinetic Model of Transient Spark: Spark Phase and NOx Formation. In Proceedings of the 22nd Symposium on Applications of Plasma Processes and 11th Eu-Japan Joint Symposium on Plasma Processing Sapp XXII, Strbske Pleso, Slovakia, 18–24 January 2019; pp. 25–32.

63. Capitelli, M.; Ferreira, C.M.; Gordiets, B.F.; Osipov, A.I. *Plasma Kinetics in Atmospheric Gases*; Springer: Berlin/Heidelberg, Germany, 2000.

64. Janda, M.; Martišovitš, V.; Hensel, K.; Dvonč, L.; Machala, Z. Measurement of the Electron Density in Transient Spark Discharge. *Plasma Sources Sci. Technol.* **2014**, *23*, 065016. [CrossRef]

65. Herron, J. Evaluated Chemical Kinetics Data for Reactions of $N(^2D)$, $N(^2P)$, and $N_2(A^3\Sigma_u^+)$ in the Gas Phase. *J. Phys. Chem. Ref. Data* **1999**, *28*, 1453–1483. [CrossRef]

66. Melton, C.E. Cross Sections and Interpretation of Dissociative Attachment Reactions Producing OH^-, O^-, and H^- in H_2O. *J. Chem. Phys.* **1972**, *57*, 4218–4225. [CrossRef]

67. Lowke, J.J.; Morrow, R. Theoretical analysis of removal of oxides of sulphur and nitrogen in pulsed operation of electrostatic precipitators. *IEEE Trans. Plasma Sci.* **1995**, *23*, 661–671. [CrossRef]

Article

Helium Atmospheric Pressure Plasma Jet Source Treatment of White Grapes Juice for Winemaking

Ramona Huzum [1] and Andrei Vasile Nastuta [2,*]

1 Integrated Center of Environmental Science Studies in the North-Eastern Development Region (CERNESIM), Department of Exact and Natural Sciences, Institute of Interdisciplinary Research, 'Alexandru Ioan Cuza' University of Iasi, 11 'Carol I' Blvd., 700506 Iasi, Romania; ramona.huzum@gmail.com

2 Physics and Biophysics Education Research Laboratory (P&B-EduResLab), Biomedical Science Department, Faculty of Medical Bioengineering, 'Grigore T. Popa' University of Medicine and Pharmacy Iasi, M. Kogalniceanu Str., No. 9-13, 700454 Iasi, Romania

* Correspondence: andrei.nastuta@gmail.com

Abstract: In the last few years, new emerging technologies to develop novel winemaking methods were reported. Most of them pointed out the need to assess the barrel aging on the wine product, fermentation process, green technologies for wine treatment for long term storage. Among these, plasma technologies at atmospheric pressure are on the way of replacing old and expensive methods for must, wine and yeast treatment, the goal being the long-term storage, aging and even decontamination of such products, and seems to meet the requirements of the winemakers. Using the principles of dielectric barrier discharge, we power up an atmospheric pressure plasma jet in helium. This plasma is used for treatment of fresh must obtained from white grapes. Our research manuscript is focused on the correlation of plasma parameters (applied voltage, plasma power, reactive species, gas temperature) with the physico-chemical properties of white must and wine (1 and 2 years old), via ultraviolet–visible and infrared spectroscopy, and colorimetry. Two types of white must were plasma treated and studied over time. The 10 W plasma source did not exceed 40 °C during treatment, the must did not suffer during thermal treatment. A higher quantity of RONS was observed during plasma-must exposure, supporting further oxidation processes. The UV-Vis and FTIR spectroscopy revealed the presence of phenols, flavones and sugar in the wine samples. Simultaneous visualization of CIE $L^*a^*b^*$ and RGB in color space charts allows easier understanding of wine changing in color parameters. These experimental results supporting the possible usability of atmospheric pressure plasma for winemaking.

Keywords: atmospheric pressure plasma jet; plasma-wine making; plasma treatment; UV-Vis spectroscopy; ATR-FTIR spectroscopy

Citation: Huzum, R.; Nastuta, A.V. Helium Atmospheric Pressure Plasma Jet Source Treatment of White Grapes Juice for Winemaking. *Appl. Sci.* **2021**, *11*, 8498. https://doi.org/10.3390/app11188498

Academic Editor: Maria Kanellaki

Received: 14 August 2021
Accepted: 10 September 2021
Published: 13 September 2021

1. Introduction

Atmospheric pressure plasma sources are rapidly gaining importance as tools for material worldwide processing, since they are easy to use, technologically simple and environmentally friendly. Applications of these plasmas include: surface modification and deposition, plasma-based synthesis of bio-medical surfaces, decontamination and sterilization, oncotherapy and wound healing [1–25]. In the last few years, new emerging technology to develop novel winemaking methods were reported. Most of them pointed out the need to assess the barrel aging on the wine product, slowing down the fermentation process and even stopping it and green technologies for wine treatment for long term storage. Among these, plasma technologies based on gas discharges, at atmospheric pressure, are on the way to replacing old and expensive methods for fruit juice, wine and yeast treatment, the goal being the long-term storage, aging and even decontamination of such products, and seems to meet most of the criteria required by the winemakers [26–44]. Depending on the utilization, the plasma source needs be tuned as to comply with the application requirements (power, electric field, reactive species). This is why it is important to characterize and monitor plasma sources from electrical and optical point of view.

To date, most of the studies regarding plasma-grapes or plasma-wine have been focused on the inactivation of microorganisms by plasma treatment [29,35], few studies approaching the impact of plasma on food components [26,45,46]. Moreover, the huge variety of existing plasma sources and treatment conditions, as well as limitless process parameters in numerous researches, makes it challenging to compare plasma effects on food by-products such as fruit juice/beverages (in our case must or wine).

Vitis vinifera sp is cultivated worldwide for grapes, juice and wine because its large adaptability for different climate and soil type. Over the world, vines grow on all kinds of soil, but individual factors of soil formation particularities of an area give rise to soil variability and different challenges of vineyard management. The terroir concept is given by a complex of variables as: soil, local climate, cultivar and winemaking technique to describe the individual character or "personality" of wine from a specific vineyard area [47,48]. *Vitis vinifera* is the most important fruit species in the world, no matter its way of consumption as fresh grapes or being processed into raisins, juice or wine. Wine can be simply defined as an alcoholic beverage made from fermented juice of grapes. In total, 95% of wine composition is represented by water and ethanol, while the remaining 5% is by other components such as glycerol, organic acids, carbohydrates, minerals, volatile and phenolic compounds [49].

Wine and winemaking history is lost in time and is closely connected with the history of agriculture, cuisine, humanity and civilization itself. It is well known that people enjoy drinking wine for its taste, flavors and for the health benefits of moderate wine consumption, nowadays being a component of the culture of many countries. A moderate amount of alcohol from wine consumption (150–300 mL/day) can protect against cardiovascular disease, dietary cancers, ischemic stroke, diabetes, hypertension and so on [50]. Additionally, the polyphenols from wine have antioxidant, antiviral, antibacterial and anti-carcinogen proprieties which imply health benefits [50–53]. Apart this, vine and wine has an important economic status, including in the world trade market as well as the agro-tourism to wineries and wine-growing areas. The vineyards became a new attraction for tourism with the thematic trips "on the way of wine" in the vineyards as well as wine tasting directly from wineries and cellars.

Romania is one of the principal producers and consumers of wine from the European Union, which account for 53% of world consumers in 2019 (according to International Organization of Vine and Wine (OIV): OIV-state of the world vitivinicultural sector in 2019 [54]). In concordance with the OIV annual report, Romania is the 5th EU country with a vineyard surface area of 191 kha after Spain (966 kha), France (794 kha), Italy (708 kha) and Portugal (195 kha). Although the weather good condition favored a potentially large 2020 harvest, Romania (3.6 mhl) had a negative variation of production in relation with 2019 and the last 5 years (−7% and −17% respectively) (in compliance with the OIV-2020 world wine production first estimates from October 2020 [55]).

The Romanian wine-growing is divided into eight regions, after Cotea et al. [56]:

- the Transylvanian Plateau (Târnave, Alba, Sebeş-Apold, Aiud, Lechinţa vineyards)
- the Moldavian Plateau (Cotnari, Iaşi, Huşi, Dealurile Fălciului, Colinele Tutovei, Zeletin, Dealu Bujorului, Nicoreşti, Iveşti, Covurlui vineyards)
- the Piedmont at the Carpathian's Curvature (Panciu, Odobeşti, Coteşti, Buzău's, Dealu Mare vineyards)
- the Getic Plateau (Ştefăneşti-Argeş, Sâmbureşti, Drăgăşani, Dealurile Craiovei, Plaiurile Drâncei, Severin Vineyards)
- The Banat-Crişana-Maramureş (Banat, Miniş-Măderat, Diosig, Valea lui Mihai, Silvaniei vineyards)
- the Sands in the South of Oltenia (Dacilor, Calafat, Sadova-Corabia vineyards)
- the Romanian Plain (Greaca, independent wine-growing centers situated in the Romanian plain)

- The Dobrogea Plateau (Sarica-Niculiţel, Istria-Babadag, Murfatlar, Ostrov vineyards)

The Moldavian Plateau, situated in the Eastern part of Romania, is the biggest wine-growing region (69,154 ha) with vines planted at 200-500 m altitude, on different types of soil. From a geological point of view the wine-growing area corresponds to the Moldavian Platform, with soil developed on Sarmatian (Bassarabian and Chersonian) sedimentary rocks consisting of clays and interlayer sand or interbanded clays and carbonates [57]. The climatic conditions of the area are characteristic of a temperate continental type.

In this report, by using the principles of dielectric barrier discharge (DBD), in a cylindrical configuration, we power up an atmospheric pressure plasma jet (appj) in helium. This plasma is used for treatment of fresh (just prepared) juice obtained from white grapes from a small family vineyard. Our research is focused on the correlation of He-appj parameters (like: applied voltage, plasma current, power, plasma excited/reactive nitrogen and oxygen species) with the physico-chemical properties of white must and the resulting white wine (1- and 2-year old), via ultraviolet–visible spectroscopy (UV-Vis) and attenuated total reflection-Fourier transform infrared (ATR-FTIR) spectroscopy, thus proving the possible usage of atmospheric pressure plasma treatment for winemaking.

2. Materials and Methods

This section contains information about the materials and methods used in conducting the experiments involved in this study. The section is divided into two parts: one related to the experimental arrangement and methods used for plasma source ignition, characterization and treatment—Section 2.1; the second includes the material (must) subject to plasma treatment, the method of obtaining it, spectroscopic and physico-chemical methods used for must and wine characterization—Section 2.2.

White grapes juice (must) and winemaking

The grapes used in these experiments were harvested at their technological maturity from the wine region of Moldova (North-Eastern Romania) as follows:

- 1. A set of grapes was collected from a small family wineyard with hybrid white Noble grapes situated in the Bârzeşti-Ştefan cel Mare region (N46°44′40,18″; E27°33′42,63″), Vaslui county. These samples were marked with '**B**'. This studied parcel of 600 m^2 was placed close to the house and no soil or leaves treatments were applied. The only agricultural practices were manual cutting and tying of vine shoots and digging.
- 2. A second set of grapes was collected from a vineyard parcel with a mixture of white grapes (Chasselas, Fetească, Busuioacă) situated in Pâhneşti-Arsura-Huşi (N46°46′54,93″; E28°02′38,21″), Vaslui county. These samples were marked with '**H**'. The studied parcel of 2300 m^2 was located in a vineyard of around 90 ha, where mechanical ploughing and different treatments for soil and plant were applied. Only cutting and tying of shoots, as well as grapes harvests, were manually made.

The *Vitis vinifera* cv. and hybrid Nobles grapes were harvested (from the above mentioned areas) manually at their optimum point of maturity in good sanitary stage, in 2018. The grapes were transported from both harvest location to the processing point, in plastic boxes, weighted and evaluated for shape, size and health. Around 4 kg of each type of grapes (around 40 bunches each) were used for the must and winemaking procedure. The white grapes were split from bunches (the de-stemming) and the berries (1–2 g each) were evaluated manually for any visible defects before hand-crushing, the resulting juice being transferred in vessels for further analysis and winemaking (alcoholic fermentation). Around 2 L of wine, for each set, was obtained and bottled for maturation.

2.1. Plasma Source and Electro-Optical Diagnosis

The discharge configuration consisted of a 100 mm long quartz tube, with inner and outer diameter of 4 and 6.1 mm respectively, and two 10 mm long copper tape electrodes. One power electrode (HV) and a grounded electrode (Gr) were wrapped on the glass tube, with an electrode gap width of 10 mm, similar to that reported by [8,11,16]. The discharge was driven by a PVM500 plasma power generator (Information Unlimited) with independent voltage, current, and frequency adjustment: voltage 1–40 kV, frequency 20–70 kHz, output power 1–300 W. The applied sinusoidal voltage Ua (up to 16 kVpp, @ 50 kHz) and the total current of the discharge Id, were monitored using voltage and current probes (Testec HVP-15HF, Pico TA131, a 50 Ω resistor) and a 200 MHz digital oscilloscope (Picoscope 2208A, two channels, 1 GS/s, 8 to 12 bits resolution, function generator + AWG 1MHz). The working gas, supplied through the discharge tube, was pure helium (spectral helium, He 5.0, Siad Romania). The gas flow rate of 2 slm was regulated using a needle valve rotameter (Platon 0–5 slm NGVS312 series). Long exposure photos of the plasma interacting with samples were captured using a Canon 600D (18Mpx, 400–750 nm spectral sensitivity) and Tamron 18–200 mm Di II lenses; IR photos of plasma-surface interaction was captured using a FLIR TG165 camera (8–14 μm spectral range, ±1.5°). The spectral emission of the discharge in the UV-Vis-NIR range (200–1100 nm) was analyzed by using a ASEQ Instruments LR1 broad range spectrometer (monochromator with a 50 μm entrance slit, a 600 gr/mm diffraction grating blazed at 300 nm, and a CCD Toshiba TCD1304DG linear array detector, calibrated for Absolute Irradiation Measurements), via a 0.4 mm diameter and 1 m long optical fiber (for 200–1400 nm, kevlar reinforced, cosine corrected adapter, Thorlabs), placed at 5 mm from the plasma.

The must samples were plasma treated (marked from now on 'PB' and 'PH') as follows: 50 mL of must was placed in Petri dishes and treated for 3 min. The discharge tube was positioned centrally to the Petri dish (25 mm height, 150 mm diameter soda glass plate), 10 mm above the must level, as in Figure 1. The procedure was repeated for 750 mL, for the two sets of must samples (the one from Bârzeşti, respectively from Huşi). The control samples (marked with 'MB' and 'MH') were also placed in identical volumes in Petri dishes and kept for 3 min each outdoors.

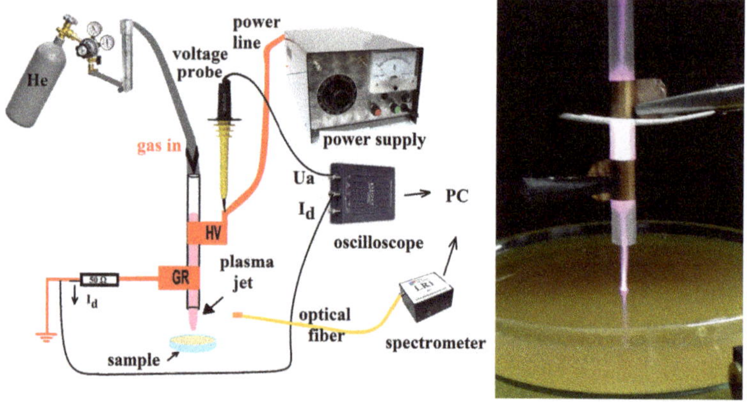

Figure 1. Plasma source set-up (**left**) and plasma-must treatment in a Petri dish (**right**).

2.2. Methods for Characterization of Must and Wine

Conventional wine analytical methods usually imply multidisciplinary approaches in order to rate wine quality and its authenticity. Just the official methods published in the OIV's Compendium of International Methods of Analysis of Wines and Musts are commonly of use for verification purposes and settling disputes upon the true origin of the wine. These analytical methods for wine investigation involve the determination of: total and volatile acidity, total and free sulfur dioxide, alcoholic strength, ethanol origin by isotope ratio mass spectrometry (IRMS), ethanol deuterium distribution by nuclear magnetic resonance (NMR), volatile compounds by gas chromatography (GC), reducing substance, principal organic acid concentration by high-pressure capillary electrophoresis (HPCE) and mineral elements by inductive plasma atomic emission spectrometry (ICP-AES) [54,55,58–60].

Regulatory authorities also assess wine for the presence of artificial sweeteners or colorants, preservatives as well as fermentation inhibitors. In any case, OIV encourages member states to continue research in the areas of interest to avoid any non-scientific evaluation of results [54,55,58,59].

Spectroscopic methods, like Ultraviolet-to-Visible spectroscopy (UV-Vis) [61–64], Fourier Transformed Infrared spectroscopy (FTIR) [48,65–71] are also applied due to their easiness of use, cost-effective and no sample preparation also in combination with numerical/statistical methods (e.g., chemometric methods) [48,66,68,72–75].

2.2.1. The Ultraviolet (UV) to Infrared (IR) Spectroscopy

The advantage of the UV-Vis method is given by the low cost due to the simple samples preparation, mainly represented by a set of dilutions, as well as saving time. The UV-VIS spectroscopy is used to identify the polyphenolic groups (non-flavoids and flavoids) that are released from solid parts of the grapes (skin, seeds, stems) into the must during the winemaking process and ageing. The polyphenols are important due to their large attribute of sensory characteristic of wine as color, taste and aroma. The UV-Vis measurements are widely used combined with multivariate regression approaches to obtain the spectroscopic calibration for prediction of phenolic compounds in grape and wine. More advanced techniques as liquid chromatography and fluorescence spectroscopy can be also used to quantify the phenolic content but UV-Vis spectroscopy remains one of higher importance due its simplicity, availability and minimal sample preparation. The UV-Vis absorption spectroscopy complies with the 'no sample pretreatment, fast and low-cost' criteria and can be easily applied to the analysis of grapes and wines and the qualitative detection of phenolic composition. The absorption spectrum in the UV-Vis region may be even used as a fingerprint in wine discrimination. Most of the studies employing UV-Vis spectroscopy are based on absorbance values in a specific wavelength region in order to quantify classes of phenolic compounds and color characteristics [61–64,66]. Recently, Minute et al. [75] proposed alternative methods for studying the pinking alteration of white wines based on spectroscopic and color properties of wine.

Another technique, the mid-range Attenuated Total Reflectance-Fourier Transform Infrared spectroscopy (ATR-FTIR) spectroscopy has become a valuable tool in wine analysis since it has been successfully used not only to detect and quantify key wine compounds but also to classify wines and monitor their fermentation or aging process, being a rapid-usage and minimum waste-producing method [65–72,76–80]. Nowadays the FTIR technique is used even for authenticity and traceability of wine [48,69,73,81]. Most of the characteristic wavenumbers associated with functional groups found in wine samples, as reported in literature, are shown in Table 1. For the presented experiments, UV-Vis absorption spectroscopy (using a Thermo Scientific Evolution 300, 190–1100 nm wavelength range, 1 nm bandwidth, quartz cuvette of 10 mm path length). Deionized water was used as the baseline in the selected wavelength range.

An ATR-FTIR FT/IR4700 spectrometer (Jasco) was used for the vibrational spectroscopy measurements. The mid-range infrared spectroscopy using the Fourier-Transform attenuated total reflection mode unveiled the followings absorbance spectra of the pristine and plasma treated must. For these measurements, 50 µL of each wine sample was positioned in contact with the diamond crystal of the attenuated total reflectance IR interferometer. All FTIR spectra were recorded in the range from 500 to 4000 cm^{-1}, at 4 cm^{-1} spectral resolution and 2 mm/s scan speed. The ATR crystal was carefully cleaned before each analysis, first with distilled water, then dried with soft tissue paper. Before each sample measurement, spectra of the clean and dry diamond against air were recorded and used as background. Each FTIR spectra was averaged from 54 scans. All measurements were performed in triplicate.

Table 1. Characteristic wavenumbers (cm^{-1}) for bending vibrations and stretching in wine samples, as reported by [67–69,72,76,77]

Wavenumber Regions (cm^{-1})	Groups	Assignment
<1000	stretching and bending vibrations	phosphates, phenolics, mono-substituted phenyl derivates, unsaturated lipids, carotenoids
1068–1065 1107–1110 1200	stretching vibration of C-O O-H stretch second overtones	sugars and organic acids
950–1250	stretching and bending vibrations	hydrolyzable and condensed tannins glucose, oligo- and polysaccharides, alcohols
1457–1288	C=O, C=C, -CH2-, C-H, -CH3, O-H	aldehydes, carboxy
1530–1600	C-N	amino acids and derivatives
1516–1519 1610–1614	C=C	aromatic compounds, flavonoids
1700	C=O	organic acids
1704–1712	C=O	esters of hydrolyzable tannins, derivatives of gallic acid and flavors
1600–1900	O-H stretching C-H3 stretch first overtone C-H2, C-H stretch first overtones	water, ethanol, glucose
2100–2300	C-H combinations vibrations and overtones	sugar and ethanol
2800–3000	C-H stretching of hydrocarbons - CH3 asymmetric stretching vibration O-H stretching of carboxylic acids	glycerol, catechins and free phenolic acids
3000–3500	-OH	alcohols, phenols, water

2.2.2. The Colour of the Wine and Its Colour Space Parameters (CIELab)

Wine is a product that can be described nowadays both as a commodity but also as a luxury, depending on its marketing price. Moreover, wine is a complex mixture of water and ethanol (as major components), and glycerol, aliphatic and aromatic alcohols, phenols, sugar, salt or organic acids (as minor components). That being said, the concentration of these minor components is of great importance for the quality and preferences classification from the industry and consumers point of view. Nonetheless, the color of wine is one of the first parameters evaluated by consumers and contributes, as well, to the wine's quality perception and acceptance. The color is mainly due to the phenols in the wine [82]. More, even the tasting of these beverages is strongly influenced by wine color, in the detriment of other sensorial parameters (temperature, aroma). Furthermore, the study of wine color is a must, bearing in mind that it is relative easy to be done. Nevertheless, many techniques are being used, even if some standardized methods are accepted by the OIV.

Through colorimetry we have the possibility to define every color as a combination of three values, mainly known as color coordinates. Moreover, in the fields of enology the determination of the color spaces of liquid samples is of great importance, therefore being widely applied. For example, researchers evaluate color using the CIE $L^*a^*b^*$ model of organic dyes and colorants and pH indicators, and enologists analyze the color of beverages (wine and spirit) samples.

The color is one of the most fundamental descriptors of wine, an attribute to which viticulturists and wine producers offer all their attention for growing grapes and winemaking. Due to different people's visual perception of wine color during tasting or organoleptic examination, a measurement of color was proposed by the Commission Internationale de l'Eclairage (CIE) and the OIV described the procedure as the one which offer the largest amount of information [83]. The CIE defines 3D graphs of all colors that humans can see, as a Cartesian coordinate system defined by three colorimeter coordinates L^* (lightness, from 0 –black to 100-white), a^* (redness/greenness, positive values for reddish and negative for greenish) and b^* (yellowness/blueness, positive for yellowish and negative for bluish) [83–86], along parameters as chroma (C_{ab}^*), angular hue or tone (H^*), color intensity (C.I.), hue, color difference (ΔE_{ab}^*) and whiteness index (WI). Furthermore, chroma is the quantitative characteristic of colorfulness that enables the characterization of differences in the grey color between samples with the same lightness for each hue. Moreover, hue is regarded as the qualitative attribute of color, being the parameter from which colors are traditionally defined, e.g., blueish, pinkish, reddish or yellowish. It is the peculiarity that specifically allows a color to be differentiated from a grey color with the same lightness. The most used mathematical relations, accepted by OIV, for the chromatic characterization of wine samples are included in the system of Equations (1).

$$
\begin{cases}
L^* = 116f(Y/Y_n) - 16 \\[2mm]
a^* = 500[f(X/X_n) - f(Y/Y_n)] \\[2mm]
b^* = 200[f(Y/Y_n) - f(Z/Z_n)] \\[2mm]
C_{ab}^* = (a^{*2} + b^{*2})^{0.5} \\[2mm]
H^* = \tan^{-1}(b^*/a^*) \\[2mm]
C.I. = A_{420\ nm} + A_{520\ nm} + A_{620\ nm} \\[2mm]
hue = A_{420\ nm}/A_{520\ nm} \\[2mm]
\Delta E_{ab}^* = [(\Delta L^*)^2 + (\Delta a^*)^2 + (\Delta b^*)^2]^{0.5} \\[2mm]
WI = 100 - [(100 - L^*)^2 + (a^*)^2 + (b^*)^2]^{0.5}
\end{cases}
\tag{1}
$$

The absorbance spectra were measured with a spectrophotometer, using a 1 cm path length rectangular cell. Measurements were taken every 1 nm between 200 and 1000 nm. Distilled water was used as the blank. From the absorption spectra, the rectangular coordinates L^* a^* b^* and the cylindrical coordinates CIE C_{ab}^* and H^* were calculated using CIE method, as well as the color intensity (C.I.), the hue, Delta E (ΔE_{ab}^*) and the whiteness index (WI) values, like in [47,49,59,82,86–88], presented in equations 1.

Once the transmittance spectra of the wine samples were recorded, the color coordinates were obtained by applying mathematical treatment (according to OIV and CIE). Based on these mathematical formulas (described previously in equation system (1)) Delgado-Gonzalez et al. [83] proposed an easy-to-use model using Microsoft Excel. These values of the color parameters obtained using the Delgado-Gonzalez et al. method were plotted in color graphs.

The wine color is given by the phenolic compounds and depending of their quantity the color can be: white, white-greenish, white-yellowish, yellow, yellow-golden, pink, pink-purplish, purple, red, red-purplish, red [82].

Usually, the white wine phenols that are responsible for browning (color changing of the wine) are the flavonoid catechins (such as: epicatechin, gallocatechin, and catechin) as well as the non-flavonoid cinnamic acid derivatives (like caftaric acid, coutaric acid or fertaric acid). The non-flavonoid phenols are known to be the primary phenols in white wine. In these beverages the oxidation followed by polymerization results in the development of a golden brown color [87].

2.2.3. The pH Value, Brix and Potential Alcohol of Must and Wine

As general rule, winemakers should keep the pH under recommended ranges and the correction by adjusting free sulfur dioxide should be made quickly because pH plays an important role in the form of SO_2 which inhibits microorganisms. The management of acidity is very important for wine stability and hence for its quality.

The wine pH values usually range from 2.8 to 3.8. Lower pH 2.8–3.2 ensures the pleasant refreshing taste of wine and the bright color, the wines are easily clarified and are resistant to microorganisms. The low-value pH wine will taste tart/sour, owing to the higher acid concentration. Higher pH values (>3.5) affect the wine quality [82]. High pH values for a grape juice make it less stable to oxidation and microbial spoilage and gives it a flat and unbalanced taste while musts with low pH (naturally occurring or after acid adjustment) require less SO_2 quantity to control the native flora and to ensure the onset of a desired fermentation [89].

The pH measurements of the pristine and plasma treated samples were performed via a PH-009(I)A Series pen pH and Temperature Tester, the measurements being taken at 20 °C.

Brix testing plays a huge role in determining when grapes are at their highest sugar content and suitable for harvest. Furthermore, Brix measurements can also be used with other forms of produce for determine the mineral, protein or amino acid concentration of the plant. Determination of carbohydrates in the samples was performed using the Abbe type refractometer, by determining the refractive index. The reading was made for the Brix index (the unit of measurement of carbohydrates in an aqueous solution), and the conversion was done with the help of a table of transformation in grams of carbohydrates / L: 1 Brix = 10.04 g/L measured at 20 °C.

The Brix determination was made by using a portable refractometer (RWN10-ATC, Czech Republic) with five scales: 5–22% VOL (±0.2%); 10–37° ČNM (±0.5°); 0–30° KMW (±0.5°); 0–150° Oe (±1°); 0–35° Bx (±0.5°).

The potential alcohol/densitometry determinations were made by using a glass glucometer (GUYOT glucometer, three thermometer bench scales: sugar 0–30, alcohol 0–20% VOL, Baume 0–18°, Enolandia) and a vinometer (0–25 % VOL, Enolandia). The pH, Brix and alcohol measurements were made in in triplicate, the values being expressed as mean± standard deviation (SD).

The formula used for estimating the potential alcohol (pa) from Brix measurements [59,87,90] of the untreated and plasma treated must samples, is usually referred to: pa (% *v/v*) = 1000(sgi-1.0)/[7.75–3.75(sgi-1.007)], where sgi is the initial specific gravity of the must.

Moreover, according to OIV International Standard for the Labelling of Wines, (Edition 2015) [58], depending on their sugar content, table and quality wines are divided into:

1. dry, with a sugar content of up to 4 g/L but not less than 2 g/L;
2. medium-dry, with a sugar content between 4.01 g/L and 12 g/L, but can go up to 18 g/L;
3. mellow or semi-sweet, with a sugar content above 18 g/L and up to 45 g/L inclusive;
4. sweet, with a sugar content of more than 45 g/L.

3. Results and Discussion

This section is divided into two subheadings (one dedicated to plasma source characterization-Section 3.1 and the second dedicated to the characterization of must and wine-Section 3.2) and provides to the reader a concise and precise description of the experimental results as well as their interpretation.

3.1. Plasma Source Electro-Optical Characterization

Plasma diagnosis methods, applied to low-temperature and atmospheric pressure, are usually related to the measurement of the applied voltage and the discharge current, the estimation of the average electric power of the discharge, but also the acquisition of the light emitted by the plasma and the identification, as well as interpretation of the excited species from the discharge.

3.1.1. Plasma Source Electrical Diagnosis

The electric characterization of the plasma was related in this study to the applied voltage, discharge current waveform visualization and power estimation. Typical waveforms are shown below in Figure 2.

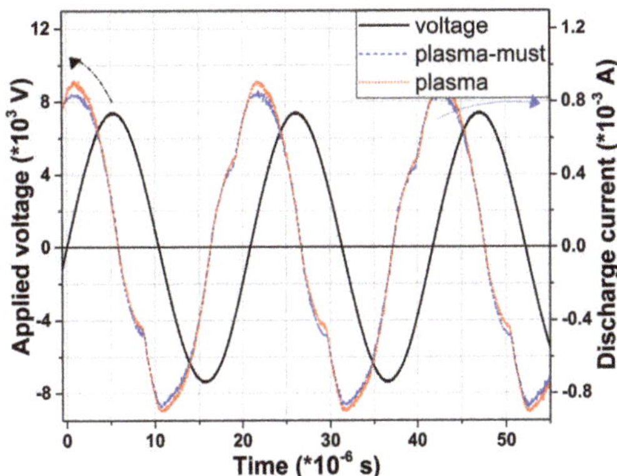

Figure 2. Typical voltage–current waveforms of the He–appj in mid air (dash line) and interacting with the withe must (short dot line).

The applied voltage on the discharge electrodes was around 16 kV peak-to-peak with a corresponding discharge current in the order of 0.8–0.9 mA (with a total charge of 1 nC per cycle), at a repetition frequency of 48 kHz. The average dissipated electrical power, estimated by integrating over one cycle the applied voltage times the discharge current [91], was in the range of 7–8 W (or a energy of up to 0.5 mJ). The estimated power density was in the 200–300 W/cm^2 range, while the energy dose (a specific energy = energy per treated area in cm^2) was between 4.6–6.6 mJ/cm^2.

Minor changes were observed in the discharge current waveform between the He-appj operating in mid air (the red-dotted line) or interacting with the white grape must (the blue-dashed line), as depicted from Figure 2. Insignificant differences were to be seen in the current waveforms of plasma-Bârzeşti-must and plasma-Huşi-must.

Our plasma source mean electrical power, around 8 W, was close to the values reported by Guo, Lukic, Nishime, Pankaj and Sarangapani (2–20 W) [26,36,92–94], but lower than 20–50 W as reported by Starek, Mujahid, Wang [38,95,96], or less than 50–200 W as reported by Bao, Pan, Zhao, Xiang, Laurita, Sainz-Garcia [29,40,41,43,97,98], and even far less than 300–750 W as reported by Ashtiani, Huang, Fan, Jambrak, Zhou [30,42,44,99,100]. The higher the dissipated plasma power got, the higher the possibility of thermal treatment of the sample was. Our appj source (up to 10 W) ensured relatively low temperature of discharge, as will be discussed in the following paragraph.

3.1.2. Plasma Source Optical Diagnosis

The Optical Emission Spectroscopy (OES) characterization of the plasma source in mid-air, as well as the OES spectra at the interface plasma-must are presented in in Figure 3.

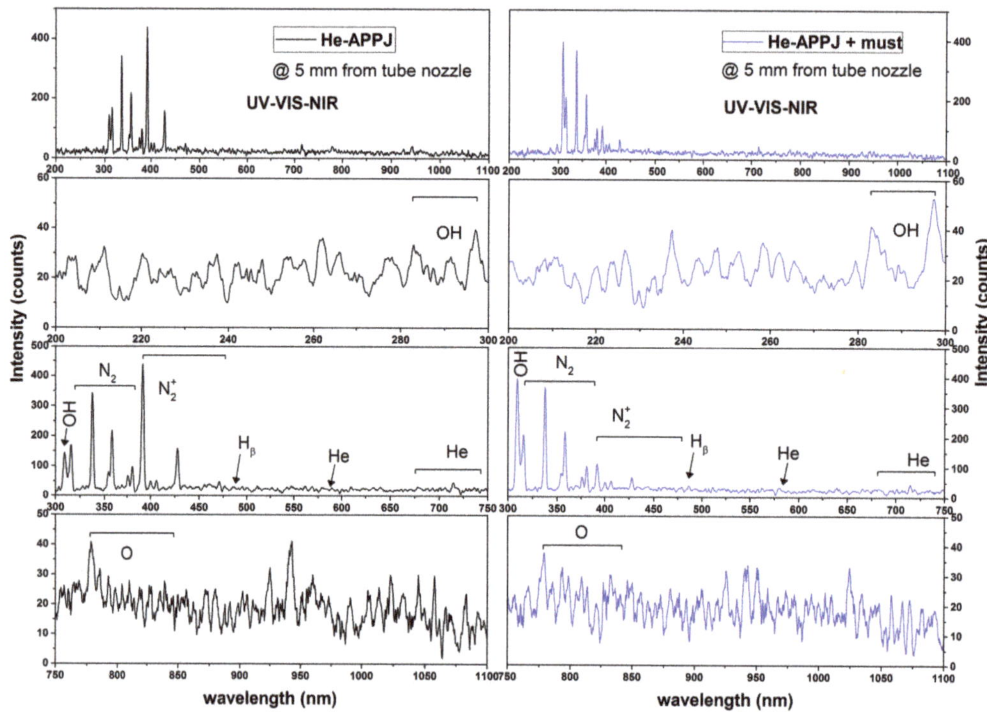

Figure 3. The emission spectra of the He-appj in mid air (**left side**) and interacting with the white must (**right side**).

Emission spectra shown in the plasma emitted light graphs, in the UV-Vis-NIR range (200–1100 nm) of the plasma jet (at 5 mm from the discharge tube nozzle) contained signatures of atomic and molecular excited species. The molecular bands, present in the low wavelength region, were assigned to hydroxyl radicals, neutral nitrogen molecules and nitrogen molecular ion.

The OH radical presented signatures between 306–310 nm. Bands of the molecular nitrogen (N_2) were seen between 315–380 nm and 399–405.9 nm. The nitrogen molecular ions (N_2^+) had bands at 391.4 nm, 427.8 nm and 470.0 nm. The generation and excitation of N_2 (the second positive nitrogen system) and N_2^+ (first negative nitrogen system) were based on the Penning effect of He metastability. Around 486 nm the atomic line corresponding to the emission of H_β was observed. Atomic lines were assigned to helium atoms (lines at 588.8 nm, 668.5 nm, 706.6 nm and 727.5 nm), as working gas, and to oxygen atoms (lines at 777.8 nm and 845.5 nm) as products of ambient O_2 and H_2O dissociation [8,11,16,21,101,102].

The presence of other excited species beside the working gas-He, suggests that these reactive oxygen and nitrogen species (RONS) can and will participate in the white grape must sample treatment. Moreover, the importance of some plasma active species changed in the case of interaction with must, as seen in the third line (300–750 nm range). More precisely, it was about the changes regarding the intensity of the hydroxyl radical bands, of the bands of nitrogen molecules as well as of the bands of molecular nitrogen ions. In the case of He-appj running in mid air, as also reported in the literature for such discharges, the bands of molecular nitrogen ions were important in intensity in the emission spectrum, often being used to normalize the entire spectrum for further spectroscopic analyses [8,11,16,21,91]. Thus, as seen in Figure 3, for appj in air, the highest intensity was attributed to the molecular nitrogen ion band, centered at 391 nm. It was followed by the intensities of the molecular bands of nitrogen (337 nm, 357 nm, 315 nm), respectively of the band corresponding to the OH radicals (308 nm). The normalized intensities can be quickly followed in Table 2. In the case of He-appj in interaction with the must, the OES spectra revealed different distribution of intensities of lines and bands. More precisely, the most intense band was now attributed to OH radicals (at 308 nm), followed by the bands of molecular nitrogen (337 and 357 nm). These intensity changes of plasma excited species can be seen and followed both in Figure 3 and Table 2.

We further used the optical emission spectroscopy, which allowed the estimation of rota-vibrational temperatures, as well as the identification of radiating species in the discharge volume. In Figure 3, an overview of the emission spectrum in the range between 200 and 1100 nm is shown. It can be seen that for the appj in mid air (Figure 3 left side graphs), nitrogen dominated the emission spectrum of the plasma as an essential element in ambient air resulting in peaks at 315, 337, and 357 nm, which were in the so called UVA region (315–400 nm). In addition, smaller peaks were found in the UVB region (280–315 nm), attributed to OH radicals bands. Below 280 nm, no radiation was observed and, thus, we can say that no UVC radiation was emitted by the studied plasma source. For the case when plasma met the must (Figure 3 right side graphs), the dominant peaks were found around 308 nm, corresponding to OH radicals, in the UVB region. Smaller peaks of molecular nitrogen and molecular nitrogen ions were found (at the same peaks wavelengths) in the UVA region. Again, no significant lines/bands were identified in the UVC region (bellow 280 nm).

Since atmospheric pressure plasma treatments are usually seen as oxidative methods, and due to the fact that phenols (the main component of white wines) are the primary support for oxidation, it is reasonable to understand the importance of knowing the plasma RONS and the changes in the UV-Vis absorption spectra of wine. Besides plasma reactive species, another important parameter in plasma treatment of matter is the plasma temperature. Since the studied plasma source is at atmospheric pressure, working in gas flow, it is easy to assume that this is a non-thermal plasma, meaning that the temperature of the electrons differs from the temperature of the ions, and that of the neutrals. In this context, of the electric discharges characteristic temperatures, it is interesting to determine these values in order to understand how these plasmas can interact with matter. Moreover, using simulation software like Lifbase [103] and Spectrum Analyzer [104], we determined, from the acquired plasma emission spectra, the characteristic plasma temperatures, such as: the rotational temperature of OH radicals, the rotational temperature of nitrogen molecular ions N_2^+, the vibrational temperature of nitrogen molecules N_2. Then, a spot thermal camera (FLIR TG165, pointed onto the surface of the must during the plasma treatment) and k-type probe thermocouple (placed bellow as well as in the Petri dish, via a PeakTech 3415 USB DMM digital multimeter) were used for monitoring the plasma gas temperature as well as the surface temperature in front of the plasma (Table 3).

Table 2. Normalized spectral lines and bands intensity observed in the discharge with energies E_k of the upper states from [105,106].

λ [nm]	Transition	Normalized Intensity in Midair	Normalized Intensity with Must	E_k [eV]
308 OH	$(A^2\Sigma^+) \leftarrow (X^2\Pi_i)$	0.333	1	4
337 N_2	$(C^3\Pi_u)_{v'=0} \leftarrow (B^3\Pi_g)_{v''=0}$	0.782	0.927	11.0
391 N_2^+	$(B^2\Sigma_u^+)_{v'=0} \leftarrow (X^2\Sigma_g^+)_{v''=0}$	1	0.299	18.7
486 H_β	$4 \leftarrow 2$	0.070	0.117	12.7
587 He	$(3d) \leftarrow (2p)$	0.067	0.078	23.1
667 He	$(3d) \leftarrow (2p)$	0.058	0.091	23.1
706 He	$(3s) \leftarrow (2p)$	0.073	0.088	22.7
728 He	$(3s) \leftarrow (2p)$	0.067	0.076	22.9
777 O	$3s\,^5S_2^0 \leftarrow 3p\,^5P_i$	0.097	0.111	10.7
844 O	$3s\,^3S_1^0 \leftarrow 3p\,^3P_i$	0.054	0.087	11.0

As can be observed from the temperature values estimated through spectroscopic measurements, the rotational temperature of hydroxyl radicals and nitrogen molecular ions, which is often attributed/equaled to the gas temperature in many plasma sources reported in literature (especially DBD based sources), was determined to be 340–355 K (66 to 80 °C). These values corresponded to a 'spectroscopic temperature' meaning energies that those plasma species (OH and N_2^+) could achieve and could be used in the plasma environment to initiate various physico-chemical reactions/processes that could take place on the surface or in the plasma volume.

Table 3. Characteristic temperatures derived from the OES spectra of the plasma source and measured via IR and k-type thermocouple (mean ± standard deviation).

Method	T_{rot} (OH) [K]	$T_{rot}(N_2^+)$ [K]	$T_v(N_2)$ [K]	T_{gas} [°C]
Spectrum Analyzer	x	x	$2265 - 2566 \pm 190$	x
Lifbase	$340 - 350 \pm 2$	$345 - 355 \pm 2$	x	x
IR camera	x	x	x	40 ± 1.5
k-type probe	x	x	x	35 ± 3.0

The gas temperature of the gas/liquid in interaction with the plasma was measured, the values being around 35–40 °C (±1.5 to 3 °C). The temperature during plasma treatment of must did not exceed 40 °C, so no thermal treatment (temperature above 80–85 °C that could induce thermal pasteurization) of the must proteins and all other must components was made [26,107].

3.2. Characterization of White Grapes Juice and Wine

This section is intended for the characterization of white grapes juice and wine. Spectroscopic methods of investigation such as UV-Vis and ATR-FTIR, as well as colorimetric and other physio-chemical methods (pH, Brix).

3.2.1. UV-Vis Spectroscopy of White Grapes Juice and Wine

As can be seen in Figure 4 the total wave range (200–1100 nm) revealed a typical waveform, as for the white wines reported in literature [61,62,66,108–110]. The spectral region from Figure 4 expressed the influence/contribution of specific organic compounds found in the wine samples. More precisely, for white wines the absorption spectra in the 240 to 400 nm region were related to the presence of esters and hydroxycinnamic acids (HCA), but correlations with pattern recognition techniques were recommended. Namely from 250 to 300 nm (around 280 nm) the absorption band was usually related to esters and different types of phenolic compounds (the absorption of benzene cycles of most of phenols), as for the 300 to 350 nm region, the band was associated with the presence of flavones and/or to nonflavonoid compounds and hydroxycinnamic acids [110]. For wines, the absorption spectrum in the UV-Vis region contained information about organic acids and phenolic compounds, including here hydroxycinnamic acids (227–245 nm, 310–332 nm), benzoic acids (235–305 nm), flavonols (250–270 nm, 350–390 nm), anthocyannins (267–275 nm, 475–545 nm) or catechins (280 nm) [109,111].

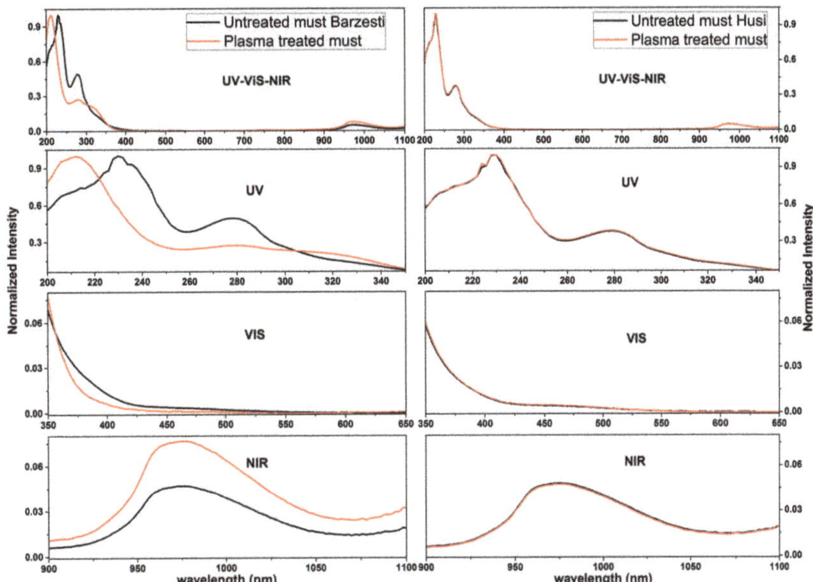

Figure 4. The absorbance spectra of untreated and treated must in the 200–1100 nm range.

Water-related absorption bands were found between 920 nm and 1050 nm. These bands were related to the third overtone of O-H stretch (around 950 nm, usually in the case of fruits, vegetables and their juices, with 70--80% of water) and to the second overtone of O-H stretch (around 990 nm, sugars and organic acid related) [111]. Therefore it was convenient to measure these components and color of wines via UV-Vis absorption spectroscopy [112]. Another aspect was related to the oxidation degree of white wines, mainly to exposure to O_2 species, that could be followed up by means of HCA changes (usually these differences are easily observed in the first or second derivative of the UV-Vis spectra [48,61–64,66,68,75].

As far as one can see, the absorption spectra of wine samples (as in Figure 4) indicated that we had notable differences only in the case of MB-PB samples. For MH-PH samples, almost imperceptible differences could hardly be traced. It can be seen from Figure 4 that the absorption spectra of the untreated MB differed from the treated one PB but also from the MH-PH. Consequently, in the UV region the maximum for MB was at 215 nm, rather than at 230 nm as for PB and MH-PH. Moreover, a second peak centered at 280 nm for all samples, seemed more broadened for the MB samples, in contrast with the narrow and well defined peak seen in PB, MH and PH samples. Furthermore, in the 350–600 nm region for the MB-PB samples a difference in the Vis absorption spectra could be observed, and also for the 900–1100 nm part of the spectra.

3.2.2. ATR-FTIR Spectroscopy of White Grapes Juice and Wine

As it can be depicted from Figure 5, the transmittance spectra of the wine samples, in the 500–4500 cm^{-1} wavenumber range, could be divided into four regions important for wine characterization, both from the chemical content as well as from the chemometric point of view.

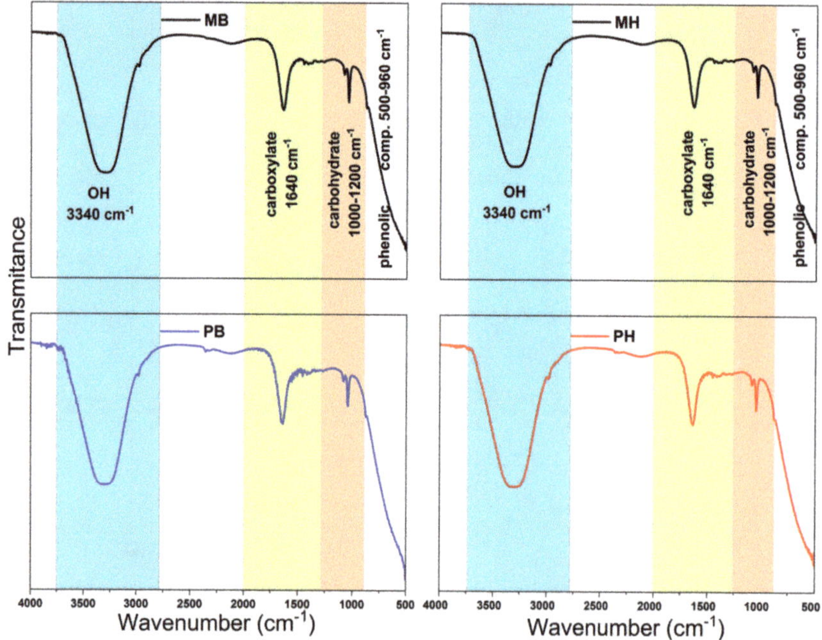

Figure 5. The IR transmittance spectra of untreated and plasma treated wine samples (1y old).

These regions of interest, as can be seen in the Figure 5, expressed the content of polyols, glycerin/glycerol and hydroxil radicals in the 2800–3500 cm^{-1}; the presence of organic acids (known also as 'total acidity') and amino acids, or carboxylates in the 1250–2000 cm^{-1}; the extant of carbohydrates, including polysaccharides (namely glucose, fructose, oligosaccharides) and polyols, in the 1000–1200 cm^{-1}, and also the presence of phenolic compounds (like phenols, esters, acetals or ketols) in the 500–960 cm^{-1} region [68,72,73,76,77,112,113]. Moreover, two clear absorption bands between 3600–3200 cm^{-1} and 1700–1560 cm^{-1} were correlated with O-H stretching vibrations. The small, yet evident, peak between 3040–2800 cm^{-1} is usually associated with the stretching vibration of C-H bond. Small differences were noticed in the FTIR spectra of all wine samples, plasma treated or untreated. These were related to the shift of the wavenumber of some observed peaks and further on listed in Table 4.

FTIR spectroscopy was involved in monitoring of wine process and prediction of its parameters (e.g., monitoring wine aging), as well as wine authenticity and traceability, or prediction of a white wine aromatic potential, as well as studies of phenolic profile and antioxidant activity during the winemaking process. All of these (and other aspects previously mentioned) make this technique very important in the vine and winemaking industry, underlining the usage of it in this study.

Table 4. Wavenumber regions measured in the studied wine samples, similar to those reported by [68,72,73,76,77,112].

Wine Samples Samples	Region 1 500–960 (cm^{-1})	Region 2 1000–1200 (cm^{-1})	Region 3 1250–2000 (cm^{-1})	Region 4 2800–3500 (cm^{-1})
MB	865, 903	1044, 1084	1318, 1385, 1636	2983, 3290
PB	876	1044, 1088	1317, 1397, 1456, 1636	2984, 3303
MH	876	1044, 1085	1274, 1392, 1455, 1636	2983, 3290
PH	875	1044, 1084	1320, 1388, 1456, 1636	2980, 3313

3.2.3. Colour of Must and Wine

The color and the hue of the white must/wine samples are presented as follows. In the case of white wines, the absorbance value read at wavelength 420 nm, in relation to distilled water, represented an estimate of the oxidation state of the color, the so called 'color closure' [82]. The color parameters of the wine samples under study, the variation of absorbance at 420, 520 and 620 nm, as well as the C.I. and hue, are shown bellow in Figure 6. As a first notable aspect, the redness/red component of the wine sample, after 2 years' increase, was seen from the absorbance at 620 nm.

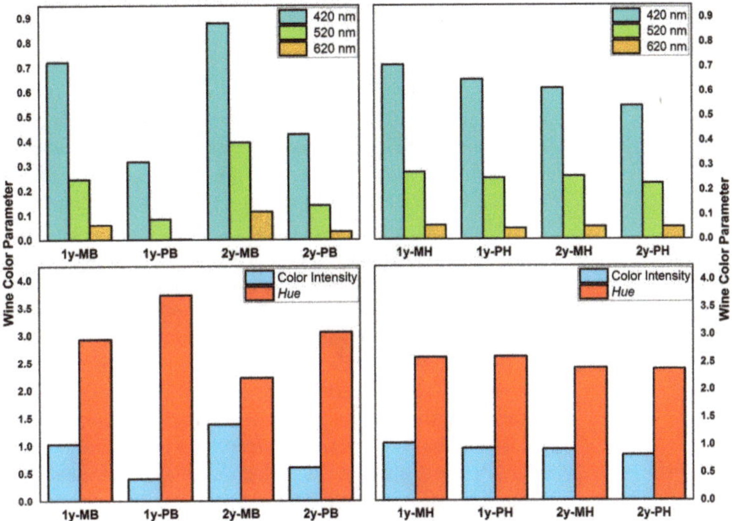

Figure 6. Colour parameters, C.I. and hue, for the untreated and plasma treated must.

The color properties/characteristics of the wine samples under study can be observed in Figure 6. Even if both studies' wines were white, the color properties were different. Moreover, for the Bârzeşti wine the color parameters, mainly the absorption values at 420, 520 and 620 nm increased with age (from 1 to 2 years) for both untreated (MB) and plasma treated samples (PB). Further on, the absorbance values of the untreated samples were greater than those for plasma treated ones. For these wine samples the color intensity (CI) increased with age and had greater value for the untreated samples unlike those treated. The Hue value depreciated over time for all wine samples, but it was with higher value for plasma treated ones. For the samples from Huşi, the 420 nm absorbance fell off significantly, while for the 520 and 620 nm the absorbance slightly reduced its value. Moreover for these Huşi wine samples the color intensity values also declined in time, as well as for the Hue parameter. However, the values were somehow bigger for plasma treated wines (PH). These aspects can be seen in Figure 6. Thus, having these values we could estimate the color differences between wine samples. As a notable fact, the Bârzeşti wine samples color were more intense unlike the Huşi wine samples, that had close values for both untreated and plasma treated samples.

The color parameters: CIELab, WI, chroma, angle hue and delta E (measured against the untreated must) of white wine are presented in the Table 5, shown below. Moreover the color parameters graphs (Figures 7 and 8) were obtained using the same method described in [83]. As reported by other researchers, a difference smaller than 3 of the wine sample color change, ΔE_{ab}^*, cannot be observed by human naked eye [107].

Table 5. Plasma effect on color parameters: CIE $L^*a^*b^*$, WI, chroma, angle hue and ΔE_{ab}^* (measured against the untreated must) of white wine and the corresponding R G B coordinates (using equation system (1) and [83]), for 1 and 2 years old wine (mean ± standard deviation).

Sample	L^*	a^*	b^*	WI	C_{ab}^*	h_{ab}	ΔE_{ab}^*	R	G	B
1y MB	85.49 ± 0.03	6.43 ± 0.02	39.19 ± 0.03	57.72 ± 0.06	39.71 ± 0.06	80.68 ± 0.09	-	250	208	138
1y PB	95.26 ± 0.02	1.20 ± 0.05	22.00 ± 0.08	77.46 ± 0.05	22.03 ± 0.09	86.88 ± 0.01	20.45 ± 0.03	255	239	198
2y MB	78.22 ± 0.07	12.18 ± 0.04	48.34 ± 0.06	45.60 ± 0.02	49.85 ± 0.01	75.85 ± 0.06	-	241	183	102
2y PB	91.13 ± 0.06	3.01 ± 0.03	25.89 ± 0.01	72.46 ± 0.09	26.07 ± 0.06	83.36 ± 0.03	27.47 ± 0.05	254	226	179
1y MH	84.75 ± 0.05	8.48 ± 0.02	42.98 ± 0.09	53.62 ± 0.05	43.80 ± 0.04	78.83 ± 0.07	-	252	204	129
1y PH	86.26 ± 0.08	7.98 ± 0.05	41.58 ± 0.04	55.49 ± 0.08	42.34 ± 0.08	79.13 ± 0.05	2.12 ± 0.02	255	209	136
2y MH	85.64 ± 0.04	8.95 ± 0.08	41.75 ± 0.07	54.95 ± 0.03	42.70 ± 0.05	77.90 ± 0.01	-	255	206	134
2y PH	87.32 ± 0.09	7.19 ± 0.06	37.83 ± 0.04	59.46 ± 0.07	38.51 ± 0.09	79.24 ± 0.03	4.62 ± 0.05	255	212	146

The effects of plasma treatment upon the color of the must samples under study are presented in the Table 5. As already seen in the UV-Vis absorption spectroscopy results, the two types of white must had different behaviour in respect to plasma treatment. The most noticeable plasma treatment results were upon the Bârzeşti samples. The L^* parameter (lightness) increased after plasma treatment compared to the control, for both Bârzeşti and Huşi samples. These changes in the values of the L^* parameter were: for the Bârzeşti wine, the L^* value depreciated by 8.5% after another one year (2y against 1y) for the control sample, while for the plasma treated one the decrease was by 4.3%. However, comparing the plasma treated wine after 1 year of storage with the control one, L^* values were 11.5% higher, and after the second year of storage the difference was about 16.5%, so an extra 5% increase of the difference between control and treated with the aging of the wine, for the Bârzeşti wine. These differences were less noticeable in the case of Huşi wine samples. More over, for the Huşi samples L^* tended to increase both over time and with plasma treatment, but with smaller percentages in comparison to Bârzeşti wines. More precisely, after 1 year of storage the L^* for the treated samples was higher with almost 1.8%, and after another year with about 1.9%, so the value of L^* remained constant over time. For the a^* and b^* coordinates (a^*-redness and b^*-yellowness), a general increase value could be observed for Bârzeşti samples, while for the plasma treated ones of the Huşi samples, a small decrease was seen (10% for a^* and b^*).

The white index (WI) had an overall tendency of increase for all samples, both with age and plasma treatment. The same was true for chroma (C_{ab}^*), with almost 25% enhancement, for the Bârzeşti control, and 18% for the plasma treated, although between control and treated samples there was a 50% difference in the control favour, while it was up to 10% for the Huşi wines. The angular hue had similar tendency of decreasing over time, in the range of around 5%. The interesting result was related to the color differences (ΔE_{ab}^*) that for the Bârzeşti plasma treated samples was 20 in the first year of aging, and up to 27 in the second year. For the Huşi wines, after 1 year $\Delta E_{ab}^* < 3$ (no visible changes), and increased a bit, 4.6 after another 1 year.

For better visualization of color parameter differences, the color space diagrams in $L^*a^*b^*$ and corresponding RGB coordinates, as resulted using Delgado-Gonzalez model [83], are presented in Figures 7 and 8.

By using this model and the absorption (converted to transmittance) spectrum of the must and wine samples we were able to not only calculate the color coordinates of CIE $L^*a^*b^*$ and RGB color space but also to visualize these parameters in a way comfortable to the reader.

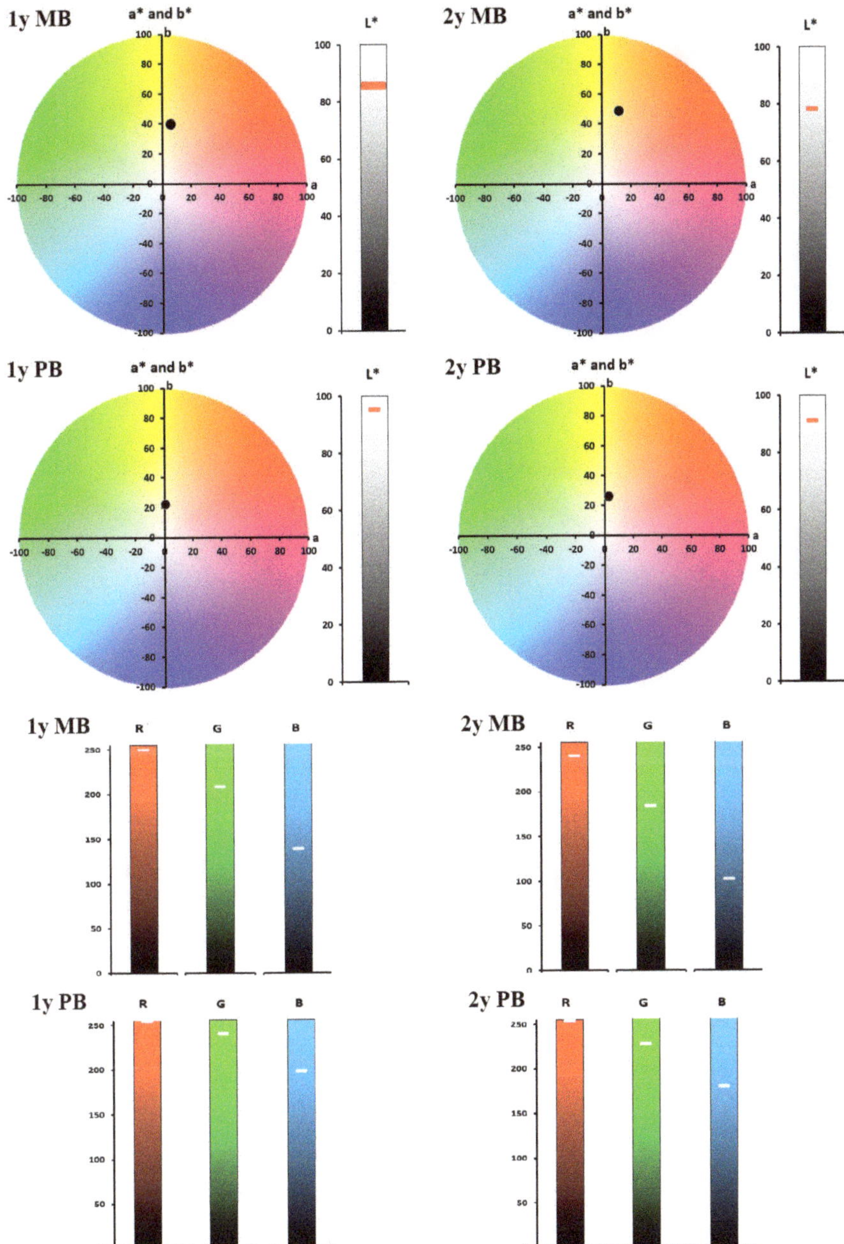

Figure 7. Color space for control (MB) and plasma treated (PB) Bârzeşti must, according to [83]. The black dots represent the points (*a**, *b**) of the samples, while the red horizontal line in the bar graphs represents the *L**, R, G and B of the 1 and 2 year old samples.

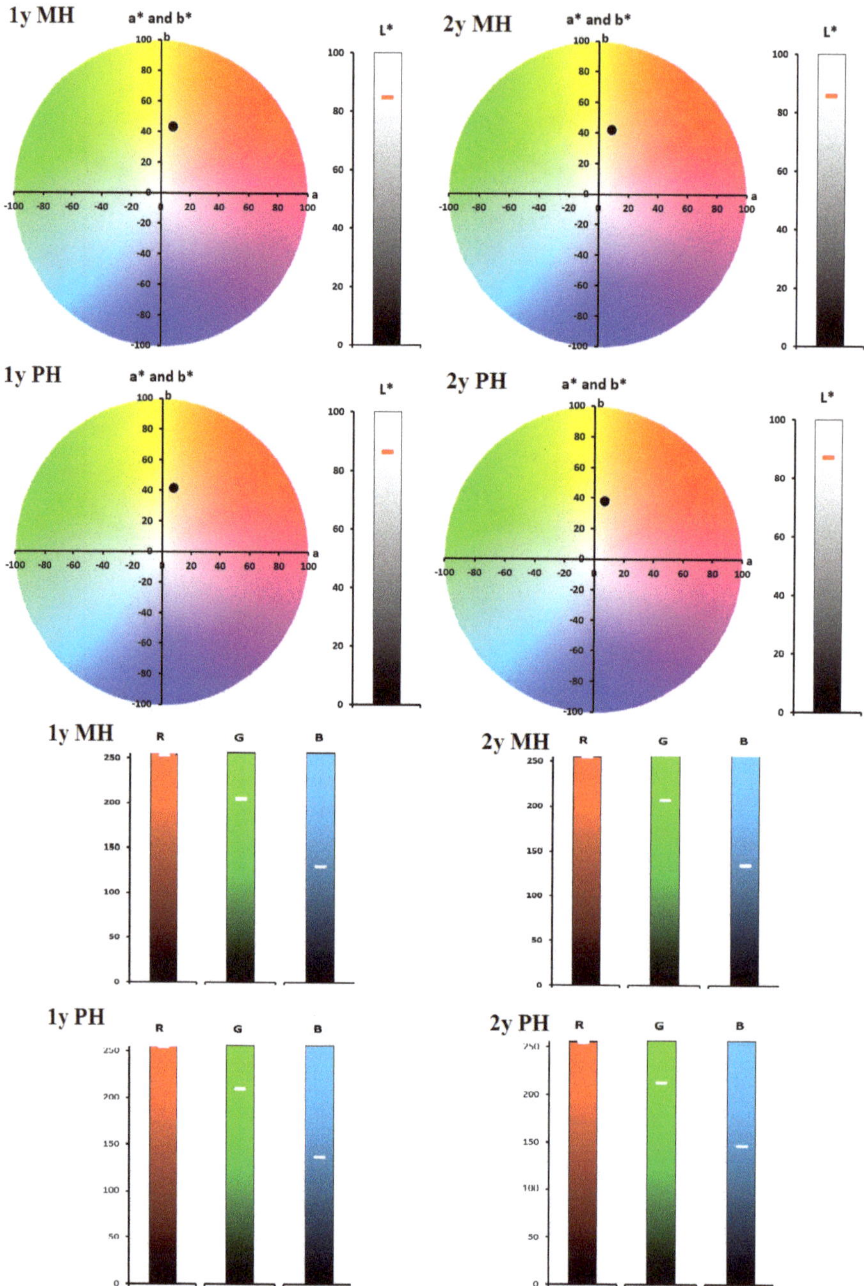

Figure 8. Color space for the control (MH) and plasma treated (PH) Huşi must, according to [83]. The black dots represent the points (*a**, *b**) of the samples, while the red horizontal line in the bar graphs represents the *L**, R, G and B of the 1 and 2 year old samples.

The parameters *L*a*b** and RGB presented in Figures 7 and 8 are easily to follow on the graphs, where the differences between the studied wine samples are visualized simultaneously.

3.2.4. pH, Brix and Densitometry Measurements of Must and Wine

Following the OIV classification (enumeration 2.2.3) and based on the estimated values of sugar (g/L) from the studied wine samples we could conclude that both Barzeşti and Huşi wines were sweet wines (>45 g/L). This aspect could be deduced even from the values obtained for the must (Table 6), so that later it could be confirmed by the data obtained for the wine samples of 1 and 2 years old respectively (Table 7).

Table 6. Physico-chemical parameters of fresh must, control and plasma treated (MB-PB, MH-PH). Measured pH, Brix, as well as the estimated (*italic*) dissolved sugar and density by using VinoCalc [90], and potential alcohol by using Equation (1) (mean ± standard deviation).

Sample	pH	Brix (%)	Dissolved Sugar (g/L)	Density (g/L)	Potential Alcohol (% v/v)
MB	3.3 ± 0.1	10.5 ± 0.5	*109.4*	*1042*	*5.55*
PB	3.3 ± 0.1	10.5 ± 0.5	*109.4*	*1042*	*5.55*
MH	3.4 ± 0.1	15.0 ± 0.5	*159.2*	*1061*	*8.13*
PH	3.4 ± 0.1	15.0 ± 0.5	*159.2*	*1061*	*8.13*

Thus, the values of the pH for the fresh must indicate that the acidity of the samples was in general agreement with those reported in literature. More, the Brix reading showed a clear differentiation between the two types of must, mainly by 5% in the favor of the Huşi sample, giving it an increased content of potential alcohol. This would further influence the properties of the wine. However, these must parameters did not guarantee the final amount of alcohol in the young or old wine.

Table 7 show the physico-chemical parameters of studied wine samples. It can be pointed out that the pH value was closed to 3.5, a value that ensured the wine's microbiological stability.

Table 7. Physico-chemical parameters of 1 and 2-year-old wine, control and plasma treated (MB-PB, MH-PH) samples. Measured pH, Brix, the estimated (*italic*) dissolved sugar and density by using VinoCalc [90], and measured alcohol by using vinometer (mean ± standard deviation).

Sample	pH	Brix (%)	Dissolved Sugar (g/L)	Density (g/L)	Alcohol (% v/v)
1y MB	3.5 ± 0.1	6.0 ± 0.5	*61.4*	*1023*	16.5 ± 1
1y PB	3.5 ± 0.1	6.0 ± 0.5	*61.4*	*1023*	15.0 ± 1
2y MB	3.6 ± 0.1	7.0 ± 0.5	*71.9*	*1027*	17.0 ± 1
2y PB	3.6 ± 0.1	6.0 ± 0.5	*61.4*	*1023*	16.0 ± 1
1y MH	3.5 ± 0.1	10.0 ± 0.5	*104.5*	*1040*	13.0 ± 1
1y PH	3.5 ± 0.1	10.0 ± 0.5	*104.5*	*1040*	11.0 ± 1
2y MH	3.4 ± 0.1	6.5 ± 0.5	*66.6*	*1025*	16.0 ± 1
2y PH	3.4 ± 0.1	6.0 ± 0.5	*61.4*	*1023*	15.5 ± 1

Comparing the wines from those two studied areas, it is obviously that after plasma treatment the alcohol content was a bit lower than for the untreated sample. In the case of Bârzeşti area, the Brix value started from 6 and rose to 7 for untreated wine, while the plasma treated samples were constant at 6. In the case of dissolved sugar a higher content was present after another aging year, but a higher content of alcohol the previously year. The same conclusions could be taken for the Huşi wine samples, that the pH values were around 3.5 and that the Brix diminished from 15% as it was in the must, to 10% and down to 6 % for the 1 and 2 year old wine. Even if the estimated dissolved sugar was higher in comparison to Bârzeşti wine samples, the measured amount of alcohol was lower. Nevertheless, after 2 years, both wines were with high alcohol amounts, around 16 % *v/v*.

As a reference, for the 1-year-old wine aged in oak barrels (200 L volume) as 'normal procedure', the measured amount of alcohol was 15 % *v/v* for MB samples and respectively 13 % *v/v* for MH, values that remained constant in the second year. These wines are usually made for domestic, non-commercial use.

Future research directions will include phenolic content and phenols identification through HPLC determinations. Additionally, studies in the first weeks of winemaking, to better capture the must-to-wine transition, using the UV-Vis / ATR-FTIR / colorimetric techniques. The 'pinking effect' as well as the 'browning' of the aged wine samples are to be considered as well as further work. Correlation with the plant-soil components is another important aspect in our future research. Nevertheless, variation of plasma parameters (higher power, increase RONS content, activated medium) in relationship to winemaking and wine preservation will be carry on as well.

4. Conclusions

This report presents the possible usability of atmospheric pressure plasma in the winemaking process. Because these processes are well-founded, there are specific ways that researchers can act to increase the quality of wine. A new method that is gaining ground is the use of plasma sources in the field of winemaking.

The electrical diagnosis revealed a stable, quiet and diffuse discharge, mainly due to helium flow and medium operating power, of up to 10 W, at 48 kHz. Moreover, the chosen working parameters ensured a cold treatment of the white must, proven also by several temperature determination methods. Through thermocouple, IR and spectroscopic techniques, the gas temperature during plasma treatment increased to a maximum 40 °C, so no pasteurization processes occurred. The spectroscopic / specific temperatures of plasma species along with their designated excitation energy provides enough information upon the energetic potential of plasma for interaction with external medium species, the white must.

The UV-to-NIR absorption spectroscopic methods showed that in must and wine there are functional groups corresponding to phenols, sugars and flavones, with differences that can be relatively easily highlighted between the control samples and plasma treated ones. The most notable differences could be seen in the Bârzeşti samples, compared to the Huşi samples.

The simultaneous visualization of CIE L*a*b* and RGB in color space charts allows easier understanding of wine changing in color parameters. Overall, the lightness parameter L^* is enhanced after plasma treatment in both PB and PH samples, and the a^* and b^* values decrease with plasma treatment. The aging influences all parameters both in control and plasma treated samples. The biggest difference in color coordinates are seen in Bârzeşti samples.

The Brix readings were used for both monitoring the must and wine as well as for estimating the dissolved sugar, density or potential alcohol. Based on the estimated sugar content, both wine types are sweet, the amount of measured alcohol being around 16 % *v/v*.

From this point of view, it can be concluded that there are many challenges involving atmospheric pressure plasma applications in winemaking. It includes optimized plasma parameters for must and wine processing, choosing the right type of wine for proper aging, targeting the proper compounds as well as the reaction mechanisms. Nonetheless, these results are the first phase towards improving our knowledge about the impact of plasma sources on wine quality, while also bringing additional helpful information in support of the subsequent optimization of plasma-assisted winemaking processes.

Author Contributions: Conceptualization: R.H. and A.V.N.; methodology and software: R.H. and A.V.N.; validation: R.H. and A.V.N.; formal analysis, investigation and data curation: R.H. and A.V.N.; writing—original draft preparation—review and editing, R.H. and A.V.N. All authors have read and agreed to the published version of the manuscript.

Funding: This research received no external funding.

Institutional Review Board Statement: Not applicable.

Informed Consent Statement: Not applicable.

Data Availability Statement: The data presented in this study are available, on reasonable request, from the corresponding author.

Acknowledgments: The Authors express their gratitude to the Iasi Plasma Advanced Research Center (IPARC), from Faculty of Physics, "Alexandru Ioan Cuza" University of Iasi, Romania, for allowing the authors to use the infrastructure (UV-Vis and FTIR spectrometers) during part of the experiment described in this manuscript.

Conflicts of Interest: The authors declare no conflict of interest.

References

1. Kogelschatz, U. Dielectric-barrier discharges: Their history, discharge physics, and industrial applications. *Plasma Chem. Plasma Process.* **2003**, *23*, 1–46. [CrossRef]
2. Kim, H.H. Nonthermal plasma processing for air-pollution control: A historical review, current issues, and future prospects. *Plasma Process. Polym.* **2004**, *1*, 91–110. [CrossRef]
3. Kogelschatz, U. Atmospheric-pressure plasma technology. *Plasma Phys. Control. Fusion* **2004**, *46*, B63. [CrossRef]
4. Fridman, G.; Friedman, G.; Gutsol, A.; Shekhter, A.B.; Vasilets, V.N.; Fridman, A. Applied plasma medicine. *Plasma Process. Polym.* **2008**, *5*, 503–533. [CrossRef]
5. Nastuta, A.; Rusu, G.; Topala, I.; Chiper, A.; Popa, G. Surface modifications of polymer induced by atmospheric DBD plasma in different configurations. *J. Optoelectron. Adv. Mater* **2008**, *10*, 2038–2042.
6. Von Woedtke, T.; Kramer, A.; Weltmann, K.D. Plasma sterilization: What are the conditions to meet this claim? *Plasma Process. Polym.* **2008**, *5*, 534–539. [CrossRef]
7. Weltmann, K.D.; Kindel, E.; Brandenburg, R.; Meyer, C.; Bussiahn, R.; Wilke, C.; Von Woedtke, T. Atmospheric pressure plasma jet for medical therapy: Plasma parameters and risk estimation. *Contrib. Plasma Phys.* **2009**, *49*, 631–640. [CrossRef]
8. Nastuta, A.V.; Topala, I.; Grigoras, C.; Pohoata, V.; Popa, G. Stimulation of wound healing by helium atmospheric pressure plasma treatment. *J. Phys. D Appl. Phys.* **2011**, *44*, 105204. [CrossRef]
9. Topala, I.; Nastuta, A.V. Helium atmospheric pressure plasma jet: Diagnostics and application for burned wounds healing. In *Plasma for Bio-Decontamination, Medicine and Food Security. NATO Science for Peace and Security Series A: Chemistry and Biology;* Machala, Z., Hensel, K., Akishev, Y., Eds.; Springer: Dordrecht, The Netherlands, 2012; pp. 335–345. [CrossRef]
10. Han, J. Review of major directions in non-equilibrium atmospheric plasma treatments in medical, biological, and bioengineering applications. *Plasma Med.* **2013**, *3*. [CrossRef]
11. Nastuta, A.V.; Pohoata, V.; Topala, I. Atmospheric pressure plasma jet—Living tissue interface: Electrical, optical, and spectral characterization. *J. App. Phys.* **2013**, *113*, 183302. [CrossRef]
12. Von Woedtke, T.; Reuter, S.; Masur, K.; Weltmann, K.D. Plasmas for medicine. *Phys. Rep.* **2013**, *530*, 291–320. [CrossRef]
13. Keidar, M. Plasma for cancer treatment. *Plasma Sources Sci. Technol.* **2015**, *24*, 033001. [CrossRef]
14. Bruggeman, P.J.; Kushner, M.J.; Locke, B.R.; Gardeniers, J.G.E.; Graham, W.G.; Graves, D.B.; Hofman-Caris, R.C.H.M.; Maric, D.; Reid, J.P.; Ceriani, E.; et al. Plasma–liquid interactions: A review and roadmap. *Plasma Sources Sci. Technol.* **2016**, *25*, 053002. [CrossRef]
15. Bekeschus, S.; Wende, K.; Hefny, M.M.; Rödder, K.; Jablonowski, H.; Schmidt, A.; von Woedtke, T.; Weltmann, K.D.; Benedikt, J. Oxygen atoms are critical in rendering THP-1 leukaemia cells susceptible to cold physical plasma-induced apoptosis. *Sci. Rep.* **2017**, *7*, 27. [CrossRef] [PubMed]
16. Nastuta, A.V.; Topala, I.; Pohoata, V.; Mihaila, I.; Agheorghiesei, C.; Dumitrascu, N. Atmospheric pressure plasma jets in inert gases: Electrical, optical and mass spectrometry diagnosis. *Rom. Rep. Phys.* **2017**, *69*, 407.
17. Dai, X.; Bazaka, K.; Richard, D.J.; Thompson, E.R.W.; Ostrikov, K.K. The emerging role of gas plasma in oncotherapy. *Trends Biotechnol.* **2018**, *36*, 1183–1198. [CrossRef] [PubMed]
18. Bekeschus, S.; Favia, P.; Robert, E.; von Woedtke, T. White paper on plasma for medicine and hygiene: Future in plasma health sciences. *Plasma Process. Polym.* **2019**, *16*, 1800033. [CrossRef]
19. Brandenburg, R.; Bogaerts, A.; Bongers, W.; Fridman, A.; Fridman, G.; Locke, B.R.; Miller, V.; Reuter, S.; Schiorlin, M.; Verreycken, T.; et al. White paper on the future of plasma science in environment, for gas conversion and agriculture. *Plasma Process. Polym.* **2019**, *16*, 1700238. [CrossRef]
20. Cvelbar, U.; Walsh, J.L.; Černák, M.; de Vries, H.W.; Reuter, S.; Belmonte, T.; Corbella, C.; Miron, C.; Hojnik, N.; Jurov, A.; et al. White paper on the future of plasma science and technology in plastics and textiles. *Plasma Process. Polym.* **2019**, *16*, 1700228. [CrossRef]
21. Nastuta, A.V.; Popa, G. Surface oxidation and enhanced hydrophilization of polyamide fiber surface after He/Ar atmospheric pressure plasma exposure. *Rom. Rep. Phys.* **2019**, *71*, 1–16.
22. Šimek, M.; Černák, M.; Kylián, O.; Foest, R.; Hegemann, D.; Martini, R. White paper on the future of plasma science for optics and glass. *Plasma Process. Polym.* **2019**, *16*, 1700250. [CrossRef]

23. Weltmann, K.D.; Kolb, J.F.; Holub, M.; Uhrlandt, D.; Šimek, M.; Ostrikov, K.; Hamaguchi, S.; Cvelbar, U.; Černák, M.; Locke, B.; et al. The future for plasma science and technology. *Plasma Process. Polym.* **2019**, *16*, 1800118. [CrossRef]

24. Braný, D.; Dvorská, D.; Halašová, E.; Škovierová, H. Cold Atmospheric Plasma: A Powerful Tool for Modern Medicine. *Int. J. Mol. Sci.* **2020**, *21*, 2932. [CrossRef]

25. Keidar, M. Introduction: Plasma for Cancer Therapy. In *Plasma Cancer Therapy*; Keidar, M., Ed.; Springer: Berlin/Heidelberg, Germany, 2020; pp. 1–13. [CrossRef]

26. Pankaj, S.K.; Wan, Z.; Colonna, W.; Keener, K. Effect of high voltage atmospheric cold plasma on white grape juice quality. *J. Sci. Food Agric.* **2017**, *97*, 4016–4021. [CrossRef] [PubMed]

27. Kalli, E.; Lappa, I.; Bouchagier, P.; Tarantilis, P.; Skotti, E. Novel application and industrial exploitation of winery by-products. *Bioresour. Bioprocess.* **2018**, *5*, 46. [CrossRef]

28. Perinban, S.; Orsat, V.; Raghavan, V. Nonthermal Plasma—Liquid Interactions in Food Processing: A Review. *Compr. Rev. Food Sci. Food Saf.* **2019**, *18*, 1985–2008. [CrossRef] [PubMed]

29. Sainz-García, E.; López-Alfaro, I.; Múgica-Vidal, R.; López, R.; Escribano-Viana, R.; Portu, J.; Alba-Elías, F.; González-Arenzana, L. Effect of the Atmospheric Pressure Cold Plasma Treatment on Tempranillo Red Wine Quality in Batch and Flow Systems. *Beverages* **2019**, *5*, 50. [CrossRef]

30. Huang, C.; Wu, J.; Wu, J.; Ting, Y. Effect of novel atmospheric-pressure jet pretreatment on the drying kinetics and quality of white grapes. *J. Sci. Food Agric.* **2019**, *99*, 5102–5111. [CrossRef]

31. Chen, Y.Q.; Cheng, J.H.; Sun, D.W. Chemical, physical and physiological quality attributes of fruit and vegetables induced by cold plasma treatment: Mechanisms and application advances. *Crit. Rev. Food Sci. Nutr.* **2020**, *60*, 2676–2690. [CrossRef]

32. Ozen, E.; Singh, R. Atmospheric cold plasma treatment of fruit juices: A review. *Trends Food Sci. Technol.* **2020**, *103*, 144–151. [CrossRef]

33. Munekata, P.E.S.; Domínguez, R.; Pateiro, M.; Lorenzo, J.M. Influence of Plasma Treatment on the Polyphenols of Food Products—A Review. *Foods* **2020**, *9*, 929. [CrossRef]

34. Santamera, A.; Escott, C.; Loira, I.; del Fresno, J.M.; González, C.; Morata, A. Pulsed Light: Challenges of a Non-Thermal Sanitation Technology in the Winemaking Industry. *Beverages* **2020**, *6*, 45. [CrossRef]

35. Sainz-García, A.; González-Marcos, A.; Múgica-Vidal, R.; Muro-Fraguas, I.; Escribano-Viana, R.; González-Arenzana, L.; López-Alfaro, I.; Alba-Elías, F.; Sainz-García, E. Application of atmospheric pressure cold plasma to sanitize oak wine barrels. *LWT* **2020**, in press. [CrossRef]

36. Sarangapani, C.; Scally, L.; Gulan, M.; Cullen, P.J. Dissipation of Pesticide Residues on Grapes and Strawberries Using Plasma-Activated Water. *Food Bioproc. Tech.* **2020**, *13*, 1728–1741. [CrossRef]

37. Song, Y.; Fan, X. Cold plasma enhances the efficacy of aerosolized hydrogen peroxide in reducing populations of Salmonella Typhimurium and Listeria innocua on grape tomatoes, apples, cantaloupe and romaine lettuce. *Food Microbiol.* **2020**, *87*, 103391. [CrossRef] [PubMed]

38. Mujahid, Z.; Tounekti, T.; Khemira, H. Cold plasma treatment to release dormancy and improve growth in grape buds: A promising alternative to natural chilling and rest breaking chemicals. *Sci. Rep.* **2020**, *10*, 2667. [CrossRef]

39. Xiang, Q.; Zhang, R.; Fan, L.; Ma, Y.; Wu, D.; Li, K.; Bai, Y. Microbial inactivation and quality of grapes treated by plasma-activated water combined with mild heat. *LWT* **2020**, 109336. [CrossRef]

40. Laurita, R.; Contaldo, N.; Zambon, Y.; Bisag, A.; Canel, A.; Gherardi, M.; Laghi, G.; Bertaccini, A.; Colombo, V. The use of plasma-activated water in viticulture: Induction of resistance and agronomic performance in greenhouse and open field. *Plasma Process. Polym.* **2020**, *18*, e2000206. [CrossRef]

41. Zhao, Y.M.; Patange, A.; Sun, D.W.; Tiwari, B. Plasma-activated water: Physicochemical properties, microbial inactivation mechanisms, factors influencing antimicrobial effectiveness, and applications in the food industry. *Compr. Rev. Food Sci. Food Saf.* **2020**, *19*, 3951–3979. [CrossRef] [PubMed]

42. Ashtiani, S.H.M.; Rafiee, M.; Morad, M.M.; Khojastehpour, M.; Khani, M.; Rohani, A.; Shokri, B.; Martynenko, A. Impact of gliding arc plasma pretreatment on drying efficiency and physicochemical properties of grape. *Innov. Food Sci. Emerg. Technol.* **2020**, *63*, 102381. [CrossRef]

43. Bao, Y.; Reddivari, L.; Huang, J. Enhancement of phenolic compounds extraction from grape pomace by high voltage atmospheric cold plasma. *LWT* **2020**, *133*, 109970. [CrossRef]

44. Jambrak, A.R.; Ojha, S.; Šeremet, D.; Nutrizio, M.; Maltar-Strmečki, N.; Valić, S.; Kljusurić, J.G.; Tiwari, B. Free radical detection in water after processing by means of high voltage electrical discharges and high power ultrasound. *J. Food Process. Preserv.* **2020**, *45*, e15176. [CrossRef]

45. Muhammad, A.I.; Xiang, Q.; Liao, X.; Liu, D.; Ding, T. Understanding the impact of nonthermal plasma on food constituents and microstructure—A review. *Food Bioprocess Technol.* **2018**, *11*, 463–486. [CrossRef]

46. López, M.; Calvo, T.; Prieto, M.; Múgica-Vidal, R.; Muro-Fraguas, I.; Alba-Elías, F.; Alvarez-Ordóñez, A. A review on non-thermal atmospheric plasma for food preservation: Mode of action, determinants of effectiveness, and applications. *Front. Microbiol.* **2019**, *10*, 622. [CrossRef]

47. White, R.E. *Understanding Vineyard Soils*, 2 nd.; Oxford University Press: Oxford, UK, 2015.

48. Basalekou, M.; Pappas, C.; Tarantilis, P.; Kallithraka, S. Wine Authenticity and Traceability with the Use of FT-IR. *Beverages* **2020**, *6*, 30. [CrossRef]

49. Aleixandre-Tudo, J.; Buica, A.; Nieuwoudt, H.; Aleixandre, J.; du Toit, W. Spectrophotometric analysis of phenolic compounds in grapes and wines. *J. Agric. Food Chem.* **2017**, *65*, 4009–4026. [CrossRef]

50. Guilford, J.M.; Pezzuto, J.M. Wine and health: A review. *Am. J. Enol. Vitic.* **2011**, *62*, 471–486. [CrossRef]

51. Golan, R.; Gepner, Y.; Shai, I. Wine and Health—New Evidence. *Eur. J. Clin. Nutr.* **2019**, *72*, 55–59. [CrossRef]

52. Albu, C.; Radu, L.E.; Radu, G.L. Assessment of Melatonin and Its Precursors Content by a HPLC-MS/MS Method from Different Romanian Wines. *ACS Omega* **2020**, *5*, 27254–27260. [CrossRef] [PubMed]

53. Ferraz da Costa, D.C.; Pereira Rangel, L.; Quarti, J.; Santos, R.A.; Silva, J.L.; Fialho, E. Bioactive Compounds and Metabolites from Grapes and Red Wine in Breast Cancer Chemoprevention and Therapy. *Molecules* **2020**, *25*, 3531. [CrossRef]

54. OIV. State of the World Vitivinicultural Sector in 2019. 2020. Available online: http://www.oiv.int/public/medias/7298/oiv-state-of-the-vitivinicultural-sector-in-2019.pdf (accessed on 1 December 2020).

55. OIV. 2020 Wine Production First Estimates 27/10/2020. 2020. Available online: http://www.oiv.int/public/medias/7541/en-oiv-2020-world-wine-production-first-estimates.pdf (accessed on 1 December 2020).

56. Cotea, D.V.; Barbu, N.; Grigorescu, C.; Cotea, V.V. *Vineyards and Wines of Romania*; Academiei Române: Bucureşti, Romania, 2005.

57. Ionesi, L.; Ionesi, B.; Lungu, A.; Rosca, V.; Ionesi, V. *Sarmaţianul mediu şi superior de pe Platforma Moldovenească*; Middle and Upper Sarmatian on Moldavian Platform; Academiei Române: Bucureşti, Romania, 2005; pp. 1–439.

58. OIV. International Standard for the Labelling of Wines-Edition 2015. Available online: http://www.oiv.int/public/medias/4776/oiv-wine-labelling-standard-en-2015.pdf (accessed on 1 December 2020).

59. OIV. Compendium of International Methods of Wine and Must Analysis. 2020. Available online: http://www.oiv.int/public/medias/7372/oiv-compendium-volume-1-2020.pdf (accessed on 1 December 2020).

60. Tzachristas, A.; Pasvanka, K.; Calokerinos, A.; Proestos, C. Polyphenols: Natural antioxidants to be used as a quality tool in wine authenticity. *Appl. Sci.* **2020**, *10*, 5908. [CrossRef]

61. Sanna, R.; Piras, C.; Marincola, F.; Lecca, V.; Maurichi, S.; Scano, P. Multivariate statistical analysis of the UV-vis profiles of wine polyphenolic extracts during vinification. *J. Agric. Sci.* **2014**, *6*, 152. [CrossRef]

62. Aleixandre-Tudo, J.L.; du Toit, W. The Role of UV-Visible Spectroscopy for Phenolic Compounds Quantification in Winemaking. In *Frontiers and New Trends in the Science of Fermented Food and Beverages*; Solís-Oviedo, R.L., de la Cruz Pech-Canul, A., Eds.; IntechOpen: Rijeka, Croatia, 2018; pp. 1–22. [CrossRef]

63. Yalçın, O.; Tekgündüz, C.; Öztürk, M.; Tekgündüz, E. Investigation of the traditional organic vinegars by UV–VIS spectroscopy and rheology techniques. *Spectrochim. Acta Part A Mol. Biomol. Spectrosc.* **2020**, *246*, 118987. [CrossRef]

64. Ríos-Reina, R.; Azcarate, S.M.; Camiña, J.; Callejón, R.M. Assessment of UV–visible spectroscopy as a useful tool for determining grape-must caramel in high-quality wine and balsamic vinegars. *Food Chem.* **2020**, *323*, 126792. [CrossRef]

65. Louw, L.; Roux, K.; Tredoux, A.; Tomic, O.; Naes, T.; Nieuwoudt, H.; Van Rensburg, P. Characterization of selected South African young cultivar wines using FTMIR spectroscopy, gas chromatography, and multivariate data analysis. *J. Agric. Food Chem.* **2009**, *57*, 2623–2632. [CrossRef]

66. Sen, I.; Ozturk, B.; Tokatli, F.; Ozen, B. Combination of visible and mid-infrared spectra for the prediction of chemical parameters of wines. *Talanta* **2016**, *161*, 130–137. [CrossRef]

67. Basalekou, M.; Kallithraka, S.; Tarantilis, P.A.; Kotseridis, Y.; Pappas, C. Ellagitannins in wines: Future prospects in methods of analysis using FT-IR spectroscopy. *LWT* **2019**, *101*, 48–53. [CrossRef]

68. Topala, C.M.; Tataru, L.D. ATR-FTIR Spectroscopy Coupled with Chemical and Chemometric Analysis to Distinguish Between Some Sweet Wines. *Rev. Chim.* **2019**, *70*, 2355–2361. [CrossRef]

69. Álvarez, Á.; Yáñez, J.; Neira, Y.; Castillo-Felices, R.; Hinrichsen, P. Simple distinction of grapevine (*Vitis vinifera* L.) genotypes by direct ATR-FTIR. *Food Chem.* **2020**, *328*, 127164. [CrossRef] [PubMed]

70. Ferrer-Gallego, R.; Rodríguez-Pulido, F.; Toci, A.; García-Estevez, I. Phenolic Composition, Quality and Authenticity of Grapes and Wines by Vibrational Spectroscopy. *Food Rev. Int.* **2020**, 1–29. [CrossRef]

71. Lucarini, M.; Durazzo, A.; Kiefer, J.; Santini, A.; Lombardi-Boccia, G.; Souto, E.; Romani, A.; Lampe, A.; Nicoli, F.; Gabrielli, P.; et al. Grape seeds: Chromatographic profile of fatty acids and phenolic compounds and qualitative analysis by FTIR-ATR spectroscopy. *Foods* **2020**, *9*, 10. [CrossRef]

72. Coimbra, M.; Gonçalves, F.; Barros, A.; Delgadillo, I. Fourier transform infrared spectroscopy and chemometric analysis of white wine polysaccharide extracts. *J. Agric. Food Chem.* **2002**, *50*, 3405–3411. [CrossRef]

73. Banc, R.; Loghin, F.; Miere, D.; Fetea, F.; Socaciu, C. Romanian wines quality and authenticity using FT-MIR spectroscopy coupled with multivariate data analysis. *Notulae Botanicae Horti Agrobotanici Cluj-Napoca* **2014**, *42*, 556–564. [CrossRef]

74. Aleixandre-Tudo, J.; Nieuwoudt, H.; Olivieri, A.; Aleixandre, J.; du Toit, W. Phenolic profiling of grapes, fermenting samples and wines using UV-Visible spectroscopy with chemometrics. *Food Control* **2018**, *85*, 11–22. [CrossRef]

75. Minute, F.; Giotto, F.; Filipe-Ribeiro, L.; Cosme, F.; Nunes, F.M. Alternative Methods for Measuring the Susceptibility of White Wines to Pinking Alteration: Derivative Spectroscopy and CIE L*a*b* Colour Analysis. *Foods* **2021**, *10*, 553. [CrossRef] [PubMed]

76. Gorinstein, S.; Moshe, R.; Deutsch, J.; Wolfed, F.; Tilis, K.; Stiller, A.; Flam, I.; Gat, Y. Determination of basic components in white wines by HPLC, FT-IR spectroscopy, and electrophoretic techniques. *J. Food Compos. Anal.* **1992**, *5*, 236–245. [CrossRef]

77. Todasca, M.C.; Chira, N.; Deleanu, C.; Rosca, S. Romanian wine study using IR spectroscopy in comparison with 1 HNMR. *UPB Sci. Bull. B Chem. Mater. Sci.* **2007**, *69*, 3–10.

78. Olejar, K.J.and Ricci, A.; Swift, S.; Zujovic, Z.; Gordon, K.; Fedrizzi, B.; Versari, A.; Kilmartin, P. Characterization of an antioxidant and antimicrobial extract from cool climate, white grape marc. *Antioxidants* **2019**, *8*, 232. [CrossRef]
79. Bañuelos, M.A.; Loira, I.; Guamis, B.; Escott, C.; Del Fresno, J.M.; Codina-Torrella, I.; Quevedo, J.M.; Gervilla, R.; Chavarría, J.M.R.; de Lamo, S.; et al. White wine processing by UHPH without SO2. Elimination of microbial populations and effect in oxidative enzymes, colloidal stability and sensory quality. *Food Chem.* **2020**, *332*, 127417. [CrossRef]
80. Dienes-Nagy, Á.; Marti, G.; Breant, L.; Lorenzini, F.; Fuchsmann, P.; Baumgartner, D.; Zufferey, V.; Spring, J.; Gindro, K.; Viret, O.; et al. Identification of putative chemical markers in white wine (Chasselas) related to nitrogen deficiencies in vineyards. *OENO One* **2020**, *54*, 583–599. [CrossRef]
81. Mendes, E.; Duarte, N. Mid-Infrared Spectroscopy as a Valuable Tool to Tackle Food Analysis: A Literature Review on Coffee, Dairies, Honey, Olive Oil and Wine. *Foods* **2021**, *10*, 477. [CrossRef]
82. Ţârdea, C. *Chimia şi Analiza Vinului [Chemistry and Wine Analysis]*; Ion Ionescu de la Brad: Iaşi, Romania, 2007.
83. Delgado-González, M.; Carmona-Jiménez, Y.; Rodríguez-Dodero, M.; García-Moreno, M. Color Space Mathematical Modeling Using Microsoft Excel. *J. Chem. Educ.* **2018**, *95*, 1885–1889. [CrossRef]
84. García-Marino, M.; Escudero-Gilete, M.; Escribano-Bailón, M.; González-Miret, M.; Rivas-Gonzalo, J.; Heredia, F. Colorimetric characteristics of the phenolic fractions obtained from Tempranillo and Graciano wines through the use of different instrumental techniques. *Anal. Chim. Acta* **2012**, *732*, 153–161. [CrossRef]
85. Rolle, L.; Giordano, M.; Giacosa, S.; Vincenzi, S.; Segade, S.; Torchio, F.; Perrone, B.; Gerbi, V. CIEL* a* b* parameters of white dehydrated grapes as quality markers according to chemical composition, volatile profile and mechanical properties. *Anal. Chim. Acta* **2012**, *732*, 105–113. [CrossRef]
86. Briones-Labarca, V.; Perez-Wom, M.; Habib, G.; Giovagnoli-Vicuña, C.; Cañas-Sarazua, R.; Tabilo-Munizaga, G.; Salazar, F.N. Oenological and quality characteristic on young white wines (sauvignon blanc): Effects of high hydrostatic pressure processing. *J. Food Qual.* **2017**, *2017*. [CrossRef]
87. Jacobson, J.L. *Introduction to Wine Laboratory Practices and Procedures*, 1st ed.; Springer: Berlin/Heidelberg, Germany, 2006; p. 135. [CrossRef]
88. Bührle, F.; Gohl, A.; Weber, F. Impact of Xanthylium Derivatives on the Color of White Wine. *Molecules* **2017**, *22*, 1376. [CrossRef] [PubMed]
89. Ewart, A., White wines. In *Fermented Beverage Production*; Lea, A.G.H., Piggott, J., Eds.; Springer: Berlin/Heidelberg, Germany, 2003; pp. 89–106. [CrossRef]
90. Musther, J. VinoCalc. 2020. Available online: https://www.vinolab.hr/calculator/gravity-density-sugar-conversions-en19 (accessed on 1 May 2020).
91. Gerber, I.; Mihaila, I.; Hein, D.; Nastuta, A.V.; Jijie, R.; Pohoata, V.; Topala, I. Time behaviour of helium atmospheric pressure plasma jet electrical and optical parameters. *Appl. Sci.* **2017**, *7*, 812. [CrossRef]
92. Guo, J.; Huang, K.; Wang, X.; Lyu, C.; Yang, N.; Li, Y.; Wang, J. Inactivation of yeast on grapes by plasma-activated water and its effects on quality attributes. *J. Food Prot.* **2017**, *80*, 225–230. [CrossRef]
93. Lukić, K.; Vukušić, T.; Tomašević, M.; Ćurko, N.; Gracin, L.; Ganić, K.K. The impact of high voltage electrical discharge plasma on the chromatic characteristics and phenolic composition of red and white wines. *Innov. Food Sci. Emerg. Technol.* **2019**, *53*, 70–77. [CrossRef]
94. Nishime, T.; Wannicke, N.; Horn, S.; Weltmann, K.D.; Brust, H. A Coaxial Dielectric Barrier Discharge Reactor for Treatment of Winter Wheat Seeds. *Appl. Sci.* **2020**, *10*, 7133. [CrossRef]
95. Starek-Wójcicka, A.; Sagan, A.; Terebun, P.; Kwiatkowski, M.; Kiczorowski, P.; Pawlat, J. Influence of a Helium–Nitrogen RF Plasma Jet on Onion Seed Germination. *Appl. Sci.* **2020**, *10*, 8973. [CrossRef]
96. Wang, J.; Han, R.; Liao, X.; Ding, T. Application of plasma-activated water (PAW) for mitigating methicillin-resistant Staphylococcus aureus (MRSA) on cooked chicken surface. *LWT* **2020**, *137*, 110465. [CrossRef]
97. Pan, Y.; Cheng, J.; Sun, D. Inhibition of fruit softening by cold plasma treatments: Affecting factors and applications. *Crit. Rev. Food Sci. Nutr.* **2020**, *61* 127164. [CrossRef] [PubMed]
98. Xiang, Q.; Fan, L.; Li, Y.; Dong, S.; Li, K.; Bai, Y. A review on recent advances in plasma-activated water for food safety: Current applications and future trends. *Crit. Rev. Food Sci. Nutr.* **2020**, 1–19. [CrossRef]
99. Fan, L.; Liu, X.; Ma, Y.; Xiang, Q. Effects of plasma-activated water treatment on seed germination and growth of mung bean sprouts. *J. Taibah Univ. Sci.* **2020**, *14*, 823–830. [CrossRef]
100. Zhou, Y.H.; Vidyarthi, S.K.; Zhong, C.S.; Zheng, Z.A.; An, Y.; Wang, J.; Wei, Q.; Xiao, H.W. Cold plasma enhances drying and color, rehydration ratio and polyphenols of wolfberry via microstructure and ultrastructure alteration. *LWT* **2020**, *134*, 110173. [CrossRef]
101. Gerling, T.; Nastuta, A.V.; Bussiahn, R.; Kindel, E.; Weltmann, K.D. Back and forth directed plasma bullets in a helium atmospheric pressure needle-to-plane discharge with oxygen admixtures. *Plasma Sources Sci. Technol.* **2012**, *21*, 034012. [CrossRef]
102. Gerling, T.; Wild, R.; Nastuta, A.V.; Wilke, C.; Weltmann, K.D.; Stollenwerk, L. Correlation of phase resolved current, emission and surface charge measurements in an atmospheric pressure helium jet. *Eur. Phys. J. Appl. Phys.* **2015**, *71*, 20808. [CrossRef]
103. Luque, J.; Crosley, D.R. LIFBASE: Database and spectral simulation program (version 1.5). *SRI Int. Rep. MP* **1999**, *99*. Available online: https://ci.nii.ac.jp/naid/10011750245 (accessed on 5 August 2021).

104. Navrátil, Z.; Trunec, D.; Šmíd, R.; Lazar, L. A software for optical emission spectroscopy-problem formulation and application to plasma diagnostics. *Czechoslov. J. Phys.* **2006**, *56*, B944–B951. [CrossRef]
105. Radzig, A.A.; Smirnov, B.M. *Reference Data on Atoms, Molecules, and Ions*; Springer: Berlin/Heidelberg, Germany, 1985; pp. 230–235. [CrossRef]
106. Kramida, A.; Ralchenko, Y.; Reader, J.; Team, N.A. NIST Atomic Spectra Database (version 5.8), 2020. Available online: https://physics.nist.gov/asd (accessed on 1 December 2020). [CrossRef]
107. Junqua, R.; Carullo, D.; Ferrari, G.; Pataro, G.; Ghidossi, R. Ohmic heating for polyphenol extraction from grape berries: An innovative prefermentary process. *OENO One* **2021**, *55*, 39–51. [CrossRef]
108. Azcarate, S.M.; Cantarelli, M.Á.; Pellerano, R.G.; Marchevsky, E.J.; Camiña, J. Classification of Argentinean Sauvignon blanc wines by UV spectroscopy and chemometric methods. *J. Food Sci.* **2013**, *78*, C432–C436. [CrossRef]
109. Kerslake, F.; Longo, R.; Dambergs, R. Discrimination of juice press fractions for sparkling base wines by a UV-Vis spectral phenolic fingerprint and chemometrics. *Beverages* **2018**, *4*, 45. [CrossRef]
110. Scutaru, I.; Balanuta, A.; Zgardan, D. The determination of oxidation behavior of white wines produced from local and european grape varieties using spectrophotometric method. *J. Eng. Sci.* **2018**, *XXV*, 82–93. [CrossRef]
111. Martelo-Vidal, M.J.; Vazquez, M. Evaluation of ultraviolet, visible, and near infrared spectroscopy for the analysis of wine compounds. *Czech J. Food Sci.* **2014**, *32*, 37–47. [CrossRef]
112. Yu, J.; Wang, H.; Zhan, J.; Huang, W. Review of recent UV–Vis and infrared spectroscopy researches on wine detection and discrimination. *Appl. Spectrosc. Rev.* **2018**, *53*, 65–86. [CrossRef]
113. Banc, R.; Loghin, F.; Miere, D.; Ranga, F.; Socaciu, C. Phenolic composition and antioxidant activity of red, rosé and white wines originating from Romanian grape cultivars. *Notulae Botanicae Horti Agrobotanici Cluj-Napoca* **2020**, *48*, 716–734. [CrossRef]

Article

Investigation of Nonthermal Plasma Jet Excitation Mode and Optical Assessment of Its Electron Concentration

Anton S. Ivankov [1,*], Anastasia M. Kozhevnikova [1], Dmitry V. Schitz [1,*] and Igor V. Alekseenko [1,2]

[1] Laboratory of Optical Radiation, Immanuel Kant Baltic Federal University, 236001 Kaliningrad, Russia; akozhevnikova1@kantiana.ru (A.M.K.); ialekseenko@kantiana.ru (I.V.A.)
[2] Institute for Laser Technology in Medicine and Measurement Technique (ILM), 89040 Ulm, Germany
* Correspondence: aivankov@kantiana.ru (A.S.I.); dschitz@kantiana.ru (D.V.S.)

Abstract: The results of a study of a plasma jet of atmospheric-pressure helium driven by a capacitive discharge using sine and pulsed modes of excitation are presented. The homogeneous discharge of a multi-channel plasma jet at gas temperature of 34 °C and helium flow rate of 0.5 L/min was achieved with short pulse excitation. A digital holography method is proposed to estimate a basic plasma parameter, i.e., its electron concentration. An automated digital holographic interferometry set-up for the observation and study of a nonthermal plasma jet in a pulse mode is developed and described. The synchronization features of recording devices with the generation of plasma pulses are considered. The electron concentration of the plasma jet is also estimated. The disadvantages of the proposed technique and its further application are discussed.

Keywords: plasma jet; bio-medicine application; cold gas-discharge plasma; digital holography; digital holographic interferometry; plasma diagnostics

Citation: Ivankov, A.S.; Kozhevnikova, A.M.; Schitz, D.V.; Alekseenko, I.V. Investigation of Nonthermal Plasma Jet Excitation Mode and Optical Assessment of Its Electron Concentration. *Appl. Sci.* **2021**, *11*, 9203. https://doi.org/10.3390/app11199203

Academic Editor: Andrei Vasile Nastuta

Received: 20 August 2021
Accepted: 29 September 2021
Published: 3 October 2021

Publisher's Note: MDPI stays neutral with regard to jurisdictional claims in published maps and institutional affiliations.

1. Introduction

Inappropriate and excessive use of antimicrobials and antibiotics creates a problem of acquired resistance of bacteria and parasites to drugs, which is now taking on dangerous proportions. In September 2016, this issue was discussed at the 71st session of the UN General Assembly. Therefore, the development of new physical therapy equipment and non-drug therapies is a very important direction of research in modern science.

Low-temperature plasma has been successfully used in medicine for the last 15 years [1]. The possibility of generating gas-discharge plasma with low gas temperature has led to the development of new and relatively safe methods of non-drug treatment accompanying the surgery of septic wounds [2]; this method is also used in the therapy of more than a dozen skin conditions, as well as in cosmetic procedures that contribute to effective facial rejuvenation, eliminating the need for invasive methods [3,4]. A plasma jet contains charged particles (electrons and ions), generates ozone, nitrogen and oxygen radicals, and produces UV radiation (within the 20–300 nm range), enabling it to destroy the membranes of pathogenic microorganisms. For this reason, treatment with a stream of low-temperature plasma in the therapy of septic wounds can be considered nonspecific, meaning that there is no acquired pathogen resistance [5].

To painlessly treat skin or open wounds, a plasma jet with a low gas temperature (lower than 40 °C) can be used. Such plasma is called *non-thermal*. The main problem with generating non-thermal plasma jet sources at atmospheric pressure is achieving a low gas temperature of the plasma jet at a low working gas flow rate. In addition, the electrode system of the plasma jets must provide a closed electrical circuit to protect patients from malicious levels of current leakage.

Sine SHF voltage is used to produce a plasma jet with low gas temperature at a low working gas flow rate (<1 L/min) [6]. Typically, SHF power supplies are bulky, expensive and inefficient. Sine HF voltage has the benefit of using power supplies that are more compact, but in this case gas temperatures lower than 40 °C can only be achieved at high flow rates of working gas (>3 L/min) [7].

Apart from the gas temperature in the jet and its spatial characteristics, the electron concentration of the plasma is the most important parameter as far as living organisms are concerned. It is obvious that the degree of ionization of working gas can mostly determine the rate of pathogen inactivation or skin regeneration efficiency.

Due to available sensors for the respective frequency ranges, it is easy to measure a radiation intensity in time by means of radiofrequency, optical, or ionizing radiation, i.e., to determine a radiation dose received by the object studied. However, the concept of *dose* is still not applied in case of treatments by plasma flow; instead, we measure the duration of exposure to plasma along with certain parameters of the electrode system, excitation mode, the type of working gas and the media in which the object is located. To compare the influence of nonthermal plasma jets on living objects, despite the conditions of plasma jet generation, we suppose that the electron concentration of the plasma flow has to be considered. Since plasma has been formed in pulse-periodic mode, probe methods cannot be applied to determine its concentration in the gas flow (gas flow being a non-stationary process). The implementation of spectral methods estimating an electron concentration in gas-discharge plasma [8], however, requires expensive equipment. On the other hand, the holographic interferometry is another method for estimating plasma electron concentration. Within the visible spectral range, it can be used to determine electron or atomic concentrations in the range of 10^{16} cm^{-3}. However, the use of multispectral holographic interferometry can increase the method's sensitivity by several orders of magnitude [9,10]. Since current holographic interferometry techniques are almost completely digital, it is possible to increase the sensitivity of holographic methods by increasing the dynamic range of digital detectors (CCD- or CMOS cameras) used for hologram acquisition [11].

In this article, we are trying to consider and solve the following problems: the first one is an investigation of the electrode system excitation regimes that can generate a nonthermal plasma jet at a low helium flow rate and low gas temperature; the second one is the development of a reliable method, based on holographic interferometry, to control and estimate the plasma parameters.

2. Materials and Methods

2.1. Testing Electrode System of the Nonthermal Plasma Jet

Tests of several electrode systems (single-barrier discharge with needle electrode, double-barrier discharge and glow discharge) to produce nonthermal plasma flows showed high efficiency and electrical safety of a double-barrier discharge (DBD) electrode system design (Figure 1). High voltage applied to the ring electrodes *2* ignites the gas discharge plasma inside the tube *1*, and the gas flow *6* pushes the plasma out in the form of a jet *5*.

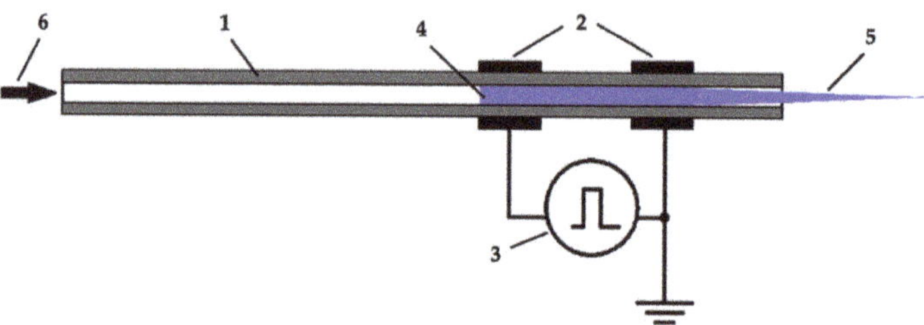

Figure 1. Electrode system of plasma-jet-based DBD: 1—glass pipe, 2—ring electrodes, 3—pulsed power supply, 4—gas discharge, 5—plasma jet, 6—gas flow.

Two power supplies were developed to compare the short-pulse and sine excitation regimes of inert gas. The short-pulse power supply was equipped with a MOSFETs-based bridge inverter and a TR1 step-up transformer (Figure 2a). The transistors switched the DC line voltage to the primary winding of TR1 so that high-voltage pulses (2 µs duration) with long pauses and alternating polarity were generated at the electrodes. Sine voltage power supply is a standard resonant generator (Figure 2b), where the L2-C6-TR2 circuit of elements works as a resonant circuit. Transistors VT5 and VT6 have a duty cycle of 50%, at a frequency close to that of the resonant circuit.

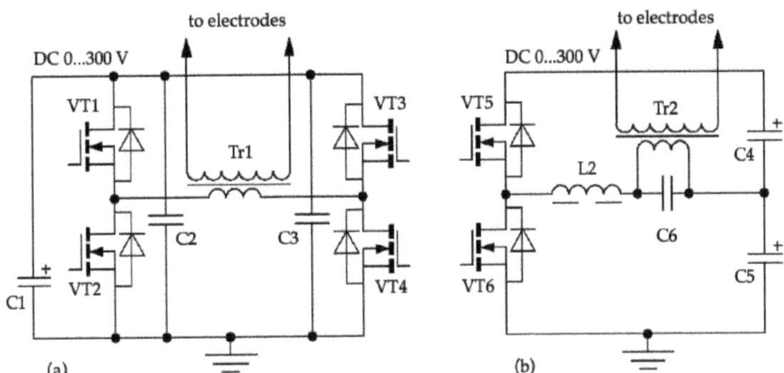

Figure 2. Electrical circuits of short pulse (**a**) and sine (**b**) voltage power supplies.

The high efficiency and long lifetime (up to 8000 h) of XeCl* and KrCl* UV radiators driven by DBD have been well described [12,13]. These results are obtained due to high-frequency short-pulse excitation [14] and a special quartz bulb processing technology [15].

The electrode voltage and gas discharge current were measured with a PPE6KV high-voltage sensor (LeCroy Corp.), a 50 Ohm shunt resistor, and a TEKTRONIX TDS2024C oscilloscope. Helium flow rate did not affect the voltage and current oscillograms. The flow rate was controlled by a PVQ13 proportional valve and PFMV530 flow sensor (SMC Corp.). Plasma gas temperature was measured with a thermocouple thermometer.

At the first stage, we tested the influence of tube diameter, ring electrode length, and interelectrode gap on gas temperature. Optimal parameters of the electrode system in terms of minimum values of plasma jet gas temperature were the same for both short pulse and sine voltages: 5.5 mm tube diameter (glass thickness 0.5 mm), 3 mm width of ring electrodes, and a 5 mm interelectrode gap.

A decrease in helium flow rate led to an increase in gas temperature of the plasma jet (Figure 3). A particularly rapid increase in gas temperature was recorded at the flow rate of less than 1 L/min. However, for excitation with short pulses, this value was much smaller. Thus, for 0.5 L/min, gas temperature was 34 °C for short-pulse excitation and 53 °C for sine excitation. The dependence shown in Figure 3 was obtained at the frequency of 33 kHz and excitation power of 4.2 W for both excitation methods. The excitation power was calculated using the technique described in [16].

Voltage and current of plasma jet at short pulse excitation and at sine voltage excitation shown in Figure 4.

With plasma excitation by microsecond pulses of alternating polarity it was possible to obtain a diffuse plasma jet without contracted channels (Figure 5a). In this case, the gas temperature of the plasma jet was 34 °C at a helium flow rate of 0.5 L/min.

Obviously, the larger the contact area of plasma with the treated surface, the less time it will take to treat it. In order to increase the diameter of the plasma flow, a multi-channel electrode system was tested. To implement this system, electrode systems absolutely identical to the construction demonstrated in Figure 1 were installed parallel to each other in one housing case (Figure 5b). The electrodes were supplied from a single power source. By means of a splitter, the helium flow was supplied to the channels at the same flow rate. The tests showed that all channels of the plasma jets could operate in parallel without affecting each other. The application of the three-channel plasma jet allowed us to increase the area of plasma contact with the surface up to 1 cm^2. To obtain an even larger diameter of the plasma jet, it was necessary to increase the number of plasma jet channels.

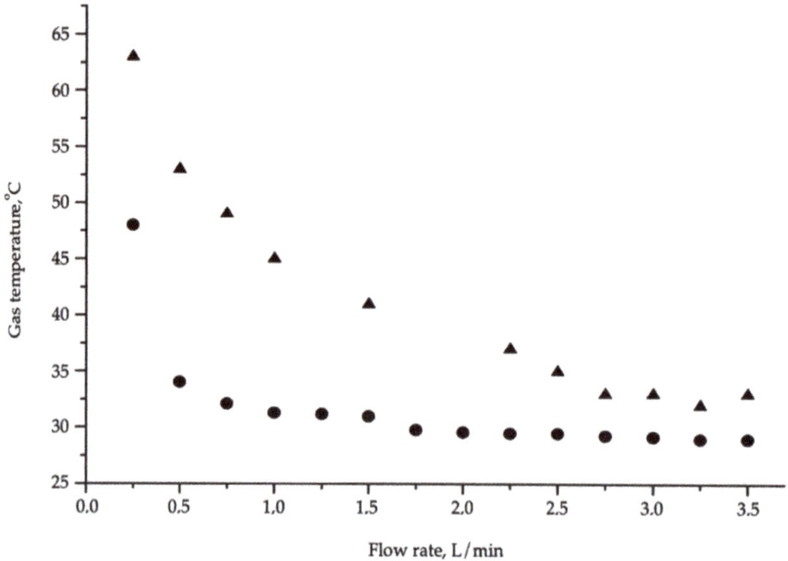

Figure 3. Dependence of plasma jet gas temperature on flow rate at short pulse (○) and sine (Δ) excitation voltage.

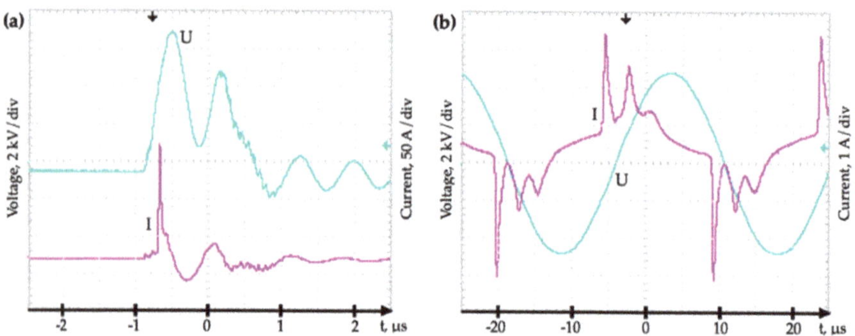

Figure 4. Voltage and current of plasma jet at short pulse excitation (**a**) and at sine voltage excitation (**b**).

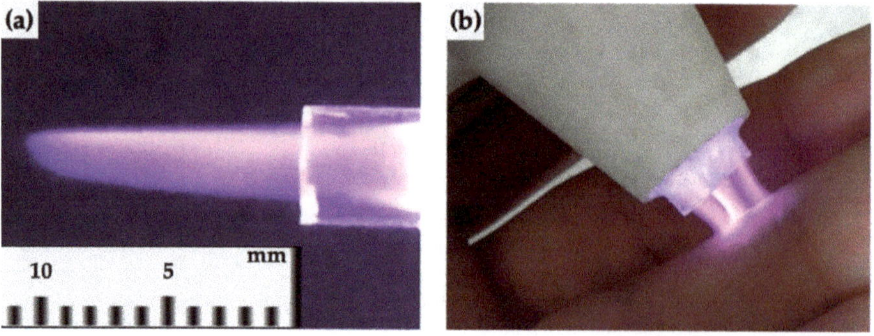

Figure 5. (**a**)—Helium plasma jet at short pulse excitation, (**b**)—three-channel plasma jet.

2.2. Evaluating of Electron Concentration in Non-thermal Plasma Jet Generated at Pulse Excitation

As we have already mentioned, the problem of plasma diagnostics can be solved by different methods. However, it would be prudent to have a sufficiently reliable and flexible method for estimating the concentration of plasma particles, and digital holographic interferometry is a proper method for this task. The results of the application of this method are preliminary and demonstrate the principal possibility of digital holographic interferometry for the investigation of a non-thermal plasma jet generated at pulse excitation.

The investigation of plasma by holographic methods is based on the study of a phase (transparent) object, in which the phase changes of the object field depend on the spatial localization of the object and its refractive index [17]. The obtained values of the plasma refractive index allow us to calculate its electron concentration [18].

It is well known in holographic interferometry that phase difference is related to the corresponding refractive index by the following expression

$$\Delta\varphi(x,y) = \frac{2\pi}{\lambda} \int_{l_1}^{l_2} [n(x,y,z) - n_0]dz \tag{1}$$

where $n(x,y,z)$ is the refractive index of the medium (plasma), and n_0 is a refractive index of the medium against which the comparison is being made [19]. The integration limits determine the length of the path along which the probing radiation propagates.

In conventional (analog) holographic interferometry, fringe analysis is carried out with the accuracy of phase detection associated with changes in the optical path up to $\lambda/10$ [20]. However, in digital holographic interferometry, holograms are recorded digitally, where the phase detection accuracy is determined by the dynamic range of the detector, with the 8-, 10- or 16-bit resolutions of an analog-to-digital converter. Thus, with 8-bit, it is possible to increase the phase detection sensitivity up to $\lambda/256$ [11].

Digital holographic interferometry is based on a comparison of the phases of at least two optical wavefronts recorded at different moments in time as digital holograms. The result of the phase comparison between the wavefronts is represented as a numerical phase difference map [21].

A digitally recorded hologram is the distribution of the interference patterns from a superposition of reference and object beams on the detector. The correct acquisition of the hologram can be provided if the condition for the sampling theorem has been met [22]. This theorem determines the angle between the reference and object beam

$$\alpha_{max} = \frac{\lambda}{2\Delta x}$$

where Δx—is the pixel size of the detector. The evaluation of the holograms and phase reconstruction were performed using the Fourier transform method [23]. If holograms are recorded at time moments t_1 and t_2, corresponding to different object states, it is possible to calculate the phase difference after the reconstruction routine.

Result intensity recorded on the camera sensor is:

$$I(x,y) = [E_R \exp(-i\varphi_R(x,y)) + E_o \exp(-i\varphi_o(x,y))] \times [E_R \exp(-i\varphi_R(x,y))E_o \exp(-i\varphi_o(x,y))]^* \tag{2}$$

where E_R E_O—are the amplitudes of the reference and object waves, respectively.

Applying double Fourier transform, followed by filtering and inverse double Fourier transform for different states of the object, allows one to obtain the phase difference function:

$$\Delta\phi = arctg[tg(\varphi_1 - \varphi_2)] = arctg\left[\frac{Im_1 \times Re_2 - Im_2 \times Re_1}{Im_1 \times Im_2 + Re_1 \times Re_2}\right] \tag{3}$$

which describes the change in the state of the object [24].

Thus, for example, if an object is subjected to mechanical loading, the phase difference will be related to the value and direction of displacements of the object's surface points [25]. When a phase object is investigated, phase difference will be determined by the integral change of the optical path related both to the size of the object and to spatial distribution of the refractive index of the medium according to the equation (1). In a simple way, the relation between the plasma refractive index and its electron concentration is determined by the following relations:

$$n - 1 = \sum_{i=1}^{k} \left(A_i + \frac{B_i}{\lambda^2} \right) N_{a_i} - 4,5 \cdot 10^{-14} \lambda^2 N_e \tag{4}$$

where, A_i and B_i are the Cauchy's constants for the medium, λ—is the laser radiation wavelength, N_e—is the plasma electron concentration, N_{a_i}—are atomic concentrations [26], and the phase difference is, respectively:

$$\Delta\varphi(x,y) = \frac{2\pi}{\lambda} l \Delta n \tag{5}$$

where Δn—is the change in the refractive index, and l—is the optical path length along the medium under study.

2.3. Experimental Setup (Experimental Verification of Pulsed Plasma Jet Registration)

The optical setup for an investigation of a non-thermal plasma jet allowed for the acquisition of the image plane hologram of the area where the phase object was localized. The setup is represented in Figure 6 and its laboratory configuration in Figure 7.

Output laser radiation was collimated to a diameter of about 15 mm and then split into the reference and object beams. Since the coherence length of the laser was 5 cm, the optical paths of the reference and object beams were aligned to an accuracy of 5 mm. The object beam, passing through a jet of nonthermal plasma, experienced a phase difference. Comparing the holographic images recorded at the moment of plasma generation and the moment of its absence made it possible to evaluate the phase changes corresponding to the refractive index of plasma.

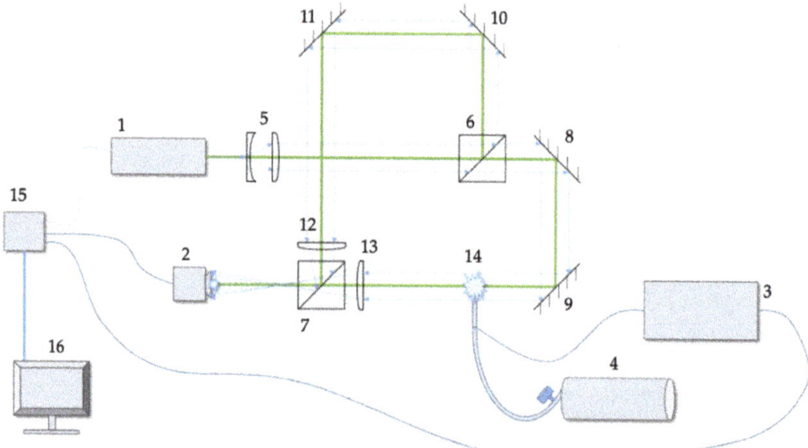

Figure 6. Configuration of the digital holographic interferometry complex for the study of plasma flow in the pulsed generation mode: (1)—Nd: YAG pulse laser, (2)—CCD Camera, (3)—pulse-plasma generator, (4)—helium container, (5)—collimator, (6, 7)—beam splitters, (8, 9, 10, 11)—mirrors, (12, 13)—lenses, (14)—plasma jet, (15)—synchronization Unit, (16)—PC.

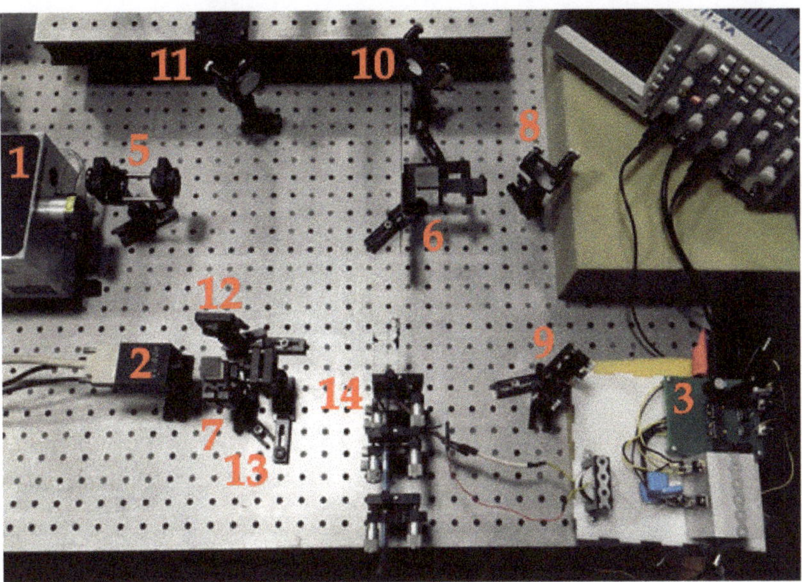

Figure 7. Configuration of the digital holographic interferometry setup for the study of plasma flow in the pulsed generation mode (photograph).

The nonthermal plasma jet was generated in pulsed mode with a frequency of 5 kHz and a pulse duration of 750 ns.

In the pulse mode of plasma generation, it is only possible to record holograms when the laser, the digital camera, and the plasma generator are synchronized with each other. Thus, we used an InnoLas "SpitLight Hybrid II" pulsed laser, with a maximum pulse repetition frequency of 50 Hz and a duration of 10 ns, and a Pulnix 1325 CL CCD camera, with a minimum exposure time of 62 μs and a frame repetition rate of 15 fps, in order to acquire holograms. A short laser pulse duration avoids the influence of random changes in the medium, which can destroy the phase information due to its averaging. Synchronization of all devices was carried out by using the NI-DAQ (National Instruments data acquisition) boards NI USB 6258 and NI SCB 68, as well as developing a hardware-software algorithm in the LabVIEW environment.

Since the frame rate of the camera and laser pulse frequency are significantly lower than the repetition rate of plasma, the recording of holograms corresponding to the plasma single-pulse is only possible when a number of plasma pulses are skipped between the hologram acquisition. Hence, the synchronization of the devices was realized in the following conditions: 15 Hz for the laser and camera repetition rates, and 4995 Hz for the plasma pulse generator. Such configuration corresponds to skipping 333 periods of plasma pulses between each moment of laser pulse "shots" and the image recording. All devices were triggered simultaneously from one external synchronizing source at the beginning of the measurement. However, due to the large difference in the frequencies of plasma pulses and the laser-camera system, the plasma pulse and trigger signal of the laser-camera had a mismatch of 60 ns in one cycle. Therefore, the trigger signal for the laser-camera shifted 60 ns along the plasma pulse. This mismatch allowed for the "scanning" of the plasma pulse that had a 750 ns duration, under the assumption that the occurring processes were identical and stationary.

3. Results

The proposed setup recorded a set of holographic images in the following sequence: holograms in the absence of helium in the observation chamber (with only air inside), holographic images when helium was supplied, images under plasma generation, and finally, the images with helium but when the plasma was turned off.

Therefore, it was possible to compare the following phase changes for refractive indices, for φ_a—phase in air, φ_h—phase in helium, and φ_{hp}—phase in helium in the presence of a plasma pulse, namely, the observation of the phase difference $\Delta\varphi_{ha} = \varphi_h - \varphi_a$ (helium—air), $\Delta\varphi_{hpa} = \varphi_{h+p} - \varphi_a$ (helium + plasma—air), and $\Delta\varphi_p = \varphi_{hp} - \varphi_h$ (helium + plasma—helium), provided that the image acquisition system was properly synchronized.

Figure 8 shows interferograms of gas flow development in air, where the white frame highlights areas in which phase changes were identified. Figure 8a depicts the phase difference with the initial state of helium flow before the appearance of plasma ($\Delta\varphi_{ha}$). The highlighted region in Figure 8b shows changes associated with the onset of plasma formation ($\Delta\varphi_{hpa}$). Further on until plasma generation was completed, the distribution of fringes corresponding to that process is shown in Figure 8c. After the completion of plasma generation, phase difference distribution returned to the initial state as shown in Figure 8d.

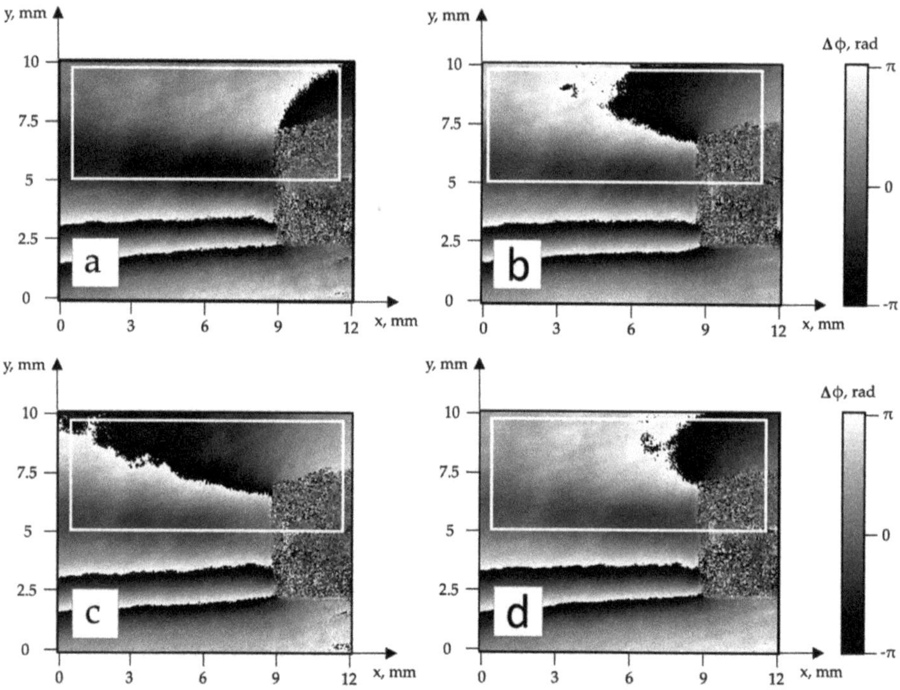

Figure 8. Interference patterns of the refraction observation in the registered sequence of holographic images: (**a**)—helium flow; (**b**)—start of plasma jet generation; (**c**)—plasma generation; (**d**)—end of plasma generation.

It is important to note that the changes in interference patterns associated with the appearance of plasma correspond to the duration of the plasma jet ($t_{plasma} = 750$ ns).

In order to show the influence of the plasma on the refractive index, the phase difference $\Delta\varphi_p$ was evaluated. Figure 9a shows the phase difference for the helium flow only. The phase distribution is uniform, and no other processes appear. Figure 9b shows the phase difference between the moments of gas with plasma and in its absence. This interferogram demonstrates the phase change that is distributed along the direction of plasma jet formation. Figure 9c shows an interferogram of the phase difference at 700 ns after the end of plasma generation. The figure shows an asymmetric intensity distribution, which may be associated with the residual transient processes in the gas after the plasma decay. Based on the phase distribution and without taking into account neutral atoms, we estimated the peak electron concentration at 0.9×10^{16} cm^{-3}.

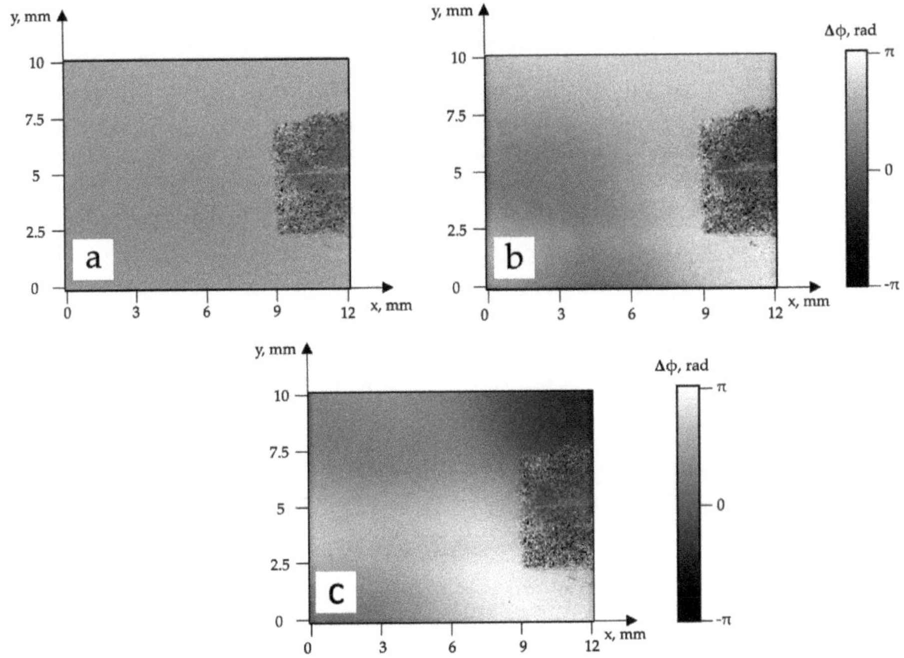

Figure 9. Phase difference: (**a**)—phase difference corresponding to the initial state; (**b**)—phase difference with plasma; (**c**)—phase difference at 700 ns after the end of plasma jet.

4. Discussion

As shown in Figure 4, gas discharge current in pulsed excitation is much shorter than in sine excitation. Short pulses of current flowing through plasma eliminate the overexcitation of gas discharge, while long pauses provide the relaxation of gas-discharge plasma. This method of excitation by short pulses makes it possible to obtain a diffuse plasma jet without contrasting channels and local plasma overheating (Figure 5). On the other hand, long excitation pulses lead to plasma contraction, plasma resistance decrease, and an increase in gas temperature.

Unlike conventional interference methods of plasma observation, the digital holographic interferometry technique makes it possible to digitally evaluate the phase difference over the entire observation field, as well as to fully automate data acquisition and processing [10].

The results do not allow us to definitely conclude that the recorded phase changes are completely related to plasma refraction, and that the neutral atoms, according to expression (4), have no effect on the phase shift. Thus, we have not yet been able to perform an accurate analysis of the electron concentration in the plasma. This method requires further improvements to increase its sensitivity, accuracy and reliability of synchronization. The sensitivity can be improved by switching to another spectral range, which will allow us to take into account the influence of neutral atoms on the resulting interferogram [25]. Increasing the accuracy of synchronization can provide precise locking of the image acquisition related to the beginning of plasma–pulse duration. It is also possible to use ultrahigh-speed digital cameras.

5. Conclusions

In this work, we show that the gas temperature of nonthermal plasma flow depends not only on the configuration of the electrode system, the excitation power level, and the rate of inert gas pumping, but also on the form of the voltage applied to the electrodes. Applying pulses of microsecond duration with long pauses leads to low values of the plasma flow temperature at a minimum level of helium flow rate.

We have also been able to demonstrate in principal the possibility of using digital holographic interferometry to register and observe refraction in pulsed plasma. We were able to see a change in the refractive index both with and without gas jet ionization. Further improvement of the technique will allow us to estimate the concentration of electrons in order to determine the dosage of plasma used in skin treatment.

Author Contributions: Conceptualization, D.V.S. and I.V.A.; methodology, I.V.A.; software, A.M.K.; validation, D.V.S., I.V.A., A.S.I. and A.M.K.; formal analysis, I.V.A.; investigation, A.S.I., A.M.K. and I.V.A.; resources, D.V.S. and I.V.A.; data curation, A.M.K. and A.S.I.; writing—original draft preparation, D.V.S. and I.V.A.; writing—review and editing, A.S.I.; visualization, A.S.I.; supervision, A.M.K. and A.S.I.; project administration, A.S.I.; funding acquisition, D.V.S. and I.V.A. All authors have read and agreed to the published version of the manuscript.

Funding: This research was partly funded by the Ministry of Science and Education of the Russian Federation under project #FZ-2020-0003, "The study of new materials and methods for plasma and phototherapy of cancer, dermatitis and suppurated complications".

Institutional Review Board Statement: Not applicable.

Informed Consent Statement: Not applicable.

Data Availability Statement: The study did not report any data.

Acknowledgments: This work was supported by the Ministry of Science and Higher Education of the Russian Federation (Grant No. FZWM-2020-0003).

Conflicts of Interest: The authors declare no conflict of interest.

References

1. Weltmann, K.D.; Kindel, E.; von Woedtke, T.; Hähnel, M.; Stieber, M.; Brandenburg, R. Atmospheric-pressure plasma sources: Prospective tools for plasma medicine. *Pure Appl. Chem.* **2010**, *82*, 223–1237. [CrossRef]
2. Weltmann, K.-D.; Metelmann, H.-R.; von Woedtke, T. Low temperature plasma applications in medicine. *Europhys. News* **2016**, *47*, 39–42. [CrossRef]
3. Heinlin, J.; Morfill, G.; Landthaler, M.; Stolz, W.; Isbary, G.; Zimmermann, J.L.; Shimizu, T.; Karrer, S. Plasma medicine: Possible applications in dermatology. *J. Dtsch Dermatol. Ges.* **2010**, *8*, 968–976. [CrossRef]
4. Choi, J.H.; Song, Y.S.; Song, K.; Lee, H.J.; Hong, J.W.; Kim, G.C. Skin renewal activity of non- thermal plasma through the activation of β-catenin in keratinocytes. *Sci. Rep.* **2017**, *7*, 6146. [CrossRef]
5. Sosnin, E.A.; Stoffels, E.; Erofeev, M.V.; Kieft, I.E.; Kunts, S.E. The effects of UV irradiation and gas plasma treatment on living mammalian cells and bacteria: A comparative approach. *IEEE Trans. Plasma Sci.* **2004**, *32*, 1544–1550. [CrossRef]
6. Ashurbekov, N.A.; Giraev, K.M.; Shakhsinov, G.S.; Israpov, E.K.; Isaeva, Z.M.; Murtazaeva, A.A.; Rabadanov, K.M. Interaction of low-temperature atmospheric pressure plasma jet mixed with argon and air with living tissues. *J. Phys. Conf. Ser.* **2016**, *1697*, 012044. [CrossRef]
7. Asenjo, J.; Mora, J.; Vargas, A.; Brenes, L.; Montiel, R.; Arrieta, J.; Vargas, V.I. Atmospheric-Pressure Non-Thermal Plasma Jet for biomedical and industrial applications. *J. Phys. Conf. Ser.* **2015**, *591*, 012049. [CrossRef]
8. Zhou, Q.; Cheng, C.; Meng, Y. Electron Density and Temperature Measurement by Stark Broadening in a Cold Argon Arc-Plasma Jet at Atmospheric Pressure. *Plasma Sci. Technol.* **2009**, *11*, 560. [CrossRef]
9. Dreiden, G.V.; Ostrovski Yu, I.; Shedova, E.N.; Zaidel, A.N. Holographic interferograms in stimulated Raman light. *Opt. Commun.* **1971**, *4*, 209–213. [CrossRef]
10. Koopman, D.R.; Sibeneck, H.J.; Jellison, G.; Niessen, W.G. Resonant holography of plasma flow phenomena. *Rev. Sci. Instrum.* **1978**, *49*, 524–525. [CrossRef] [PubMed]
11. Alexeenko, I.V.; Gusev, M.E.; Redkorechev, V.I.; Zyubin, A.Y.; Samusev, I.G. Digital holographic interferometry for the nanodisplacement measurement. In Proceedings of the 2017 Progress In Electromagnetics Research Symposium-Spring (PIERS), St. Petersburg, Russia, 22–25 May 2017; pp. 610–612. [CrossRef]
12. Lomaev, M.I.; Skakun, V.S.; Sosnin, E.A.; Tarasenko, V.F.; Schitz, D.V.; Erofeev, M.V. Excilamps: Efficient sources of spontaneous UV and VUV radiation. *Phys. Uspekhi.* **2003**, *46*, 193–209. [CrossRef]

13. Sosnin, E.A.; Avdeev, S.M.; Tarasenko, V.F.; Skakun, V.S.; Schitz, D.V. KrCl Barrier-Discharge Excilamps: Energy Characteristics and Applications (Review). *Instrum. Exp. Tech.* **2015**, *58*, 309–318. [CrossRef]
14. Lomaev, M.I.; Schitz, D.V.; Skakun, V.S.; Tarasenko, V.F. Influence of excitation pulse form on barrier discharge excilamp efficiency. In *Selected Research Papers on Spectroscopy of Nonequilibrium Plasma at Elevated Pressures*; International Society for Optics and Photonics: Bellingham, WA, USA, 2002; Volume 4460, pp. 38–45. [CrossRef]
15. Erofeev, M.V.; Skakun, V.S.; Sosnin, E.A.; Tarasenko, V.F.; Chernov, E.B. Lifetime of working mixtures of XeCl and KrCl excilamps. *Atmos. Ocean. Opt.* **2000**, *13*, 286–288.
16. Lomaev, M. Determination of energy input in barrier discharge excilamps. *Atmos. Ocean. Opt.* **2001**, *14*, 1005–1008.
17. Vest, C.M. *Holographic Interferometry*; John Wiley&Sons: Hoboken, NJ, USA, 1979; pp. 387–396.
18. Khanzadeh, M.; Jamal, F.; Shariat, M. Experimental investigation of gas flow rate and electric field effect on refractive index and electron density distribution of cold atmospheric pressure-plasma by optical method, Moiré deflectometry. *Phys. Plasmas* **2018**, *25*, 043516. [CrossRef]
19. Kreis, T. *Handbook of Holographic Interferometry: Optical and Digital Methods*; WILEY-VCH Verlag GmbH & Co.: Hoboken, NJ, USA; Volume 370, 2005.
20. Zaidel', A.N. Application of holographic interferometry for plasma diagnostics. *Sov. Phys. Uspekhi* **1986**, *29*, 447. [CrossRef]
21. Schedin, S.; Pedrini, G.; Tiziani, H.J.; Aggarwal, A.K.; Gusev, M.E. High sensitive pulsed digital holography for defect analysis using an in-built laser excitation. *Appl. Opt.* **2001**, *40*, 100–103. [CrossRef]
22. Juptner, W. *Digital Holography*; Springer: Berlin/Heidelberg, Germany; Volume 64, 2005.
23. Pedrini, G.; Zou, Y.L.; Tiziani, H.J. Simultaneous quantitative evaluation of in-plane and out-of-plane deformations by use of a multidirectional spatial carrier. *Appl. Opt.* **1997**, *36*, 786–792. [CrossRef]
24. Takeda, M.; Ina, H.; Kobayashi, S. Fourier-transform method of fringe-pattern analysis for computer-based topography and interferometry. *JosA* **1982**, *72*, 156–160. [CrossRef]
25. Schnars, U. Direct phase determination in hologram interferometry with use of digitally recorded holograms. *JosA* **1994**, *11*, 2011–2015. [CrossRef]
26. Ostrovskaya, G.V.; Ostrovsky, Y.I. IV Holographic Methods of Plasma Diagnostics. *Prog. Opt.* **1985**, *22*, 197–270. [CrossRef]

Article

Cold Atmospheric Pressure Plasma Jet Operated in Ar and He: From Basic Plasma Properties to Vacuum Ultraviolet, Electric Field and Safety Thresholds Measurements in Plasma Medicine

Andrei Vasile Nastuta [1,*] and Torsten Gerling [2,3,*]

[1] Physics and Biophysics Education Research Laboratory (P&B-EduResLab), Biomedical Science Department, Faculty of Medical Bioengineering, 'Grigore T. Popa' University of Medicine and Pharmacy Iasi, Str. M. Kogalniceanu No. 9-13, 700454 Iași, Romania

[2] ZIK *plasmatis*, Leibniz Institute for Plasma Science and Technology (INP), Felix-Hausdorff-Str. 2, 17489 Greifswald, Germany

[3] Diabetes Competence Centre Karlsburg (KDK), Leibniz Institute for Plasma Science and Technology (INP), 17495 Karlsburg, Germany

* Correspondence: andrei.nastuta@gmail.com (A.V.N.); gerling@inp-greifswald.de (T.G.)

Abstract: Application desired functionality as well as operation expenses of cold atmospheric pressure plasma (CAP) devices scale with properties like gas selection. The present contribution provides a comparative investigation for a CAP system operated in argon or helium at different operation voltages and distance to the surface. Comparison of power dissipation, electrical field strength and optical emission spectroscopy from vacuum ultraviolet over visible up to near infrared ((V)UV-VIS-NIR) spectral range is carried out. This study is extended to safety relevant investigation of patient leakage current, induced surface temperature and species density for ozone (O_3) and nitrogen oxides (NO_x). It is found that in identical operation conditions (applied voltage, distance to surface and gas flow rate) the dissipated plasma power is about equal (up to 10 W), but the electrical field strength differs, having peak values of 320 kV/m for Ar and up to 300 kV/m for He. However, only for Ar CAP could we measure O_3 up to 2 ppm and NO_x up to 7 ppm. The surface temperature and leakage values of both systems showed different slopes, with the biggest surprise being a constant leakage current over distance for argon. These findings may open a new direction in the plasma source development for Plasma Medicine.

Keywords: CAP; electric diagnosis; E-field measurements; vacuum-ultraviolet spectroscopy; patient leakage current; power measurement; voltage-charge plot; OES

1. Introduction

Cold atmospheric pressure plasma (CAP) sources are rapidly gaining importance as tools for material processing worldwide, since they are easy to use, technologically simple and environmentally friendly. Applications of these plasmas include: surface modification and deposition, plasma-based synthesis of bio-medical surfaces, decontamination and sterilization, oncotherapy and wound healing. Depending on the application, the plasma source must be tuned as to fulfill the application requirements. This is why it is important to characterize and monitor plasma sources from electrical and optical point of view. CAP's are nowadays versatile tools involved in many applications, starting from basic surface cleaning to Plasma Medicine (covering also plasma pharmaceutics, plasma oncotherapy, plasma bioengineering, etc.) [1–16]; plasma for environment, gas conversion and agriculture [17–20]; plasma for plastics and textiles [21]; plasma for optics and glass [22]; and the future for plasma science and technology [23–25]; plasma for aerospace and automotive [26–28].

The variety of fields of application for cold atmospheric pressure plasmas each has different implications on the application conditions. While a surface treatment in industry

Citation: Nastuta, A.V.; Gerling, T. Cold Atmospheric Pressure Plasma Jet Operated in Ar and He: From Basic Plasma Properties to Vacuum Ultraviolet, Electric Field and Safety Thresholds Measurements in Plasma Medicine. *Appl. Sci.* **2022**, *12*, 644. https://doi.org/10.3390/app12020644

Academic Editor: Emilio Martines

Received: 21 December 2021
Accepted: 6 January 2022
Published: 10 January 2022

Publisher's Note: MDPI stays neutral with regard to jurisdictional claims in published maps and institutional affiliations.

is more flexible with operation temperatures, the reproducibility and long term stability are important conditions. For plasma medicine, it was recently found that the treatment distance impacts the effectivity and the fundamental plasma development [29]. While a close contact treatment produces a conductive operation with the plasma visibly interacting with the surface, at higher distances, the interaction is separated and a non-conductive mode is established. While both modes produce a significant level of reactive species, the composition of species as well as the efficiency of the device change. Furthermore, the literature presents a variety of e.g., plasma jet devices, while some operate in argon and others in helium with no clear comparison between both.

In the past few years, new diagnostic techniques were developed and used for CAP's characterization, from which few have a significant role, due to the importance of the information that it can bring upfront for better understanding of plasma properties in correlation to future applications. The usage of vacuum-ultra-violet spectroscopy [30–39] and of electrical field determination through electro-optical crystals (Pockels electro-optic effect based sensors) [40–44] are such diagnostics. While these diagnostics are of a rather complex nature, the measurement of mainly safety relevant and basic characterization plasma values is a necessity when discussing medical usability. In order to evaluate any potential plasma source for a biomedical application with a potential patient contact, an initiative generated a German industrial norm proposal called DIN-specification 91315 [45]. This approach summarizes potential physical, chemical and biological risk aspects as well as performance criteria from plasma devices in application, and it is frequently used for such testings [46–50]. These measurements allow a standard-compliant assessment of the medical potential.

In this study, we report the results obtained from the diagnosis of a cold atmospheric pressure plasma jet running both in helium and argon. Using sinusoidal high voltage excitation, the plasma is generated using the principle of a dielectric barrier discharge inside a capillary. After leaving the dielectric tube, in air, plasma has a jet shape, with a length of up to few centimeters. The combination of the results retrieved from power measurements, safety characterization (leakage current, temperature, toxic species generation), the electrical field determinations and optical studies revealed new insights on the plasma jets properties and its potential applications (from industrial to biomedical ones).

2. Materials and Methods

2.1. Plasma Source Configuration

The cold atmospheric pressure plasma jet (CAP) is generated using a 100 mm long quartz tube (inner diameter 4 mm, outer diameter 6.1 mm), having two copper tape electrodes (10 mm width, a power (HV) and a grounded (G1) electrode) wrapped on the external surface, with 10 mm gap, and 6 mm till the tube exit nozzle. Depending on the intended measurement, either a metallic (copper plate) or dielectric (alumina covered probe, quartz glass or magnesium fluoride MF_2 window) surface act as the second electrode (G2) placed at distance d and connected to the ground (Figure 1).

Two types of working gas were supplied through the discharge tube: pure helium (He 4.6) or a pure argon (Ar 4.8). The constant gas flow rate of 2 slm (ensuring a laminar flow regime) was regulated through a mass flow controller (MKS Multi Gas Controller 647C or Bronkhorst E-8412 with a F-201CV). In this way, an atmospheric pressure helium or argon plasma jet is generated and operates freely in air. For the comparative nature of the paper, the following abbreviations will be used: He-CAP for the helium jet and Ar-CAP for the argon jet. Long exposure photos were captured using a Nikon D80 camera (CCD: 23.6 × 15.8 mm, 10.2 Mpx, ISO: 100–3200, shutter speed: 30–1/4000 s, 3 fps) equipped with Tamron SP 180 mm f/3.5 Di LD Macro lenses.

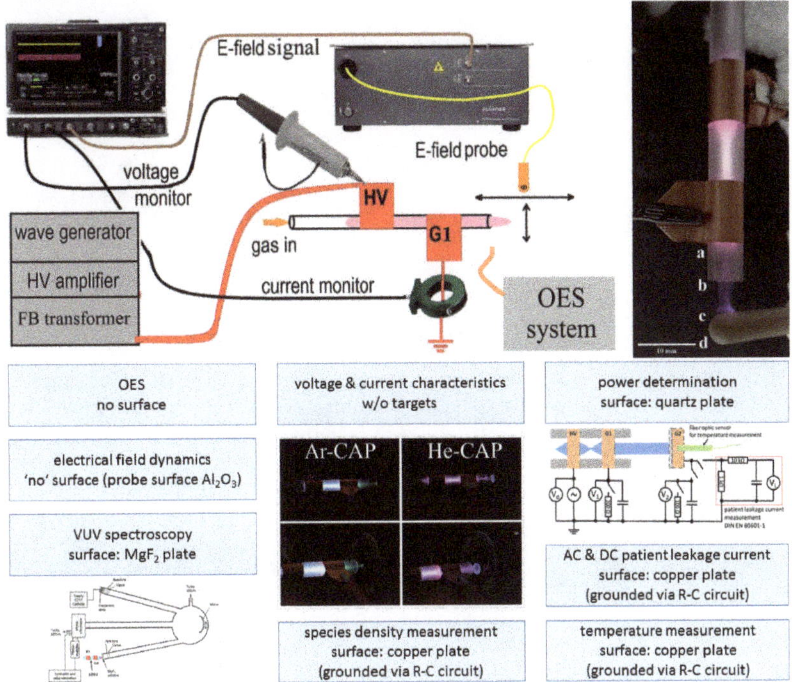

Figure 1. Overview of the experimental arrangements: (**top**) CAP, E-field and VUV set-up with a photo of the CAP and E-field probe; (**bottom**) CAP facing the VUV MgF₂ window with a list of performed diagnostics throughout the manuscript and the description of w/o surface. Leakage current circuit in accordance with [51].

The discharge was driven by a high voltage fly-back transformer (SAXAR MC-FB-001) fed by an amplifier (Amplifier T&C Power Conversion, Inc. AG Amplifier, NY, USA) which was controlled by a function generator (Tektronix AFG 3101). The applied sinusoidal voltage U_a (up to 20 kVpp, frequency 18 kHz) and the total current of the discharge I_d were monitored using voltage and current probes (Tektronix 6015A and CT2, Pearson 6585 current monitor) and a 4-channel digital oscilloscope (Tektronix TDS 5104—1 GHz with 5 Gs/s, and Lecroy Waverunner 640 Zi—4 GHz with 40 Gs/s).

2.2. Characterization of Plasma Source: E-Field Measurement Set-Up

For the electrical field (E-field) measurements, we used a 1 transverse electrical field component optical probe (Kapteos, EoProbe ET5-air, low permittivity ∼3.6, sensitivity 250 mV/m, up to 10 MV$_{rms}$/m, bandwidth 10–12 GHz, spatial resolution of 1 mm, 5 m optical fiber cord) coupled to an electromagnetic field measurement system (Kapteos, EoSense LF 100U-1). The ET5-air is an electrooptic (EO) probe based on an isotropic EO crystal that acts as a transducer (due to the Pockels effect) converting the E-field to be determined into an optical modulation of a laser beam that can be measured via a polarizer and a photodiode [40,44,52–55].

The EO probe was moved axially to the CAP tube and measurements were made at the nozzle exit (marked as 'b', 0 mm), at 5 mm away from nozzle ('c'), at 10 mm away from nozzle ('d') and near the ground electrode, on the tube at −5 mm from the nozzle exit (positioned instead of the letter 'a'), exactly as in Figure 1 and further on presented in the electrical field characterization paragraph.

2.3. Characterization of Plasma Source: Power Measurement Setup

The power dissipation at different locations inside the discharge were determined from the area underneath the voltage-charge plot according to [56,57]. A dielectric quartz plate was placed at a distance d from 0 mm to 20 mm towards the tube edge. Behind the quartz, a copper electrode was connected via a capacity of 4.7 nF towards ground. The capacity was determined by measuring the setup capacity as being between 0.2 pF and 0.6 pF with a LCR meter (GW Instek LCR-8110G), rounded towards 1 pF to correct for uncertainty and multiplying with 1000 ("rule of thumb") [58]. An identical capacity was connected between the second ring electrode and ground. A voltage probe (PP023, Teledyne Lecroy) was placed each to determine the charge accumulation. Together with the high voltage measurement (P 6015A, Tektronix) and an oscilloscope (Waverunner 640 Zi, Teledyne Lecroy), a setup for power determination at different settings of working gas, distance and applied voltage was realized. Data acquisition was set on 50 averages to account for variances due to the wide range of operation voltage. The evaluation of the acquired data sets was performed via a Matlab script (R2012a Matlab, Mathworks Inc., Carlsbad, CA, USA) performing the polyarea function. A comparison with our previous Python scripts [57] were consistent.

2.4. Characterization of Plasma Source: Basic Safety Properties Setup According to DIN SPEC 91315

The temperature of the plasma source was measured at a distance d from 0 mm to 20 mm towards the tube edge. A 10×10 mm^2 copper disc acted as the surface with an extension to host a fiber optic temperature sensor (FOT Labor Kit, LumaSense Technologies, Inc. GmbH, CA, USA). To simultaneously determine the patient leakage current (PLC), an RC-circuit according to DIN 60601-1-6 [51] via UNIMET® 800ST (Bender GmbH & Co. KG, Gruenberg, Germany) was connected to the copper surface [50]. The combined setup was mounted via an *x-y-z*-linear stage system (Qioptiq Photonics GmbH & Co. KG, Gö ttingen, Germany) with 10 μm precision for ideal placement. Data acquisition was performed with a home made Python software PlaDinSpec (made with Python 3.8, Python Software Foundation, Wilmington, DE, USA) [59], measuring and averaging 100 values.

Densities of reactive oxygen and nitrogen species (ROS and RNS) in ambient air were further measured at practical distances (50 mm to 500 mm) along a copper surface with a focus on long living species, namely by using ozone and nitrogen oxides monitors (O_3: APOA-360, NO_x: APNA-370, Horiba Europe GmbH, Oberursel, Germany).

2.5. Characterization of Plasma Source: Spectroscopic VUV to UV-VIS-NIR Measurement Set-Up

The spectral emission of the discharge in the ultra-violet, visible and near infrared spectral range (UV-VIS-NIR) was analyzed by using a fiber optic monochromator (Avantes, AvaSpec 3648, 200–1100 nm range, ~2 nm resolution, 300 lines/mm grating, blazed at 300 nm, deep-UV-detector coated CCD linear array) with a 0.4 mm diameter orifice and a 0.5 m long optical fiber placed at 5 mm from the plasma. The optical emission spectroscopy (OES) of the discharge in the vacuum ultra-violet (V-UV) range was measured by using a calibrated 0.5 m VUV-monochromator (Acton Research Corp., VM-505, 110–280 nm range, ~0.1 nm accuracy, 1200 lines/mm grating, Thorn/EMI 9635 QB detector), described also in [31,48,60]. For these measurements, the CAP was placed in front of the MgF_2 entrance window (cut off wavelength of 115 nm) of the monochromator, at a distance d = 4 mm, as depicted in Figure 1.

3. Results and Discussion

3.1. Basic Electrical and E-Field Characterization of Plasma Source

The voltage and current curves for helium and argon operation show a periodic signal with a dominant symmetry between the positive and negative voltage period (Figure 2). For identical applied voltage, the current shape under argon generates a main discharge current up to 4.2 mA with a duration of about 6 μs and a second discharge current of 4.5 mA and a duration of 3 to 6 μs. The helium cap shows discharge conditions of 2 mA amplitudes and a discharge duration of 12 μs. The estimated electrical power for the Ar-CAP, using the traditional applied voltage multiply by discharge current, revealed a mean value of 0.7 to 1 W, while for He-CAP 0.4 to 0.96 W.

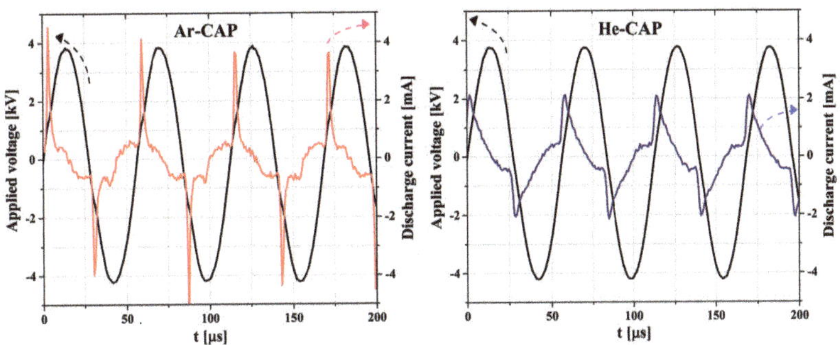

Figure 2. The applied voltage and discharge current for the Ar (**left**) and He (**right**) CAP in mid air (no surface).

While the geometry is rather simple, the operation in argon especially is scarce in literature and comparison hence not easy. Considering the works in a similar frequency range for argon, an investigation on argon metastable density distribution with a capillary wrapped by the grounded electrode and an internal powered electrode correlated phase resolved imaging with the main current peak with the discharge inside the capillary between both electrodes and the second peak with the generation of the guided streamer exiting the capillary [61,62]. Another recent work with a close geometry focused on the atomic auroral oxygen line emission rather than the discharge current [63]. The described phase should be similar, while the geometric properties and the flow range differ. However, the current was only measured on the input line and an overall high displacement current is included.

The helium signals for voltage and current are comparable to characteristics in other studies indicating the jet current signal relates to the guided streamer propagation [64,65]. Our previous works with a similar basic geometry showed discharge peaks of shorter duration [57,66,67]. When comparing images of the different systems, the effluent differs from the previous system with a filamented effluent and the present system with a rather diffuse discharge (as in discharge photos from Figure 1).

Electric field diagnosis

The presented voltage and current characteristics in Figure 2 are for the case without a surface. Meanwhile, the signals measured when placing the electrical field probe represent the operation with a dielectric target placed in front (Figure 3). As it is observed, the basic characteristics are similar, with comparable time scales and slightly increased amplitudes. The symmetry between negative and positive half phase of the voltage period remains as well. The main change is, however, the occurrence of more consecutive discharge peaks after the first, increasing in amplitude and number in accordance with the applied voltage. Operation in argon generates up to three peaks per half cycle while helium goes up to two peaks, respectively. Consecutive discharge peaks like these were already reported to correlate with weaker but dominantly repeating discharge dynamics [31].

In a previous study for an in principle identical electrode setting, we reported the generation of an inverse current pulse following the primary discharge pulse which correlated with the bullet propagation and charge exchange processes [66,67]. While voltage settings and electrode geometries equal the present setup, changes in capillary dimension, implying flow condition changes, dielectric properties and electrode dimensions are present.

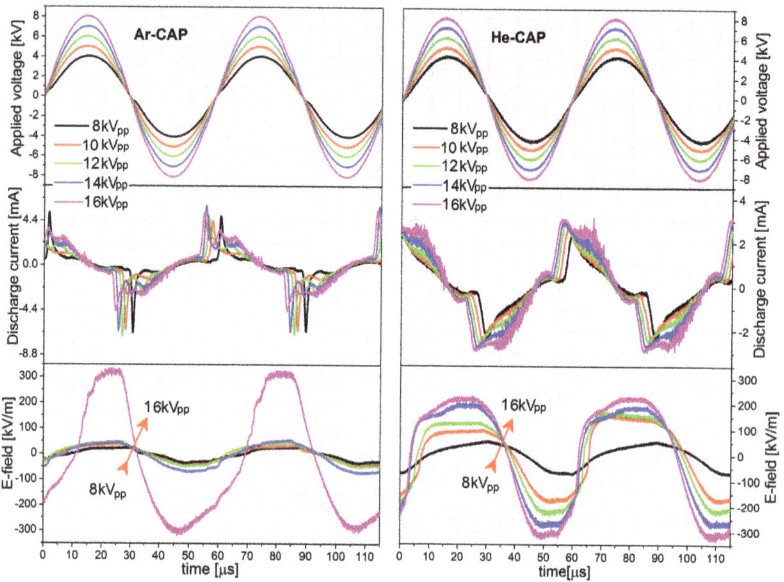

Figure 3. The applied voltage, discharge current and electrical field data for the Ar (**left**) or He (**right**) CAP facing a dielectric surface at a distance d = 10 mm (probe head).

The electrical field probe shows an increase of electrical field amplitudes with increasing applied voltage. The signal correlates with events observed on the current signal. An enhanced electrical field amplitude from 25 kV/m up to 320 kV/m in argon and from 50 kV/m up to 300 kV/m in helium is achieved (see Figure 4). For Ar-CAP and He-CAP operation, a plateau dominates each discharge cycle. While the discharge peak is still proven to result from the discharge between the two ring electrodes around the capillary, the guided streamers leaving the system follow shortly after [67]. This correlates with the observed peaks in the electrical field strength measurements that are delayed by few microseconds from the main discharge peak.

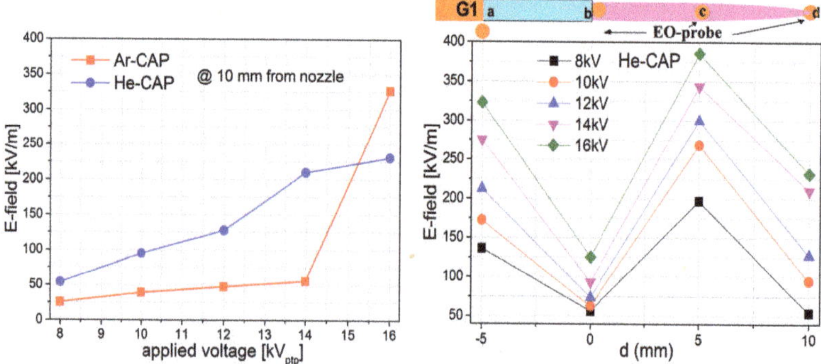

Figure 4. Electrical field strength as function of applied voltage for Ar-CAP and He-CAP (**left**) and for He-CAP as function of distance from the G1 electrode to 10 mm in front of the exit nozzle (**right**). The EO-probe was moved axially to the CAP tube and measurements were made at the nozzle exit ('b' = 0 mm), at 5 mm away from nozzle ('c'), at 10 mm away from nozzle ('d') and near the ground electrode, on the tube at −5 mm from the nozzle exit (positioned instead of the letter 'a'), exactly as in the right figure (orange circles = EO-probe).

The probe is actively placed into the discharge channel and the dielectric surface acts as a third electrode, including charging and discharging through the plasma channel as measured previously [66]. While the maximal value correlates with measured and simulated values from other groups in helium or argon discharges [64,68–70], the high plateau is interpreted to result more from charge accumulation than from ongoing discharge dynamics.

3.2. Determination of Dissipated Power

The representative acquired voltage charge plots (Q-V plots) with a quartz substrate placed 5 mm away are shown in Figure 5. For the grounded ring electrode around the capillary, the voltage-charge curve does not resemble a parallelogram but is similar to the shape of another system [57]. The shape remains similar with increased applied voltage, but the charge amplitude increases for argon and helium operation. The quartz surface is placed on top of a grounded copper plate and the measured voltage-charge curve is mostly rounded with one more or less pronounced jump on the rising slope.

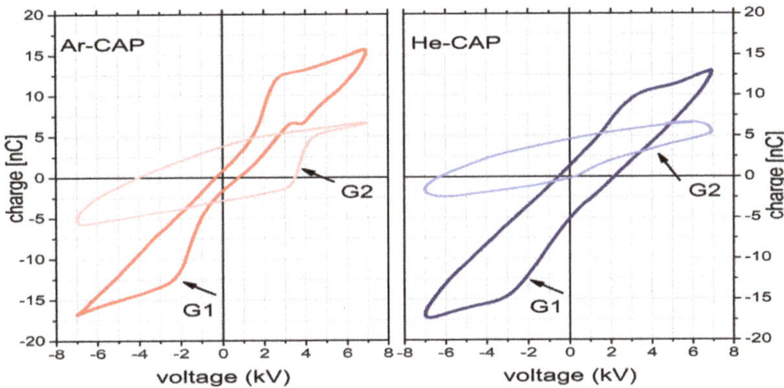

Figure 5. Charge-voltage plot for device operation in argon (**left**) and helium (**right**), each for an applied voltage of 14 kV$_{pp}$ and a quartz plate at 5 mm distance.

From the enclosed loop of the Q-V plot, one can determine the energy of a cycle and by multiplying with the frequency the determination of power is possible [71]. The respective energy values are 56 µJ (G1), 63 µJ (G2) for argon and 76 µJ (G1), 54 µJ (G2) for helium operation, representing a power dissipation of 1.00 W (G1), 1.15 W (G2) for argon and 1.38 W (G1), 0.97 W (G2) for helium. Previous investigation already states that plasma jet systems are not closed and energy is consumed by further reactions of the plasma inside as well as outside the capillary like light emission, heat production, and reactive species generation [56,72].

In Figure 6a–c, the determined power is presented at given applied voltage over different distances as well as for given distance over increased applied voltage. When changing the distance at constant voltage, a constant power is observed for helium and argon while an increase in power is found when the voltage amplitude is increased.

Between G1 and G2 in each case, a tendency is observed for an increase in power dissipation for G1 in exchange for a decrease in power dissipation for G2. Increasing the applied voltage at a fixed distance leads to a continuous increase of dissipated power in G1 and G2. For the highest investigated voltage settings, G2 surpasses G1 in dissipated power. The power range overall between helium and argon shows comparable values for G2, while G1 in helium surpasses argon by a total amount of 0.7 W with a final value around 2.1 W. The total power dissipation achieved within this measurements is 3.9 W for argon and 4.7 W for helium operation. For further investigations, the applied voltage was set to either 8 kV$_{pp}$ with identical power dissipation values or 14 kV$_{pp}$ with comparable G2 values yet already increased G1 value for helium.

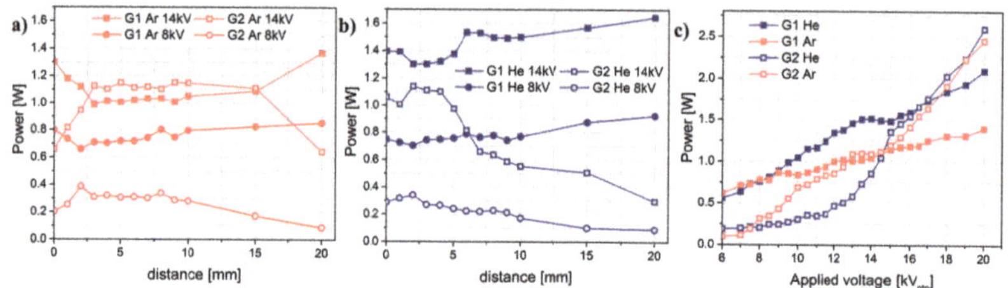

Figure 6. Dissipated power as a function of (**a**,**b**) distance and (**c**) applied voltage (at d = 5 mm) for helium and argon operation. G1 and G2 describe the measurement position of either grounded ring electrode at the capillary (G1) and ground surface (G2).

3.3. Basic Safety Properties

When considering dissipated power values of up to 5 W in the system, the range of safety relevant parameters should be considered to qualify potential application relevant settings. The results on patient leakage current and temperature of a treated copper surface are shown in Figure 7. According to DIN SPEC 91315 [45] and the norms within the leakage current, the safety threshold is set to 0.1 mA in AC and 0.01 mA in DC mode. Temperature should not cross 40 °C to avoid denaturation of proteins.

Results within a certain range of these thresholds can be adjusted by further adjustment of system parameters, while significant overvalues cast doubts on a direct medical application. In addition, these values are correlated with efficacy parameters, so that a device is always both safe and effective and not just either. Here, that value is accessed via reactive oxygen and nitrogen species (ROS and RNS) measurements. While intuitively a higher power value implies an improved effectivity, the present system will show for the first time that this correlation is not that simple. Concerning the AC patient leakage current at different voltage amplitudes, the overall range for helium and argon are comparable, with no DC leakage current detectable in either He-CAP and Ar-CAP. While the He-CAP delivers AC leakage current values of 10 μA at 7 kV$_{pp}$, the Ar-CAP delivers 80 μA at identical settings.

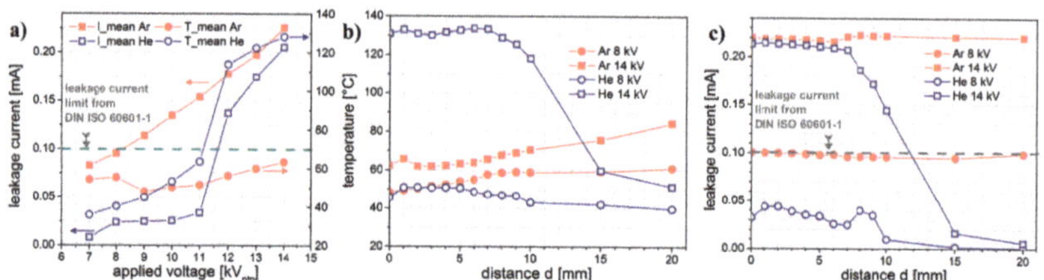

Figure 7. Measured values of patient leakage current and temperature for different voltage and distance settings in helium and argon. (**a**) shows the AC-leakage current for various applied voltages and the temperature for d = 5 mm to a copper surface (the DC-leakage current is below the detection limit of 1 μA for all conditions); (**b**) shows the temperature as a function of distance for two applied voltages (8 and 14 kV$_{pp}$); (**c**) shows the AC-leakage current for different distances for two investigated voltages (8 and 14 kV$_{pp}$).

In the Ar-CAP, the values rise continuously with increased voltage up to 250 μA. In contrast, the He-CAP provides a constant leakage current up to 11 kV$_{pp}$ and a jump with a further on increase up to 200 μA. The surface temperature rises in dependence on the leakage current curve for He-CAP with values from 35 °C up to 130 °C. For Ar-CAP, the temperature ranges between 40 °C and 60 °C. In case of Ar-CAP, the temperature could be reduced by increasing the gas flow to optimize for a potential medical setting with compliant leakage current values at lower amplitudes. For the He-CAP, the range of operation voltage with a compliant set of leakage current and temperature is increased with a more pronounced threshold transgression.

A variation of the distance for these measurements considers application relevant uncertainties since a fixed treatment distance in complex geometries is neither possible in hand-held operation nor in case of complex geometries with robotic operation. For the temperature values in He-CAP, a drop is observed for longer distance of over 10 mm, while prior to it a mostly constant value is detected (Figure 7b). In Ar-CAP, interestingly, the value rises slightly over the whole distance for both investigated voltages. On the contrary, for Ar-CAP, the cooling effect of the working gas is reduced the further away the treated surface is moved. Hence, here again an adjustment of gas flow range could impact the temperature amplitude. For He-CAP, it seems that most of the heat is produced inside the capillary with a volume discharge while outside mainly diffusion and buoyancy cools the output gas from the capillary [72,73].

When the temperature drops for the He-CAP, a correlating drop in leakage current is observed in Figure 7c as described previously for the voltage variation (Figure 7a). For the Ar-CAP, however, a constant leakage current is measured over the whole investigated range of 20 mm with a filament reaching the surface continuously. A constant leakage value over the whole range is not reported in literature yet and could neutralize distance uncertainties in applications caused by user errors or complex geometries.

The correlation of He-CAP and Ar-CAP operation at different applied voltage amplitudes and ROS and RNS density production is presented in Figure 8 with ozone and nitrogen oxides as semi stable products from the chemistry between plasma and neutral gas [74].

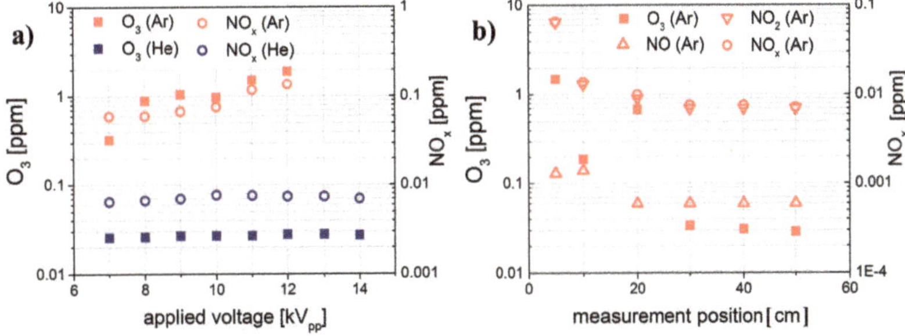

Figure 8. Distribution of ROS and RNS production for argon and helium operation for a conductive surface at d = 5 mm for different (**a**) applied voltages and (**b**) different measurement positions of the gas inlet at 8 kV$_{pp}$.

For Ar-CAP, the density is exponentially increasing with the increased voltage and hence scales with the dissipated power. He-CAP, on the other hand, is not affected by the increased applied voltage or the correlating increase in dissipated power. The measurements for He-CAP were within range of the environmental offset and hence the species density could not be detected. A further investigation considered the ROS and RNS species density over the distance to the device. For He-CAP, the densities remained below the detection limit and the results are hence not shown. For Ar-CAP, a strong decrease over distance with a local maximum at about 20 cm is measured and in accordance with previous measurements before the values drop below the thresholds [46,50]. For both cases, He-CAP and Ar-CAP, the ROS and RNS content will be further evaluated qualitatively with OES in the next chapter.

3.4. Optical Diagnosis of the Plasma Source

Using OES on the He-CAP and Ar-CAP allows a relative access to the generation and excitation of reactive oxygen and nitrogen species within the plasma. If OES is used in the UV, VIS to NIR spectral range, the existence of atomic lines from hydrogen, nitrogen and oxygen as well as molecular bands of nitrogen oxide NO, hydroxyl OH, molecular nitrogen (N_2) and molecular nitrogen ion (N_2^+) can be detected [31,75]. The measurement of radiation in the VUV further provides information on atomic lines from nitrogen, hydrogen and oxygen as well as the molecular argon excimer continuum and even ozone [30,76,77].

3.4.1. VUV-Radiance Measurements

The emission spectra in the 110–200 nm range using a VUV scanning monochromator are presented in Figure 9. A set of atomic lines of hydrogen (121.5 nm), oxygen (130.5 nm) and nitrogen (149.5 nm and 174.5 nm) is detected, emitting powerful VUV radiation with energies from 7 eV to 9.5 eV. In addition, the continuous profile of the molecular argon excimer is detected for the Ar-CAP from 110 nm up to 140 nm. These species are of interest due to their energy that can be transferred to the surface under treatment. These values of dissociation energies under consideration are between 4.3 eV (for atomic H and Ar) to 9.8 eV (for atomic N and O) [78,79]. It has to be noted that the helium excimer is located below the detection limit of the applied system and hence no information on it can be stated here [37].

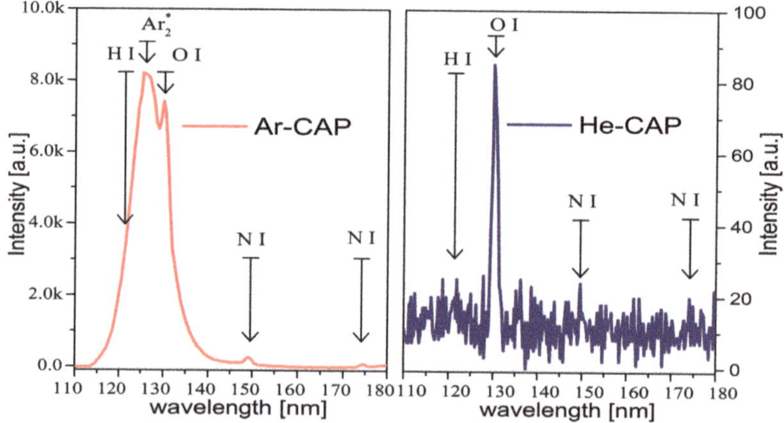

Figure 9. VUV spectra of the discharge in argon (**left**) and helium (**right**) for an applied voltage of 8 kV$_{pp}$ and a distance to the MgF$_2$ window of 4 mm. Ar-CAP generated a spectral radiance of up to 350 mW/(nm mm^2 sr) at around 125 nm.

When comparing the intensities of He-CAP and Ar-CAP, a much more intense VUV emission in argon could be found. While Ar-CAP radiance could be quantified to reach up to 350 mW/(nm mm^2 sr), no radiance level could be determined for He-CAP. The VUV radiance results present another indication that the RONS production is better observed for the Ar-CAP.

3.4.2. OES Measurements

Emission spectra in the 200–900 nm spectral range of the plasma jets show signatures of molecular and atomic excited species (Figure 10). The molecular bands are assigned to nitrogen oxide (NO$_\gamma$ system), hydroxyl radicals (OH), neutral nitrogen molecules (N_2) and nitrogen molecular ion (N_2^+). The NO$_\gamma$ system is observed in the 200–280 nm region with characteristic bands, as for the OH radical, their signature is between 306–310 nm. Bands of N_2 are observable in the spectral range of 315–380 nm and 399–405.9 nm. The N_2^+ have bands at 391.4 nm, 427.8 nm and 470.0 nm. The atomic lines form the emitted spectra are assigned to helium atoms (lines at 588.8 nm, 668.5 nm, 706.6 nm and 727.5 nm) or to argon atoms (lines between 680 to 850 nm) as working gas, but also to oxygen atoms (lines at 777.8 nm and 845.5 nm) as products of ambient O_2 and H_2O dissociation [31,65,67,80–82].

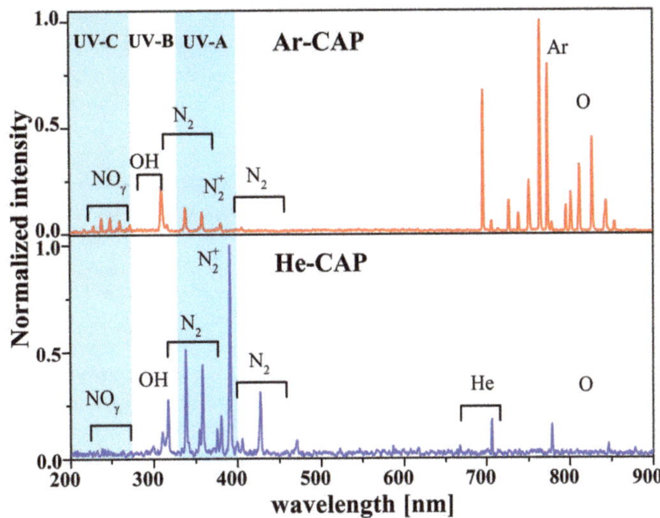

Figure 10. The emitted light in the UV-Vis-NIR from CAP in Ar (red line) and He (blue line), acquired at 5 mm from the nozzle exit, in mid air, 8 kV$_{pp}$ applied voltage.

In the 220–280 nm spectral region (the UV-C), we observed only for the Ar-CAP the emission signatures of the nitric oxide molecular band (NO$_\gamma$, in Figure 10), well known for their role as signaling molecules involved in wound healing [83]. The OH and O lines are observed for He-CAP and Ar-CAP, while the line intensity was higher in Ar-CAP. This observation supplements the data acquired from VUV measurements as well as the species density measurements. Another characteristic of the OES spectra of the He-CAP is that the presence of excited nitrogen species is dominantly observed for both N$_2$ and N$_2^+$ compared to Ar-CAP due to the discharge mechanisms, mainly based on penning effect for helium. In particular, the nitrogen molecular ion is observed in helium CAP as the most intense line. Nonetheless, in the Ar discharge, the NO bands are observed at a lower temperature than in He discharges.

Variations of the atomic lines and bands emitted intensities for the plasma jet operating in air, depending on the working gas, can be easily observed in Figure 10. Moreover, the importance of such plasma active/reactive species is to be considered also from the energetic point of view of these species. Thus, energy values from 4 eV up to 23 eV are attributed to OH radicals, N$_2$ and N$_2^+$ molecules, hydrogen He, Ar and O atoms, as reported by [78,84].

The results until now show that both devices operated at comparable power values, identical flow and voltage settings generate species with a different efficiency. One way to explain this tendency is via the basic plasma properties of rotational temperature and by considering the given temperatures involved in the chemistry. The ratio between O$_3$ and NO$_x$ is known to depend on the temperature [85]. However, our measurement of the surface temperature, via a temperature optical fiber, in Figure 7 shows a temperature range from 35 °C up to 135 °C for the He-CAP and a more or less constant temperature for the Ar-CAP in between the helium range. According to neutral gas temperature alone, this effect of species densities can not be explained. Since the studied plasma source is at atmospheric pressure, a non-thermal plasma is assumed. Therefore, the temperature of the electrons is significantly higher than the temperature of the ions and of the neutrals. A frequently used assumption considers the rotational temperature of neutral molecules to be close to the neutral temperature. Therefore, a further investigation of the rotational and vibrational temperatures in He-CAP and Ar-CAP will be performed. Moreover, using simulation software like *Lifbase* [86] and *Spectrum Analyzer* [87], we determined from the acquired plasma emission spectra the characteristic plasma temperatures, such as the rotational temperature of OH radicals (T_{rot}(OH)), the rotational temperature of nitrogen molecular ions N_2^+ ($T_{rot}(N_2^+)$) and the vibrational temperature of nitrogen molecules N_2 ($T_v(N_2)$).

As a result, we estimated for Ar-CAP a T_{rot}(OH) around 420 ± 25 K ($\sim$146 °C), $T_{rot}(N_2^+)$ 450 ± 50 K ($\sim$176 °C), and $T_v(N_2)$ 1766 ± 167 K. Moreover, for He-CAP, the T_{rot}(OH) around 420 ± 20 K ($\sim$146 °C), $T_{rot}(N_2^+)$ 425 ± 25 K ($\sim$151 °C), and $T_v(N_2)$ 2381 ± 603 K. In most cases, cold atmospheric pressure plasmas are considered non-thermal plasmas, and thus the rotational temperatures are often equivalent to those of the working gas. These estimated values of the rotational temperatures of OH radicals and molecular ions of N_2^+ are slightly higher than those measured using FOT optical fiber. No significant rotational temperature difference was to be observed between the Ar and He-CAP, this leading to the idea that these two plasmas are at the same temperature, when ignited at the same parameters.

However, it must be borne in mind that these values correspond to '*spectroscopic temperatures*' referring to the energies that those plasma species (OH, N_2 and N_2^+) can be achieved by various processes (dissociation, Penning, etc.) and can be further used in the plasma environment to initiate various physico-chemical processes that can take place on the surface or in the plasma volume. The noticeable difference is within the vibration temperature, when He-CAP T_v is almost 35% higher than that estimated for Ar-CAP. In addition, it might be that a higher $T_v(N_2)$ found for He-CAP to be to blamed for the small or even non-existent production of NO_x and O_3.

4. Conclusions

The present study shows for the first time a direct comparison of an identical setup within an identical range of operation settings with helium and argon operation. The range of applied voltage ranged from 8 kV_{pp} up to 20 kV_{pp} for both He-CAP and Ar-CAP, while the resulting dissipated power rises linearly up to 5 W for both systems. An investigation of the electrical field amplitude for He-CAP and Ar-CAP showed peak values of 300 kV/m and 320 kV/m, respectively.

As expected of such high power consumption, the temperature induced on a conductive surface ranges from 40 °C up to 130 °C for He-CAP and from 50 °C up to 85 °C for Ar-CAP. The temperature slope over distance is constant and further reduction can be achieved by parameter adjustment of e.g., gas flow. The patient leakage current for He-CAP ranges from 10 μA up to 200 μA over applied voltage and drops over distance. The leakage current measurement for Ar-CAP, however, shows a unique slope with a linear increase over applied voltage and a constant slope over distance. This was never observed in literature before and offers unique conditions for surface treatments with constant temperature and leakage values at changing treatment distance over a range of up to the investigated 20 mm.

The production of reactive species was investigated by emission spectroscopy in the (V)UV-VIS-NIR range. Atomic nitrogen and oxygen could be detected in VUV for both systems; however, Ar-CAP showed much stronger signals in all spectral ranges. In addition, the nitrogen oxide emission in the UV range was only measured for argon operation. In accordance with the OES results, species densities of O_3 and NO_x could only be measured for Ar-CAP, while, for He-CAP, they were below the detection limits.

Plasma 'spectroscopic temperature' values, rotational temperatures of OH and N_2, were estimated to be the same for Ar-CAP and He-CAP, around 420 K (160 °C), near the maximum values measured via FOT (130 °C). The vibrational temperatures N_2^+ revealed different values (higher for He-CAP than Ar-CAP), mainly due to different plasma ignition mechanisms. In addition, the higher T_v in the case of helium plasma might be associated with lower production of NO_x and O_3.

At identical CAP settings of geometry and operation conditions, overall, the Ar-CAP shows the more promising parameters, a constant leakage current over distance and a significant species production. Meanwhile, He-CAP generates comparable values in power, electrical field, temperature and leakage current with no detectable reactive species density.

It is an interesting observation and further investigations are clearly required to trace the pathways of the helium power consumption leading to the presented observations. In the meantime, the Ar-CAP provides the interesting property of constant leakage current over distance that indicates a voltage dependent current limitation of the conductive channel that is generated by the discharge. The He-CAP might be an interesting approach for a high electrical field application with no toxic species as byproducts. These findings may open a new direction in the plasma source development for Plasma Medicine.

Nevertheless, more experiments should be made, like surface charge production, ultra-fast photography or mass spectrometry in order to have the whole understanding of the plasma source.

Author Contributions: Conceptualization, A.V.N. and T.G.; methodology, A.V.N. and T.G.; validation, A.V.N. and T.G.; formal analysis, A.V.N. and T.G.; investigation, A.V.N. and T.G.; data curation, A.V.N. and T.G.; writing—review and editing, A.V.N. and T.G. All authors have read and agreed to the published version of the manuscript.

Funding: The research of A.V.N. was funded by the UEFISCDI, PNCDI III, project PN-III-P1-1.1-MC-2017-1098 and by 'Gr. T. Popa' University of Medicine and Pharmacy Iasi, project GI No. 30339/28.12.2017. T.G. research was funded by the German Ministry of Education and Research (BMBF 13N13960) and by the Ministry of Education, Science and Culture of the State of Mecklenburg-Vorpommern (AU 15 001).

Institutional Review Board Statement: Not applicable.

Informed Consent Statement: Not applicable.

Data Availability Statement: Raw data may be available, on reasonable request, from the authors.

Acknowledgments: The authors would like to acknowledge *Laura Vilardell Scholten* for writing the customized code *"PlaDinSpec"* for CAP temperature and leakage current analysis. Furthermore, *Peter Druckrey, Peter Holtz, Christiane Meyer* and *Rüdiger Titze* are acknowledged for technical assistance.

Conflicts of Interest: The authors declare no conflict of interest.

References

1. Han, J. Review of major directions in non-equilibrium atmospheric plasma treatments in medical, biological, and bioengineering applications. *Plasma Med.* **2013**, *3*, 175–243. [CrossRef]
2. Von Woedtke, T.; Reuter, S.; Masur, K.; Weltmann, K.D. Plasmas for medicine. *Phys. Rep.* **2013**, *530*, 291–320. [CrossRef]
3. Topala, I.; Nastuta, A. Helium atmospheric pressure plasma jet: Diagnostics and application for burned wounds healing. In *Plasma for Bio-Decontamination, Medicine and Food Security*; Machala Z., Hensel K., Akishev Y., Ed.; Springer: Berlin/Heidelberg, Germany, 2012; pp. 335–345. [CrossRef]
4. Bruggeman, P.J.; Kushner, M.J.; Locke, B.R.; Gardeniers, J.G.; Graham, W.; Graves, D.B.; Hofman-Caris, R.; Maric, D.; Reid, J.P.; Ceriani, E.; et al. Plasma–liquid interactions: A review and roadmap. *Plasma Sources Sci. Technol.* **2016**, *25*, 053002. [CrossRef]
5. Bekeschus, S.; Mueller, A.; Miller, V.; Gaipl, U.; Weltmann, K.D. Physical plasma elicits immunogenic cancer cell death and mitochondrial singlet oxygen. *IEEE Trans. Radiat. Plasma Med. Sci.* **2017**, *2*, 138–146. [CrossRef]
6. Bekeschus, S.; Wende, K.; Hefny, M.M.; Rödder, K.; Jablonowski, H.; Schmidt, A.; von Woedtke, T.; Weltmann, K.D.; Benedikt, J. Oxygen atoms are critical in rendering THP-1 leukaemia cells susceptible to cold physical plasma-induced apoptosis. *Sci. Rep.* **2017**, *7*, 2791. [CrossRef] [PubMed]
7. Gerber, I.; Mihaila, I.; Hein, D.; Nastuta, A.; Jijie, R.; Pohoata, V.; Topala, I. Time Behaviour of Helium Atmospheric Pressure Plasma Jet Electrical and Optical Parameters. *Appl. Sci.* **2017**, *7*, 812. [CrossRef]
8. Laroussi, M.; Lu, X.; Keidar, M. Perspective: The physics, diagnostics, and applications of atmospheric pressure low temperature plasma sources used in plasma medicine. *J. Appl. Phys.* **2017**, *122*, 020901. [CrossRef]
9. Dai, X.; Bazaka, K.; Richard, D.J.; Thompson, E.R.W.; Ostrikov, K.K. The emerging role of gas plasma in oncotherapy. *Trends Biotechnol.* **2018**, *36*, 1183–1198. [CrossRef] [PubMed]
10. Bekeschus, S.; Favia, P.; Robert, E.; von Woedtke, T. White paper on plasma for medicine and hygiene: Future in plasma health sciences. *Plasma Process. Polym.* **2019**, *16*, 1800033. [CrossRef]
11. Braný, D.; Dvorská, D.; Halašová, E.; Škovierová, H. Cold Atmospheric Plasma: A Powerful Tool for Modern Medicine. *Int. J. Mol. Sci.* **2020**, *21*, 2932. [CrossRef] [PubMed]
12. Busco, G.; Robert, E.; Chettouh-Hammas, N.; Pouvesle, J.M.; Grillon, C. The emerging potential of cold atmospheric plasma in skin biology. *Free. Radic. Biol. Med.* **2020**, *161*, 290–304. [CrossRef] [PubMed]

13. Hahn, V.; Grollmisch, D.; Bendt, H.; von Woedtke, T.; Nestler, B.; Weltmann, K.D.; Gerling, T. Concept for Improved Handling Ensures Effective Contactless Plasma Treatment of Patients with kINPen® MED. *Appl. Sci.* **2020**, *10*, 6133. [CrossRef]
14. Keidar, M. *Introduction: Plasma for Cancer Therapy*; Springer: Berlin/Heidelberg, Germany, 2020; pp. 1–13. [CrossRef]
15. Liu, D.; Zhang, Y.; Xu, M.; Chen, H.; Lu, X.; Ostrikov, K. Cold atmospheric pressure plasmas in dermatology: Sources, reactive agents, and therapeutic effects. *Plasma Process. Polym.* **2020**, *17*, 1900218. [CrossRef]
16. von Woedtke, T.; Emmert, S.; Metelmann, H.R.; Rupf, S.; Weltmann, K.D. Perspectives on cold atmospheric plasma (CAP) applications in medicine. *Phys. Plasmas* **2020**, *27*, 070601. [CrossRef]
17. Brandenburg, R.; Bogaerts, A.; Bongers, W.; Fridman, A.; Fridman, G.; Locke, B.R.; Miller, V.; Reuter, S.; Schiorlin, M.; Verreycken, T.; et al. White paper on the future of plasma science in environment, for gas conversion and agriculture. *Plasma Process. Polym.* **2019**, *16*, 1700238. [CrossRef]
18. Adhikari, B.; Adhikari, M.; Park, G. The Effects of Plasma on Plant Growth, Development, and Sustainability. *Appl. Sci.* **2020**, *10*, 6045. [CrossRef]
19. Varilla, C.; Marcone, M.; Annor, G.A. Potential of Cold Plasma Technology in Ensuring the Safety of Foods and Agricultural Produce: A Review. *Foods* **2020**, *9*, 1435. [CrossRef]
20. Huzum, R.; Nastuta, A.V. Helium Atmospheric Pressure Plasma Jet Source Treatment of White Grapes Juice for Winemaking. *Appl. Sci.* **2021**, *11*, 8498. [CrossRef]
21. Cvelbar, U.; Walsh, J.L.; Černák, M.; de Vries, H.W.; Reuter, S.; Belmonte, T.; Corbella, C.; Miron, C.; Hojnik, N.; Jurov, A.; et al. White paper on the future of plasma science and technology in plastics and textiles. *Plasma Process. Polym.* **2019**, *16*, 1700228. [CrossRef]
22. Šimek, M.; Černák, M.; Kylián, O.; Foest, R.; Hegemann, D.; Martini, R. White paper on the future of plasma science for optics and glass. *Plasma Process. Polym.* **2019**, *16*, 1700250. [CrossRef]
23. Weltmann, K.D.; Kolb, J.F.; Holub, M.; Uhrlandt, D.; Šimek, M.; Ostrikov, K.; Hamaguchi, S.; Cvelbar, U.; Černák, M.; Locke, B.; et al. The future for plasma science and technology. *Plasma Process. Polym.* **2019**, *16*, 1800118. [CrossRef]
24. Lietz, A.M.; Damany, X.; Robert, E.; Pouvesle, J.M.; Kushner, M.J. Ionization wave propagation in an atmospheric pressure plasma multi-jet. *Plasma Sources Sci. Technol.* **2019**, *28*, 125009. [CrossRef]
25. Slikboer, E.; Sobota, A.; Garcia-Caurel, E.; Guaitella, O. In-situ monitoring of an organic sample with electric field determination during cold plasma jet exposure. *Sci. Rep.* **2020**, *10*, 13580. [CrossRef] [PubMed]
26. Keidar, M.; Beilis, I. *Plasma Engineering: Applications from Aerospace to Bio and Nanotechnology*; Academic Press: London, UK, 2013. [CrossRef]
27. Sim, K.B.; Baek, D.; Shin, J.H.; Shim, G.S.; Jang, S.W.; Kim, H.J.; Hwang, J.W.; Roh, J.U. Enhanced Surface Properties of Carbon Fiber Reinforced Plastic by Epoxy Modified Primer with Plasma for Automotive Applications. *Polymers* **2020**, *12*, 556. [CrossRef] [PubMed]
28. GmbH, P. Plasma processes reduce costs in automotive manufacturing. *IST Int. Surf. Technol.* **2020**, *13*, 28–29. [CrossRef]
29. Miebach, L.; Freund, E.; Clemen, R.; Weltmann, K.D.; Metelmann, H.; von Woedtke, T.; Gerling, T.; Wende, K.; Bekeschus, S. Conductivity augments ROS and RNS delivery and tumor toxicity of an argon plasma jet. *Free Radic. Biol. Med.*, in press.
30. Brandenburg, R.; Lange, H.; von Woedtke, T.; Stieber, M.; Kindel, E.; Ehlbeck, J.; Weltmann, K.D. Antimicrobial effects of UV and VUV radiation of nonthermal plasma jets. *IEEE Trans. Plasma Sci.* **2009**, *37*, 877–883. [CrossRef]
31. Gerling, T.; Nastuta, A.; Bussiahn, R.; Kindel, E.; Weltmann, K. Back and forth directed plasma bullets in a helium atmospheric pressure needle-to-plane discharge with oxygen admixtures. *Plasma Sources Sci. Technol.* **2012**, *21*, 034012. [CrossRef]
32. Schneider, S.; Lackmann, J.W.; Ellerweg, D.; Denis, B.; Narberhaus, F.; Bandow, J.E.; Benedikt, J. The role of VUV radiation in the inactivation of bacteria with an atmospheric pressure plasma jet. *Plasma Process. Polym.* **2012**, *9*, 561–568. [CrossRef]
33. Reuter, S.; Sousa, J.S.; Stancu, G.D.; van Helden, J.P.H. Review on VUV to MIR absorption spectroscopy of atmospheric pressure plasma jets. *Plasma Sources Sci. Technol.* **2015**, *24*, 054001. [CrossRef]
34. Jablonowski, H.; Bussiahn, R.; Hammer, M.; Weltmann, K.D.; von Woedtke, T.; Reuter, S. Impact of plasma jet vacuum ultraviolet radiation on reactive oxygen species generation in bio-relevant liquids. *Phys. Plasmas* **2015**, *22*, 122008. [CrossRef]
35. Es-sebbar, E.t.; Bénilan, Y.; Fray, N.; Cottin, H.; Jolly, A.; Gazeau, M.C. VUV Spectral Irradiance Measurements in H2/He/Ar Microwave Plasmas and Comparison with Solar Data. *Astrophys. J. Suppl. Ser.* **2019**, *240*, 7. [CrossRef]
36. Zhang, Y.; Ishikawa, K.; Mozetič, M.; Tsutsumi, T.; Kondo, H.; Sekine, M.; Hori, M. Polyethylene terephthalate (PET) surface modification by VUV and neutral active species in remote oxygen or hydrogen plasmas. *Plasma Process. Polym.* **2019**, *16*, 1800175. [CrossRef]
37. Golda, J.; Biskup, B.; Layes, V.; Winzer, T.; Benedikt, J. Vacuum ultraviolet spectroscopy of cold atmospheric pressure plasma jets. *Plasma Process. Polym.* **2020**, *17*, 1900216. [CrossRef]
38. Liu, F.; Nie, L.; Lu, X.; Stephens, J.; Ostrikov, K. Atmospheric plasma VUV photon emission. *Plasma Sources Sci. Technol.* **2020**, *29*, 065001. [CrossRef]
39. Zaplotnik, R.; Vesel, A. Effect of VUV Radiation on Surface Modification of Polystyrene Exposed to Atmospheric Pressure Plasma Jet. *Polymers* **2020**, *12*, 1136. [CrossRef] [PubMed]
40. Gaborit, G.; Reuter, S.; Iseni, S.; Duvillaret, L. Cold plasma diagnostic using vectorial electrooptic probe. In Proceedings of the 5th International Conference on Plasma Medicine (ICPM 5), Nara, Japan, 18–23 May 2014.

41. Darny, T.; Pouvesle, J.M.; Puech, V.; Douat, C.; Dozias, S.; Robert, E. Analysis of conductive target influence in plasma jet experiments through helium metastable and electric field measurements. *Plasma Sources Sci. Technol.* **2017**, *26*, 045008. [CrossRef]

42. Lu, X.P.; Reuter, S.; Laroussi, M.; Liu, D.W. *Nonequilibrium Atmospheric Pressure Plasma Jets: Fundamentals, Diagnostics, and Medical Applications*, 1st ed.; CRC Press: Boca Raton, FL, USA, 2019. [CrossRef]

43. Iséni, S. Mapping the electric field vector of guided ionization waves at atmospheric pressure. *Plasma Res. Express* **2020**, *2*, 025014. [CrossRef]

44. Aljammal, F.; Gaborit, G.; Bernier, M.; Iséni, S.; Galtier, L.; Revillod, G.; Duvillaret, L. Pigtailed Electrooptic Sensor for Time-and Space-Resolved Dielectric Barrier Discharges Analysis. *IEEE Trans. Instrum. Meas.* **2021**, *70*, 9512609. [CrossRef]

45. DIN SPEC 91315:2014-06. General Requirements for Plasma Sources in Medicine. 2014. Available online: https://www.beuth. de/en/technical-rule/din-spec-91315/203493369 (accessed on 1 November 2021).

46. Mann, M.S.; Tiede, R.; Gavenis, K.; Daeschlein, G.; Bussiahn, R.; Weltmann, K.D.; Emmert, S.; von Woedtke, T.; Ahmed, R. Introduction to DIN-specification 91315 based on the characterization of the plasma jet kINPen® MED. *Clin. Plasma Med.* **2016**, *4*, 35–45. [CrossRef]

47. Lehmann, A.; Pietag, F.; Arnold, T. Human health risk evaluation of a microwave-driven atmospheric plasma jet as medical device. *Clin. Plasma Med.* **2017**, *7*, 16–23. [CrossRef]

48. Xaubet, M.; Baudler, J.S.; Gerling, T.; Giuliani, L.; Minotti, F.; Grondona, D.; Von Woedtke, T.; Weltmann, K.D. Design optimization of an air atmospheric pressure plasma-jet device intended for medical use. *Plasma Process. Polym.* **2018**, *15*, 1700211. [CrossRef]

49. Thana, P.; Wijaikhum, A.; Poramapijitwat, P.; Kuensaen, C.; Meerak, J.; Ngamjarurojana, A.; Sarapirom, S.; Boonyawan, D. A compact pulse-modulation cold air plasma jet for the inactivation of chronic wound bacteria: Development and characterization. *Heliyon* **2019**, *5*, e02455. [CrossRef] [PubMed]

50. Timmermann, E.; Bansemer, R.; Gerling, T.; Hahn, V.; Weltmann, K.D.; Nettesheim, S.; Puff, M. Piezoelectric-driven plasma pen with multiple nozzles used as a medical device: Risk estimation and antimicrobial efficacy. *J. Phys. D Appl. Phys.* **2020**, *54*, 025201. [CrossRef]

51. DIN EN 60601-1:2013-12 VDE 0750-1:2013-12. Medical Electrical Equipment—Part 1: General Requirements for Basic Safety and Essential Performance (IEC 60601-1:2005 + Cor.:2006 + Cor.:2007 + A1:2012); German Version EN 60601-1:2006 + Cor.:2010 + A1:2013. 2013. Available online: https://www.beuth.de/en/standard/din-en-60601-1/193923032 (accessed on 1 November 2021).

52. Gaborit, G.; Dahdah, J.; Lecoche, F.; Jarrige, P.; Gaeremynck, Y.; Duraz, E.; Duvillaret, L. A nonperturbative electrooptic sensor for in situ electric discharge characterization. *IEEE Trans. Plasma Sci.* **2013**, *41*, 2851–2857. [CrossRef]

53. Gaborit, G.; Jarrige, P.; Lecoche, F.; Dahdah, J.; Duraz, E.; Volat, C.; Duvillaret, L. Single shot and vectorial characterization of intense electric field in various environments with pigtailed electrooptic probe. *IEEE Trans. Plasma Sci.* **2014**, *42*, 1265–1273. [CrossRef]

54. Gaborit, G.; Dahdah, J.; Lecoche, F.; Treve, T.; Jarrige, P.; Gillette, L.; Piquet, J.; Duvillaret, L. Optical sensor for the vectorial analysis of the plasma induced electric field. In Proceedings of the 2015 IEEE International Conference on Plasma Sciences (ICOPS), Antalya, Turkey, 24–28 May 2015; p. 15360911. [CrossRef]

55. Aljammal, F.; Gaborit, G.; Revillod, G.; Iséni, S.; Duvillaret, L. Optical Probe for the Real Time and Vectorial Analysis of the Electric Field Induced by Ionized Gazes. In Proceedings of the 24th International Symposium on Plasma Chemistry (ISPC), Naples, Italy, 9–14 June 2019; pp. 1–3.

56. Gerling, T.; Brandenburg, R.; Wilke, C.; Weltmann, K.D. Power measurement for an atmospheric pressure plasma jet at different frequencies: Distribution in the core plasma and the effluent. *Eur. Phys. J. Appl. Phys.* **2017**, *78*, 10801. [CrossRef]

57. Teschner, T.; Bansemer, R.; Weltmann, K.D.; Gerling, T. Investigation of power transmission of a helium plasma jet to different dielectric targets considering operating modes. *Plasma* **2019**, *2*, 348–359. [CrossRef]

58. Brandenburg, R. Dielectric barrier discharges: Progress on plasma sources and on the understanding of regimes and single filaments. *Plasma Sources Sci. Technol.* **2017**, *26*, 053001. [CrossRef]

59. Scholten, L.V. PlaDinSpec. 2021. Available online: https://pm-gitlab.intranet.inp-greifswald.de/vilardell/pladinspec (accessed on 1 November 2021).

60. Foest, R.; Kindel, E.; Lange, H.; Ohl, A.; Stieber, M.; Weltmann, K.D. RF capillary jet-a tool for localized surface treatment. *Contrib. Plasma Phys.* **2007**, *47*, 119–128. [CrossRef]

61. Bussiahn, R.; Kindel, E.; Lange, H.; Weltmann, K. Spatially and temporally resolved measurements of argon metastable atoms in the effluent of a cold atmospheric pressure plasma jet. *J. Phys. D Appl. Phys.* **2010**, *43*, 165201. [CrossRef]

62. Boeuf, J.; Yang, L.; Pitchford, L. Dynamics of a guided streamer ('plasma bullet') in a helium jet in air at atmospheric pressure. *J. Phys. D Appl. Phys.* **2012**, *46*, 015201. [CrossRef]

63. Jaiswal, S.; Aguirre, E.; Prakash, G.V. A KHz frequency cold atmospheric pressure argon plasma jet for the emission of O (1 S) auroral lines in ambient air. *Sci. Rep.* **2021**, *11*, 1893. [CrossRef]

64. Sretenović, G.B.; Krstić, I.B.; Kovačević, V.V.; Obradović, B.M.; Kuraica, M.M. Spatio-temporally resolved electric field measurements in helium plasma jet. *J. Phys. D Appl. Phys.* **2014**, *47*, 102001. [CrossRef]

65. Nastuta, A.V.; Pohoata, V.; Topala, I. Atmospheric pressure plasma jet—Living tissue interface: Electrical, optical, and spectral characterization. *J. Appl. Phys.* **2013**, *113*, 183302. [CrossRef]

66. Wild, R.; Gerling, T.; Bussiahn, R.; Weltmann, K.; Stollenwerk, L. Phase-resolved measurement of electric charge deposited by an atmospheric pressure plasma jet on a dielectric surface. *J. Phys. D Appl. Phys.* **2013**, *47*, 042001. [CrossRef]

67. Gerling, T.; Wild, R.; Nastuta, A.V.; Wilke, C.; Weltmann, K.D.; Stollenwerk, L. Correlation of phase resolved current, emission and surface charge measurements in an atmospheric pressure helium jet. *Eur. Phys. J. Appl. Phys.* **2015**, *71*, 20808. [CrossRef]

68. Sobota, A.; Guaitella, O.; Garcia-Caurel, E. Experimentally obtained values of electric field of an atmospheric pressure plasma jet impinging on a dielectric surface. *J. Phys. D Appl. Phys.* **2013**, *46*, 372001. [CrossRef]

69. Norberg, S.A.; Johnsen, E.; Kushner, M.J. Helium atmospheric pressure plasma jets touching dielectric and metal surfaces. *J. Appl. Phys.* **2015**, *118*, 013301. [CrossRef]

70. Goldberg, B.M.; Reuter, S.; Dogariu, A.; Miles, R.B. 1D time evolving electric field profile measurements with sub-ns resolution using the E-FISH method. *Opt. Lett.* **2019**, *44*, 3853–3856. [CrossRef] [PubMed]

71. Kettlitz, M.; Höft, H.; Hoder, T.; Reuter, S.; Weltmann, K.; Brandenburg, R. On the spatio-temporal development of pulsed barrier discharges: Influence of duty cycle variation. *J. Phys. D Appl. Phys.* **2012**, *45*, 245201. [CrossRef]

72. Schmidt-Bleker, A.; Reuter, S.; Weltmann, K. Quantitative schlieren diagnostics for the determination of ambient species density, gas temperature and calorimetric power of cold atmospheric plasma jets. *J. Phys. D Appl. Phys.* **2015**, *48*, 175202. [CrossRef]

73. Jiang, N.; Yang, J.L.; He, F.; Cao, Z. Interplay of discharge and gas flow in atmospheric pressure plasma jets. *J. Appl. Phys.* **2011**, *109*, 093305. [CrossRef]

74. Schmidt-Bleker, A.; Winter, J.; Bösel, A.; Reuter, S.; Weltmann, K.D. On the plasma chemistry of a cold atmospheric argon plasma jet with shielding gas device. *Plasma Sources Sci. Technol.* **2015**, *25*, 015005. [CrossRef]

75. Gerling, T.; Hoder, T.; Bussiahn, R.; Brandenburg, R.; Weltmann, K. On the spatio-temporal dynamics of a self-pulsed nanosecond transient spark discharge: A spectroscopic and electrical analysis. *Plasma Sources Sci. Technol.* **2013**, *22*, 065012. [CrossRef]

76. Reuter, S.; Winter, J.; Schmidt-Bleker, A.; Schroeder, D.; Lange, H.; Knake, N.; Schulz-Von Der Gathen, V.; Weltmann, K. Atomic oxygen in a cold argon plasma jet: TALIF spectroscopy in ambient air with modelling and measurements of ambient species diffusion. *Plasma Sources Sci. Technol.* **2012**, *21*, 024005. [CrossRef]

77. Bruno, G.; Wenske, S.; Mahdikia, H.; Gerling, T.; von Woedtke, T.; Wende, K. Radiation Driven Chemistry in Biomolecules—Is (V)UV Involved in the Bioactivity of Argon Jet Plasmas? *Front. Phys.* **2021**, *9*, 708. [CrossRef]

78. Radzig, A.A.; Smirnov, B.M. (Eds.) *Reference Data on Atoms, Molecules, and Ions*; Springer: Berlin/Heidelberg, Germany, 1985. [CrossRef]

79. Capitelli, M.; Ferreira, C.M.; Osipov, A.I.; Gordiets, B.F. *Plasma Kinetics in Atmospheric Gases*; Springer: Berlin/Heidelberg, Germany, 2000.

80. Nastuta, A.V.; Topala, I.; Grigoras, C.; Pohoata, V.; Popa, G. Stimulation of wound healing by helium atmospheric pressure plasma treatment. *J. Phys. D Appl. Phys.* **2011**, *44*, 105204. [CrossRef]

81. Nastuta, A.; Topala, I.; Pohoata, V.; Mihaila, I.; Agheorghiesei, C.; Dumitrascu, N. Atmospheric pressure plasma jets in inert gases: Electrical, optical and mass spectrometry diagnosis. *Rom. Rep. Phys.* **2017**, *69*, 407.

82. Nastuta, A.; Popa, G. Surface oxidation and enhanced hydrophilization of polyamide fiber surface after He/Ar atmospheric pressure plasma exposure. *Rom. Rep. Phys.* **2019**, *71*, 413.

83. Dobrynin, D.; Arjunan, K.; Fridman, A.; Friedman, G.; Clyne, A.M. Direct and controllable nitric oxide delivery into biological media and living cells by a pin-to-hole spark discharge (PHD) plasma. *J. Phys. D Appl. Phys.* **2011**, *44*, 075201. [CrossRef]

84. Kramida, A.; Ralchenko, Y.; Reader, J.; Team, N.A. NIST Atomic Spectra Database; Version 5.8. 2020. Available online: https://physics.nist.gov/asd (accessed on 1 December 2020).

85. Bansemer, R.; Schmidt-Bleker, A.; van Rienen, U.; Weltmann, K.D. Investigation and control of the-to-transition in a novel sub-atmospheric pressure dielectric barrier discharge. *Plasma Sources Sci. Technol.* **2017**, *26*, 065005. [CrossRef]

86. Luque, J.; Crosley, D. *LIFBASE: Database and Spectral Simulation Program*; SRI International Report; Technical Report MP 99-009; Scientific Research Publishing: Wuhan, China, 1999.

87. Navrátil, Z.; Trunec, D.; Šmíd, R.; Lazar, L. A software for optical emission spectroscopy-problem formulation and application to plasma diagnostics. *Czechoslov. J. Phys.* **2006**, *56*, B944–B951. [CrossRef]

 **applied sciences**

Article

Characteristics of 2.45 GHz Surface-Wave-Sustained Argon Discharge for Bio-Medical Applications

Evgenia Benova [1,*], Plamena Marinova [2], Radka Tafradjiiska-Hadjiolova [3], Zafer Sabit [3], Dimitar Bakalov [3], Nikolay Valchev [3], Lubomir Traikov [4], Todor Hikov [4], Ivan Tsonev [5,6] and Todor Bogdanov [4]

[1] Plasma Technology Laboratory, Clean & Circle Center of Competence, Sofia University, 1164 Sofia, Bulgaria
[2] Faculty of Forest Industry, University of Forestry, 1756 Sofia, Bulgaria; plamena_am@abv.bg
[3] Department of Physiology and Pathophysiology, Faculty of Medicine, Medical University of Sofia, 1431 Sofia, Bulgaria; rhadjiolova@medfac.mu-sofia.bg (R.T.-H.); zsabit@medfac.mu-sofia.bg (Z.S.); dbakalov@medfac.mu-sofia.bg (D.B.); nvalchev@medfac.mu-sofia.bg (N.V.)
[4] Department of Medical Physics and Biophysics, Faculty of Medicine, Medical University of Sofia, 1431 Sofia, Bulgaria; lltraikov@medfac.mu-sofia.bg (L.T.); thikov@medfac.mu-sofia.bg (T.H.); tbogdanov@medfac.mu-sofia.bg (T.B.)
[5] Research Group PLASMANT, Department of Chemistry, University of Antwerp, 2610 Antwerp, Belgium; Ivan.Tsonev@uantwerpen.be
[6] Faculty of Physics, Sofia University, 1164 Sofia, Bulgaria
* Correspondence: ebenova@uni-sofia.bg

Abstract: Cold atmospheric plasma (CAP) applications in various fields, such as biology, medicine and agriculture, have significantly grown during recent years. Many new types of plasma sources operating at atmospheric pressure in open air were developed. In order to use such plasmas for the treatment of biological systems, plasma properties should fulfil strong requirements. One of the most important is the prevention from heating damage. That is why in many cases, the post-discharge region is used for treatment, but the short living particles in the active discharge zone and reactions with them are missed in that case. We use the active region of surface-wave-sustained argon plasma for biological systems treatment. The previous investigations showed good bactericidal, virucidal, seeds germination and decontamination effects at a short treatment time, but the discharge conditions for bio-medical applications need specific adjustment. A detailed theoretical and experimental investigation of the plasma characteristics and their possible optimization in order to meet the requirements for bio-medical applications are presented in this paper. The length of the plasma torch, the temperature at the treatment sample position and the microwave radiation there are estimated and optimized by the appropriate choice of discharge tube size, argon flow rate and microwave power.

Keywords: bio-medical plasma applications; surface-wave-sustained discharge; microwave discharge; cold atmospheric plasma; microwave plasma torch

Citation: Benova, E.; Marinova, P.; Tafradjiiska-Hadjiolova, R.; Sabit, Z.; Bakalov, D.; Valchev, N.; Traikov, L.; Hikov, T.; Tsonev, I.; Bogdanov, T. Characteristics of 2.45 GHz Surface-Wave-Sustained Argon Discharge for Bio-Medical Applications. *Appl. Sci.* **2022**, *12*, 969. https://doi.org/10.3390/app12030969

Academic Editor: Andrei Vasile Nastuta

Received: 20 December 2021
Accepted: 14 January 2022
Published: 18 January 2022

Publisher's Note: MDPI stays neutral with regard to jurisdictional claims in published maps and institutional affiliations.

1. Introduction

The low-temperature, non-equilibrium atmospheric pressure plasmas have attracted increasing interest as simple and less expensive plasma sources operating in open space for applications in biology, medicine, agriculture, and the food industry [1–10]. Various types of atmospheric pressure plasma sources were developed for such applications. Initially, the plasma produced at atmospheric pressure was with a temperature much higher than 40 °C, and mainly the thermal effects of plasma were used, as in the argon plasma coagulation device [11,12]. Later on, the investigations were focused on developing plasma sources producing plasma at atmospheric pressure with a gas temperature low enough (below 40 °C, so-called cold atmospheric plasma—CAP) in order to avoid the destructive effects and damaging of heat-sensitive materials and biological systems. Two main requirements are formulated in [1] for the direct application of plasma "on or in the human (or animal) body" and they are the same for treatment in vivo or in vitro of any biological system, including seeds, plants, fresh fruits, etc.: (i) good stability and reproducibility of the plasma source operating at atmospheric pressure in the open space; and (ii) low temperature (<40 °C) at the tissue (sample) contact zone to avoid thermal destruction.

Various types of plasma devices for these purposes were developed in recent years. The most widely studied and used plasma sources for bio-medical applications in various configurations and designs are based on the dielectric barrier discharge (DBD) and the atmospheric pressure plasma jets (APPJ) [13–15]. In [16], the plasma sources (assuming that the plasma is produced between electrodes) are categorized in three types: (i) direct plasma sources, when one of the electrodes is the sample (human body); (ii) indirect plasma sources, when the plasma is produced between two electrodes and the active plasma components are transported to the treated sample; and (iii) hybrid plasma sources—surface micro discharge (SMD) technology. A detailed review of non-thermal atmospheric pressure plasma sources for various applications and their plasma characteristics is presented in [17], and in [18] the focus is on plasma sources for medical purposes.

The strategies for controlling the physical properties of the plasma for bio-medical applications and for keeping the plasma parameters bellow the damaging limits at the treatment area include the following: (i) using low electrical power; (ii) high frequency (kHz to MHz) or pulsed regime of operation; and (iii) appropriate choice of gas or gas mixture and the gas flow rate for plasma production. One of the CE certified as a medical device plasma source is the kINPen [19]. Various modifications exist under this name, and they operate in different regimes (power from 1.9 W up to 120 W; frequency from 0.8 MHz to 27.12 MHz; gases Ar, He + 2% molecular gas admixture, air with flow rates usually 3–5 slm) and one of them (the kINPen MED) is actually CE certified as a medical device. It needs about 10 years to decrease the operating temperature from about 80 °C [20–22] to 35–38 °C. In [23], the kINPen is compared with another plasma source, MicroPlaSter, which is also CE certified for medical use. The MicroPlaSter operates at 2.45 GHz and 80–110 W electrical power. Both devices use plasma, which is transported out of the ignition region by the gas flow. For MicroPlaSter, the gas flow rate is 2–5 L/min, and the distance between the device nozzle and the treated sample is fixed to 2 cm by a plastic spacer. The diameter of the treated area is 1 cm^2 with kINPen and 4–5 cm^2 with MicroPlaSter. The treatment time is also a very important parameter, and it is shown in [23] that the treatment time with kINPen can vary from 20 s to 5 min, and with MicroPlaSter, it is usually 2–5 min.

Another type of plasma source operating at atmospheric pressure and a microwave frequency of 2.45 GHz, the surface-wave-sustained discharge (SWD) was investigated for possible applications in biology, medicine, agriculture, and the food industry [24–30]. The specific way of plasma produced and sustained by an electromagnetic wave traveling along the plasma–dielectric interface (surface wave) determines the features of these discharges.

Surface-wave-sustained discharges operating at low and intermediate pressure have already a successful history after several types of wave launcher were introduced by Moisan in 1974 [31–35]. The good stability and reproducibility of these electrodeless discharges in a wide range of discharge conditions are very important characteristics for applications and the reason for intensive investigation during the past decades. A detailed review of these investigations can be found in [36,37]. For these discharges, the operating frequency can vary from MHz to GHz. The discharge tube diameter, thickness and dielectric permittivity play a very important role for the characteristics of the plasma sustained inside it because the tube is not only a container for sustaining the plasma, but is an important part of the waveguide structure for the electromagnetic wave propagation. The plasma itself is also a part of the same waveguide, and its properties can be easily controlled by appropriate selection of the discharge tube geometric parameters and permittivity. The plasma density (electron number density) decreases almost linearly from the wave launcher to the plasma column end. The plasma can be much longer that the size of the wave launcher, and its length depends on the wave power. The theoretical and experimental investigations of SWD at this period are at low and intermediate gas pressure [37,38]. It is shown that at these discharge conditions, the plasma is a strongly non-equilibrium one: the electron energy distribution function (EEDF) is non-Maxwellian; because of this, the electron temperature T_e is defined as 2/3 of the mean electron energy $<u>$ obtained from the calculated EEDF ($T_e = 2<u>/3$); and the electron temperature is much higher than the gas temperature (the temperature of the heavy particles in the plasma), $T_e >> T_g$.

The situation is more complicated for SWD operating at atmospheric pressure. A stable atmospheric pressure plasma inside the discharge tube or going outside it (plasma torch) can be sustained by high wave power, which leads to high gas temperature. In [39], stable argon plasma with an electron density of about 3×10^{20} m^{-3} and gas temperature of 2500 K is produced by 70 W wave power at 915 MHz and gas flow from 0.2 to 17 L/min. It is mentioned there that at a wave power above 200 W, the surfatron wave exciter needs external water cooling. The further surface wave plasma torch development is directed toward devices operating at high power and with high gas temperature (4000–8000 K) for applications in cutting, welding, material processing [40], carbon dioxide elimination [41] and many others [42]. Being electrodeless, the corrosion of electrodes and plasma contamination is avoided, and a precise, clean composition of working gases and gas mixtures can be achieved.

Producing a stable and well-reproducible surface-wave plasma torch at low wave power with low gas temperature is a tricky challenge. The filamentation in pure argon is one of the problems at atmospheric pressure SWD [43]. By adding a small amount of nitrogen, the filamentation can be eliminated [25,44] but the gas temperature of the Ar + N plasma is much higher than is appropriate for bio-medical applications. It is possible to obtain a single filament discharge in pure argon when the discharge tube is with a small radius, but this requires a higher wave frequency in the microwave range, e.g., 2.45 GHz. The electromagnetic waves in the microwave region may also have unacceptable biological effects, which need further investigation. For the direct treatment of biological samples, the plasma must come out of the discharge tube, i.e., we need a plasma torch (Figure 1). The surface-wave argon plasma torch can be produced in a gas flow regime by increasing of the wave power so that the electromagnetic surface wave propagation along the plasma–dielectric interface continues along the plasma–air boundary. The part of the plasma outside the tube is not an afterglow, but an active discharge region with high enough wave power for sustaining plasma, which leads to a higher temperature. At the same time, for bio-medical application, it is important to keep the gas temperature of the sustained plasma below 40 °C.

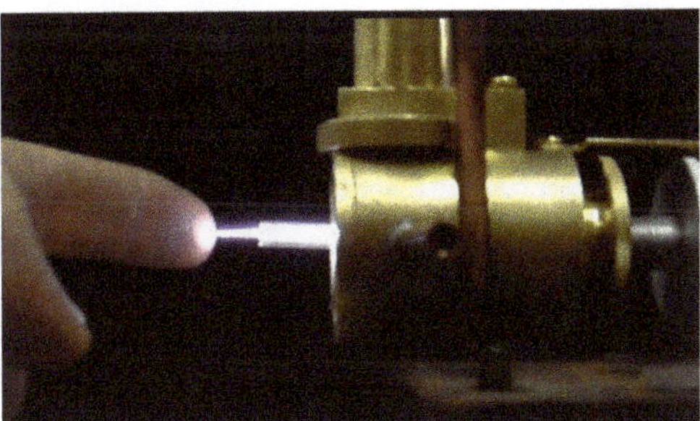

Figure 1. Plasma torch produced by a surfatron wave launcher.

We found some regimes of operation allowing a microwave plasma torch to be applied for the treatment of biological systems without thermal damage. The investigations show high bactericidal [24] and virucidal [26] effects at very short treatment time (5–30 s). This treatment time is much shorter than the typical 2–5 min of MicroPlaSter [23]. Promising results are also obtained for the decontamination of seeds, plants, and fruits at a similarly short treatment time by a microwave plasma torch [28–30]. Some preliminary experiments on wound treatment on mice [27] also stimulate the investigations in these directions. The reactive oxygen and nitrogen (RONS) species important for bio-medical plasma effects are produced, even when the working gas is pure argon as a result of the interaction of the plasma torch with ambient air [25].

The biggest problem of the microwave plasma torches sustained by a traveling wave that needs systematic investigation is that by using the same (low) wave power and gas flow, in some cases, the temperature at the plasma torch tip is low enough to be touched (Figure 1) and in other cases, can be above 100 °C [25]. In some conditions, only one stable filament exists, while at the same wave power and gas flow, a second filament can appear. Another important difference between the surface wave plasma torch and the MicroPlaSter device [23] is connected with the microwave field used for the plasma ignition. The microwave electric field in MicroPlaSter is applied between electrodes inside a metal container, and the plasma is transported outside the active region by the gas flow. In this way, the treated sample is prevented from the microwave radiation, but only long-living active plasma particles can be used for the treatment process. The surface wave plasma torch is produced by an electromagnetic wave traveling along the plasma with the wave power decreasing along the torch. At low and intermediate pressures, the end of the plasma column is at a position where the wave power becomes zero and the wave cannot propagate anymore. At high and atmospheric pressures, the wave power for sustaining plasma is much higher, and the plasma column ends when the wave power is not enough for sustaining plasma, but usually is not zero. The plasma used for the samples treatment is a part of the active discharge region with all short- and long-living particles produced in the plasma (which can be the reason for obtaining good results at very short treatment time), but also the microwave field there may be too high. It is important to investigate the microwave radiation at the position of the treated sample in order to avoid any possible damage during the treatment.

The purpose of this work is to investigate the parameters of the surface-wave-sustained plasma torch and to find the well-reproducible discharge conditions at which the plasma is appropriate for bio-medical applications. The dependance of the plasma torch length, the temperature of the treated samples, the microwave radiation at the samples on the wave power and discharge tube geometrical parameters are investigated and presented in the paper.

2. Materials and Methods

2.1. Experimental

The experimental set-up is schematically presented in Figure 2a. A simple surfatron-type wave launcher is used for plasma sustaining. The surfatron (Sairem, SURFATRON 80) is connected with a solid-state microwave generator at 2.45 GHz (Sairem, GMS 200 W) by a coaxial cable. The plasma is produced inside the discharge tube situated at the surfatron axis. Various quarts tubes with different outer and inner diameters (Table 1) are used. For all of them, the real and imaginary parts of the dielectric permittivity are $\varepsilon_r = 5.58$ and $\varepsilon_i = 0.036$, respectively. The discharge tubes end is fixed at 2 mm out of the surfatron, and the plasma torch is formed out of the discharge tube in the air. The working gas is argon 5.0 (purity of 99.999%), and its mass flow is controlled by Omega FMA-A2408 mass flow controller. The system is vertically installed with the gas flow from top to down.

Table 1. Outer and inner diameters of quarts tubes used.

Tube Notation	Outer Diameter	Inner Diameter
8/2	8 mm	2 mm
8/3	8 mm	3 mm
8/4	8 mm	4 mm

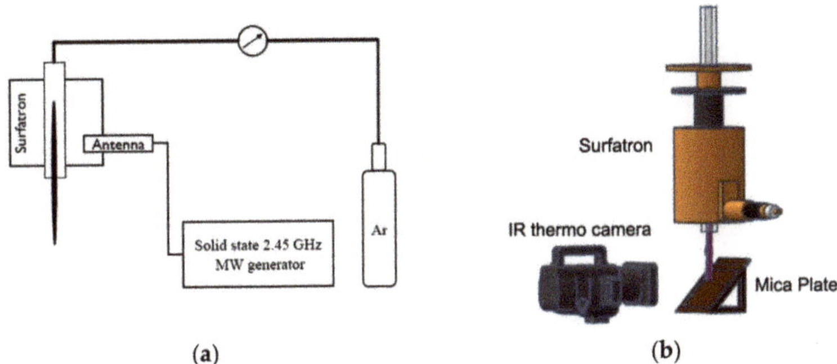

Figure 2. Experimental set-up: (**a**) surface-wave sustained argon plasma torch; (**b**) Mica Plate and infrared thermo camera system.

As it was mentioned above, the propagating surface wave produces strongly non-equilibrium plasma not only at low and intermediate, but also at atmospheric pressure. Under such conditions, the assumption that the translational (gas) temperature is equal to the rotational temperature cannot be used a priori. This is shown for other types of non-equilibrium discharges in [45] and discussed in detail in [46]. Approaching carefully this problem, it is shown in [25] that the rotational temperatures calculated from the nitrogen second positive 0–2 band is in the interval 1500–3000 K, while it is between 600 K and 1000 K when using the OH (A → X) 0–0 band. No information about the gas temperature of the plasma torch is obtained in this way. In [25], the maximum temperature on the treated sample (fresh pork skin, 10 s treatment) measured by a thermo camera when the plasma torch tip touches the sample is about 110–120 °C. Although the area on the sample surface with such a temperature is very small and the average temperature is lower, such conditions are not acceptable for bio-medical applications.

It is also observed that around the plasma filament in the discharge tube, there exists hot argon gas, which flows outside close to the plasma torch. Its temperature is usually higher than that of the plasma torch, but it does not radiate in the visible region and is not registered by optical emission spectroscopy.

Because of the above reasons, the plasma temperature is obtained by two independent methods: contact thermometry at the plasma tip and via thermal emission measured by Mica Plate and an infrared thermo camera system. (Figure 2b). The used contact thermometer is a mercury thermometer with fused quartz with a range of up to 400 °C. The temperature determined in this way is the average temperature of the plasma column and of the gas surrounding the plasma at the position where the treated sample usually is placed.

We used a Testo 865 thermal imager for the infrared thermography of a standard Mica Plate inclined at 45 degrees to the plasma torch axis and the camera-to-plate center direction. The distance between the plasma column end and Mica Plate was 29 cm and distance of the camera–plate was 55 cm. The ambient temperature was 21 °C, and the relative humidity was 48%. The method of infrared thermography determined the average plasma jet temperature, ignoring the carrying gas temperature.

Microwave radiation measurements from the Surfatron device were done by half-wave dipolar antenna attached to HF ANALYSER HF59B by Gigahertz Solutions® or wattmeter 2M-66. Measurements were done with use of 20 dB attenuator and antenna at the level of the plasma column end.

2.2. Modeling

The surface-wave discharge plasma is sustained by an electromagnetic wave propagating along the plasma–dielectric interface in a waveguide structure, including also the sustained plasma. This requires a self-consistent model for describing the wave–plasma system. The self-consistent model includes a kinetic part describing the elementary processes inside the plasma and an electrodynamic part for the wave propagation based on Maxwell's equations. These two parts are linked in a self-consistent way by the energy balance equations of the electromagnetic wave and the electrons in each point along the plasma torch because the plasma is axially inhomogeneous. The self-consistent approach used in this paper is schematically presented in Figure 3.

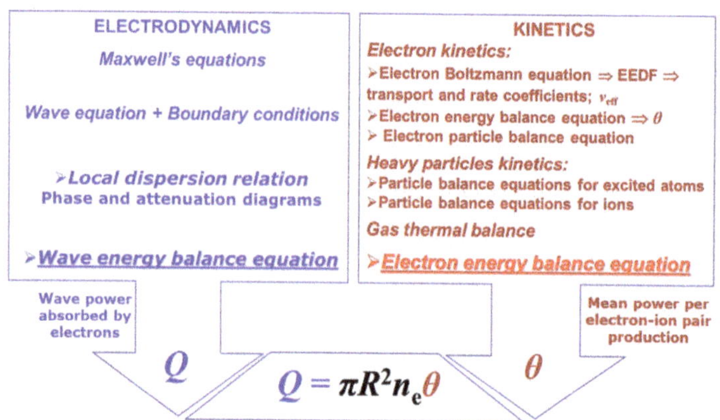

Figure 3. Schematic of self-consistent modeling approach.

The main equations in the electrodynamic part are the local dispersion equation and the wave energy balance equation obtained from Maxwell's equations. The kinetic part is based on the electron Boltzmann equation together with the electron energy balance equation and the particle balance equations for electrons, ions and excited atoms. The kinetic model of the surface-wave-sustained argon plasma at atmospheric pressure is presented in detail in [47].

3. Results

The modeling allows obtaining the plasma and wave characteristics by varying the discharge condition parameters, such as the tube size, its dielectric permittivity and the wave frequency. The plasma goes out of the discharge tube, forming the plasma torch in the open space, where the electromagnetic wave continues its propagation along the plasma–air interface. In this way, the region of the plasma torch is a region of active plasma sustained by the electromagnetic field of the wave.

3.1. Effects of the Dielectric Tube Permittivity and the Wave Frequency

The dielectric permittivity is one of the important characteristics of the discharge tube, which is a part of the waveguide structure for the wave propagation. Using different materials for the discharge tube, one can obtain higher or lower plasma density and temperature at the same geometry parameters and wave power. Figure 4 illustrates the changes in the plasma density (electron number density) n_e and the wave power axial distributions at two different dielectric permittivity values ($\varepsilon_d = 4$ and 5.58) of the same size discharge tube (8 mm/2 mm, see Table 1). For easily comparing the results, the end of the plasma torch is at position $z = 0$, and the surfatron position is on the left side of the graph. The well-known almost linear axial distribution of the plasma density is confirmed for all discharge conditions. The effect of the dielectric permittivity increasing is in the increasing in the plasma density, but it is not very significant.

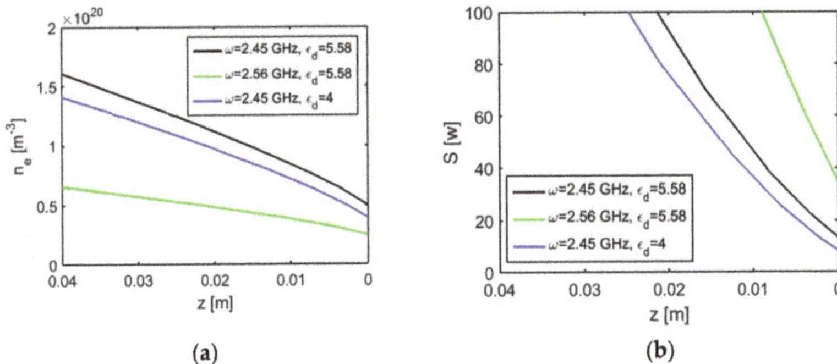

(a)   **(b)**

Figure 4. Theoretical axial distribution for the discharge tube 8 mm/2 mm at different discharge conditions of (**a**) plasma density; (**b**) wave power.

The increase in the wave frequency turns out to produce a much more visible result in decreasing the plasma density combined with increasing the wave power necessary for its sustaining. This is seen in comparison with the black and green curves in Figure 4.

3.2. Effects of the Dielectric Tube Inner Diameter on the Plasma Column Size

The axial distribution of the plasma density n_e obtained theoretically for the three discharge tubes 8/2, 8/3 and 8/4 (see Table 1) is presented in Figure 5a—solid lines. The plasma torch where plasma is surrounded by air is presented by dashed lines. The corresponding wave power can be seen in Figure 5b. The length of the plasma torch also increases with the wave power. The torch length produced with the same wave power (100 W) is marked by dots in Figure 5a. One can see that the theoretical prediction obtains a longer plasma column with a higher plasma density at the same wave power in the discharge tube with the smaller inner radius. Respectively, the plasma column with the same length is produced in the discharge tube with a smaller inner radius by a lower wave power. The plasma torch is longer, but with a lower electron density compared to the plasma inside the dielectric tube with the same wave power. From Figure 5b, one can see also that at the end of the plasma column inside the dielectric tube, the microwave power is not completely absorbed, and the residual power is higher at a smaller internal tube radius. At the end of the plasma torch (dashed lines), the residual microwave power is much lower.

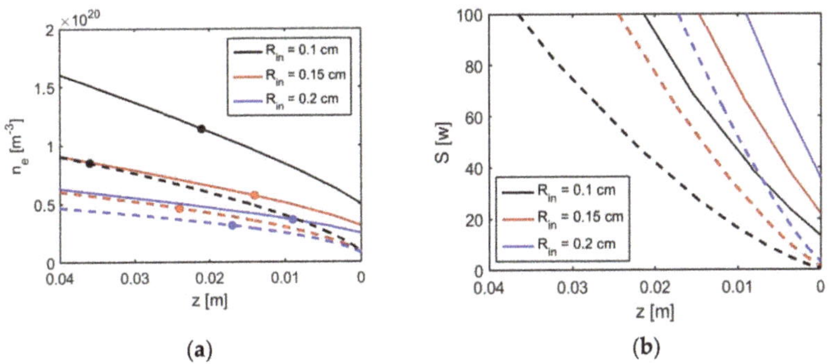

(a)   **(b)**

Figure 5. Theoretical axial distribution for three discharge tubes with different internal radii of (**a**) plasma density; (**b**) wave power. The solid lines correspond to plasma inside dielectric tube and the dashed lines are for plasma torch (plasma surrounded by air).

The experimentally measured plasma torch length produced by using the three discharge tubes shows different behavior. In Figure 6a, the fast camera imaging of the plasma torches in the three tubes are presented. From such images, the plasma torch length is measured at a different wave power, and the graphs are presented in Figure 6b. At the same wave power, the longer plasma torch is obtained in the discharge tube with the bigger internal radius, which is just the opposite to the theoretical results. There are two reasons for such discrepancy: (i) in the modeling, the gas flow is neglected, while all experiments are in the gas flow regime, and the gas glow varies from 2 L/min to 15 L/min. The experimental results presented in Figure 6b are at a gas flow rate of 5 L/min. (ii) The model results are obtained under the assumptions that the discharge tube is completely filled with plasma (or the plasma torch radius is equal to the discharge tube radius), while the experiment shows that inside the tube, there exist one or more filaments. Initially, at a low wave power, the filament is only one in each tube, but does not fill the tube completely in the radial direction. With wave power, the increasing of the length of the filament is almost linear, as it is theoretically predicted, but at some power, depending on the tube inner radius, a second filament appears, and the length of the first one stops increasing. This is illustrated in Figure 6c for the 8 mm/3 mm discharge tube. Our investigation shows that the appearance of the second filament happens at a lower wave power when the inner diameter of the discharge tube is smaller. It is shown by experimental [48] and radial modeling [49] investigations that the molecular argon ions play a critical role in the radial contraction of argon microwave discharges at atmospheric pressure. Our modeling results show the axial distribution of argon atomic (Ar^+) and two molecular (Ar_2^+ and Ar_3^+) ions along the plasma column in the three discharge tubes (Figure 7). As it can be seen, the molecular Ar_2^+ ion has a much higher concentration than the other two ions in all tubes, which is in good agreement with the prediction in [48,49]. The highest concentration of Ar_2^+ ion is in the tube with the smaller inner diameter (8 mm/2 mm). This can be the reason for the fast formation of the second filament at a lower wave power in this tube, but further investigations would be needed to explain it.

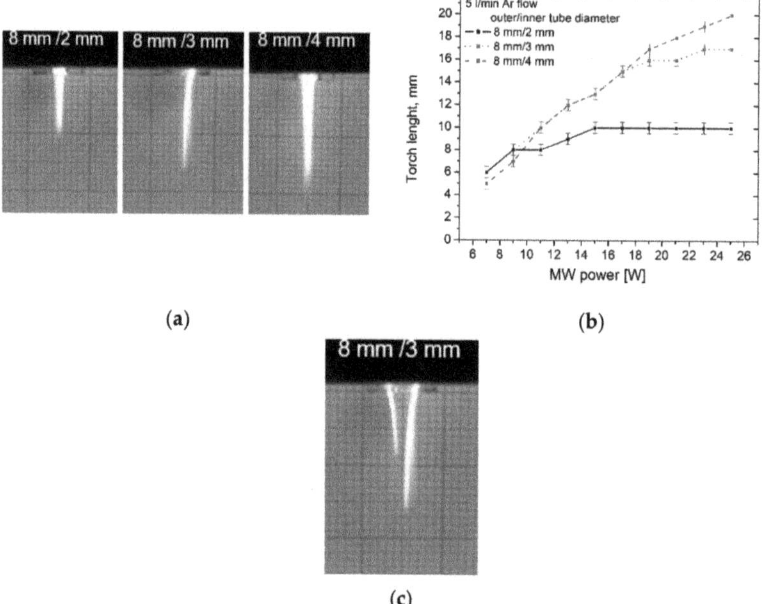

(a)

(b)

(c)

Figure 6. (**a**) Fast camera images of the plasma torch in the three tubes. (**b**) Experimental results for the plasma torch length in the three tubes. (**c**) Appearance of a second filament at higher wave power for 8 mm/3 mm tube.

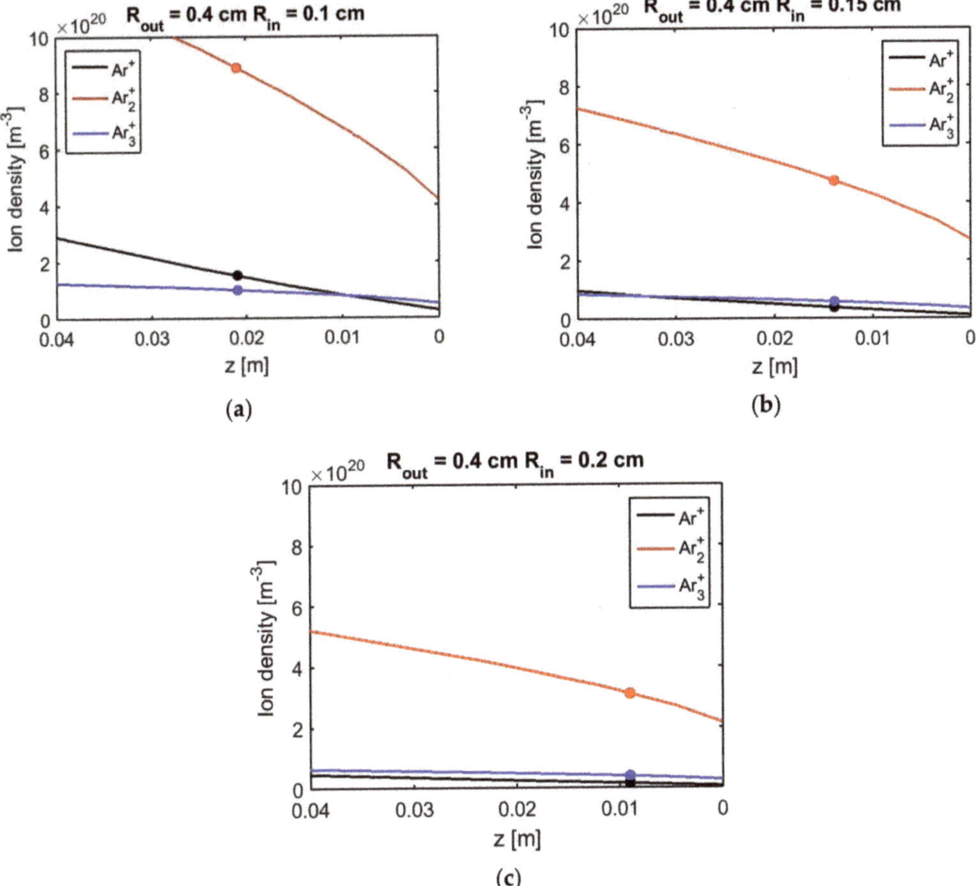

Figure 7. Axial distribution of argon atomic ion Ar⁺ (balck lines) and two molecular ions Ar_2^+ (red lines) and Ar_3^+ (blue lines) along the plasma column in the three discharge tubes: (**a**) 8 mm/2 mm; (**b**) 8 mm/3 mm; (**c**) 8 mm/4 mm. The dots correspond to the same wave power as in Figure 5a.

These experimental results show also that even at a low wave power, the length of the plasma torches in all tubes is enough for the direct treatment of the samples. The important question for bio-medical applications is what the temperature at the position of the samples is so that thermal damage is avoided.

3.3. Temperature at the Sample Position

For this investigation, the argon flow rate is varied from 2 L/min to 15 L/min for the three tubes. The temperature is measured by a contact thermometer and by the Mica Plate and infrared thermo camera system. Figure 8 illustrates the latter experiments and obtained images used for the temperature estimation at an argon flow rate of 10 L/min. Similar images are obtained at all the other discharge conditions.

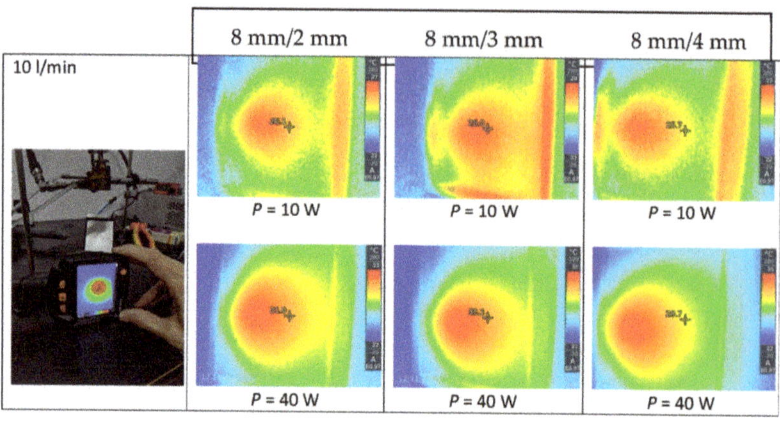

Figure 8. Images from Mica Plate and infrared thermo camera system.

Figure 9 demonstrates the effect of the discharge tube inner diameter on the temperature at fixed gas flow rate. In Figure 9a, the gas flow rate is 5 L/min and in Figure 9b it is twice higher, 10 L/min. The temperature measured by the contact thermometer is always higher than that obtained by the IR camera. We observe that the temperature of the plasma is lower than the temperature of the not ionized Argon in the tube and ambient gas outside it. By the contact thermometer, the average temperature of the plasma and ambient gas is measured, while infrared thermography determines only the average plasma temperature, ignoring the carrying gas temperature.

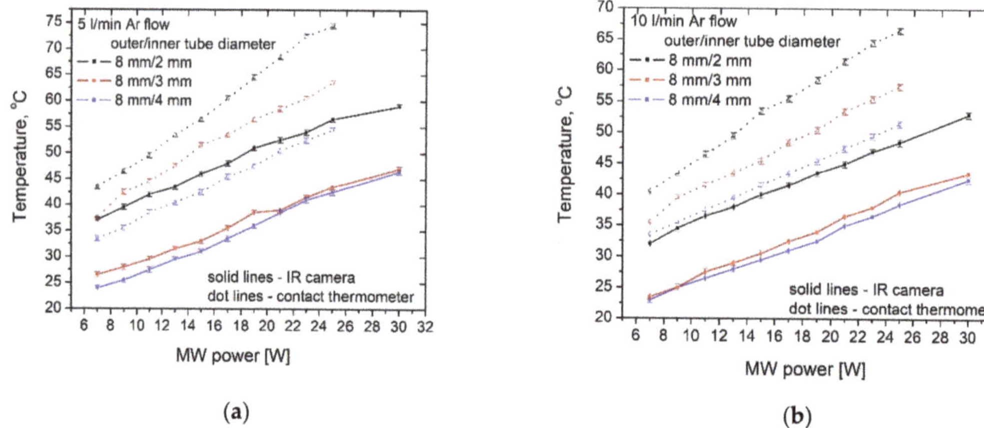

(a)

(b)

Figure 9. Temperature measured at the position of the sample as function of the wave power for three inner diameters of the discharge tubes. Solid symbols and lines are the results obtained by the IR camera and the empty symbols and dot lines by contact thermometer. The argon flow rate is fixed at (**a**) 5 L/min; (**b**) 10 L/min.

Both methods show that the lower temperature is obtained by using the discharge tube with a bigger inner diameter, i.e., 8 mm/4 mm. For this tube, even at 20 W wave power, the temperature does not exceed 45 °C, measured by the contact thermometer. Similar results are obtained for all the other gas flow rates, i.e., the best performance is shown for the 8 mm/4 mm discharge tube. More detailed dependance of the temperature on the gas flow rate for the 8/4 discharge tube is shown in Figure 10.

At a higher gas flow rate, the temperature is lower, but this effect is less than 5 °C at a low wave power and less than 10 °C at 30 W. Comparing the results presented in Figures 9 and 10, one can conclude that for decreasing the temperature at the sample, it is more important to choose a discharge tube with a bigger inner diameter than to increase the gas flow and gas consumption. In addition, if the sample is solid, its surface is not much affected by the high gas flow, but this is not the case when the treated sample is liquid. In the latter case, the gas flow cannot be too high. For the discharge tube of 8 mm/4 mm, the temperature below 45 °C can be achieved by a gas flow rate of 5 L/min or 8 L/min at a wave power of up to 20 W, which could be the optimal operational regime.

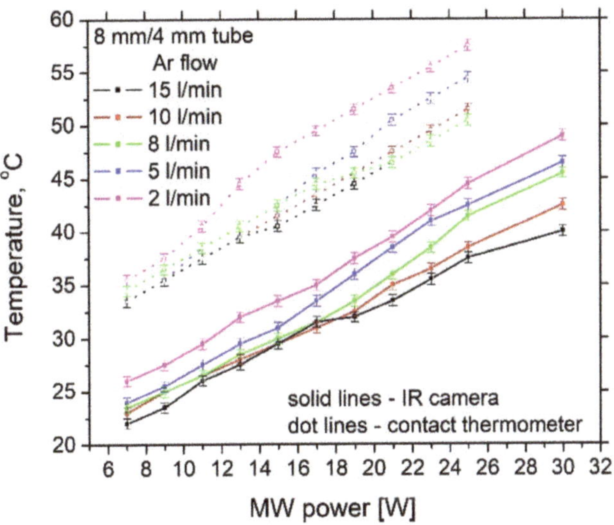

Figure 10. Effect of argon flow rate on the temperature for 8 mm/4 mm discharge tube.

3.4. Microwave Radiation at the Sample Position

One of the reasons for not widely using the SWD at atmospheric pressure for biomedical purposes is its operation at 2.45 GHz and the possible residual microwave power at the torch end and treatment sample position. The possible microwave radiation on the biological systems can produce unwanted and unacceptable biological effects. Although for SWD, the wave power is well absorbed by the plasma and usually the reflected power is less than 1 W, this problem requires special attention. The microwave radiation distribution around the sample position is illustrated in Figure 11 at typical discharge conditions. The results show that at all discharge conditions, the microwave radiation at the sample position is from 0.3 mW to 0.45 mW. Thus, we cannot expect any negative effect from the MW radiation on the biological systems due to the plasma treatment. The highest level of microwave radiation (4–5 mW) is at the connection of the surfatron antenna to the coaxial cable. This problem can be easily solved by shielding this region, which will not disturb the sample treatment.

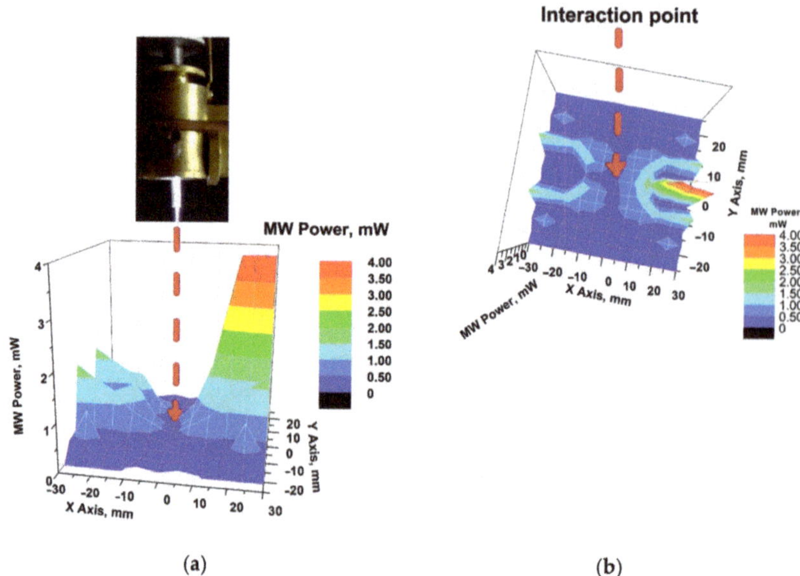

Figure 11. Microwave radiation distribution around the sample position at wave frequency 2.45 GHz, wave power 9 W, reflected power less than 1 W, discharge tube 8/3, argon flow 7 L/min: (**a**) side view; (**b**) top view.

4. Discussion

The results obtained show that the plasma characteristics of the surface-wave-sustained argon plasma torch operating at 2.45 GHz can be easily optimized to meet the requirements for bio-medical applications. For such applications, the operation regime is at a low wave power. With increasing the wave power, the length of the plasma torch increases. It is usually assumed for SWD that the increase in the wave power leads to the addition of a new part of the plasma column close to the wave exciter but without changing the plasma properties already produced by the lower power part near the plasma column end. This is correct for SWD at low pressure, but this investigation shows that at atmospheric pressure, the temperature at the same distance from the column end increases with the wave power. In order to keep the temperature below 45 °C, it is necessary to work at a low wave power (to about 30 W) regime of operation. It is good to keep the wave power as low as possible in order to avoid thermal damage to the sample.

Another important parameter than can be varied for decreasing the temperature is the discharge tube size. The temperature is lower when the inner diameter of the tube is bigger. For our tubes with an outer diameter of 8 mm, the temperatures lower than 45 °C are obtained at inner diameters of 4 mm and 3 mm at the wave power that can be increased up to 30 W. The temperature at the same position of the plasma in the tube with an inner diameter of 2 mm can be below 45 °C only at a wave power lower than 10 W.

The argon flow rate is another parameter that can be varied for decreasing the temperature. With increasing the gas flow rate, the temperature decreases. However, this decrease is not as significant as expected. Increasing the argon flow rate from 2 L/min to 15 L/min leads to a decrease in the temperature of about 10 °C at high wave power and even less at low wave power. Additional problems are the high gas consumption, affecting the soft surfaces during the treatment by the high gas flow. That is why the optimal operational regime is at a gas flow rate from 5 L/min to 8 L/min at the given discharge tubes of 8 mm/4 mm and 8 mm/3 mm.

The measurements show that problem with microwave radiation on the treated sample is not significant under the low wave power operational conditions. At the sample position, the microwave 2.45 GHz radiation does not exceed 0.4 mW. The radiation up to 4 mW that appears at the connection of the coaxial cable with the surfatron antenna can be easily shielded without disturbing the sample treatment.

5. Conclusions

The results obtained show that the surface-wave-sustained argon plasma torch operating at atmospheric pressure can keep the temperature at the treated sample surface below 45 °C, which is one of the main requirements for using this plasma for in vivo and in vitro treatment in biology, medicine and agriculture. By choosing a discharge tube with bigger inner diameter, varying the working gas flow rate and keeping low enough microwave power, the optimal regime of operation with low temperature can be achieved. At a low wave power regime, the microwave radiation at the sample position does not exceed 0.4 mW and does not affect the treated sample. All obtained results show that the microwave plasma torch fulfills the requirements to be successfully used for bio-medical applications.

Author Contributions: The investigations concept was proposed during personal discussion between all authors. The modeling approach and conditions were proposed by E.B. and P.M. The modeling results were prepared by P.M. Contact measurement of plasma column temperature, T.B., I.T., R.T.-H., Z.S. and N.V.; infrared thermography, T.B., I.T., T.H., D.B. and L.T.; measurement of radiated electromagnetic field, T.B., T.H. and L.T. The manuscript writing was done by E.B., T.B. and P.M. All authors have read and agreed to the published version of the manuscript.

Funding: This research was funded by the Grant No. BG05M2OP001-1.002-0019: "Clean Technologies for Sustainable Environment-Waters, Waste, Energy for a Circular Economy", financed by the Science and Education for Smart Growth Operational Program (2014-2020) and co-financed by the EU through the ESIF.

Institutional Review Board Statement: Not applicable.

Informed Consent Statement: Not applicable.

Data Availability Statement: The data that support the presented results of this study are available from the corresponding author upon reasonable request.

Conflicts of Interest: The authors declare no conflict of interest.

References

1. Weltmann, K.-D.; von Woedtke, T. Plasma medicine—Current state of research and medical application. *Plasma Phys. Control. Fusion* **2017**, *59*, 014031. [CrossRef]
2. Park, G.Y.; Park, S.J.; Choi, M.Y.; Koo, I.G.; Byun, J.H.; Hong, J.W.; Sim, J.Y.; Collins, G.J.; Lee, J.K. Atmospheric-pressure plasma sources for biomedical applications. *Plasma Sources Sci. Technol.* **2012**, *21*, 043001. [CrossRef]
3. Isbary, G.; Shimizu, T.; Li, Y.-F.; Stolz, W.; Thomas, H.M.; Morfill, G.E.; Zimmermann, J.L. Cold atmospheric plasma devices for medical issues. *Expert Rev. Med. Devices* **2013**, *10*, 367–377. [CrossRef] [PubMed]
4. Kong, M.G.; Kroesen, G.; Morfill, G.; Nosenko, T.; Shimizu, T.; van Dijk, J.; Zimmermann, J.L. Plasma medicine: An introductory review. *New. J. Phys.* **2009**, *11*, 115012. [CrossRef]
5. Laroussi, M. Low-Temperature Plasma Jet for Biomedical Applications: A Review. *IEEE Trans. Plasma Sci.* **2015**, *43*, 703–712. [CrossRef]
6. Laroussi, M.; Bekeschus, S.; Keidar, M.; Bogaerts, A.; Fridman, A.; Lu, X.-P.; Ostrikov, K.; Hori, M.; Stapelmann, K.; Miller, V.; et al. Low Temperature Plasma for Biology, Hygiene, and Medicine: Perspective and Roadmap. *IEEE Trans. Radiat. Plasma Med. Sci.* **2021**. [CrossRef]
7. Laroussi, M. Cold Plasma in Medicine and Healthcare: The New Frontier in Low Temperature Plasma Applications. *Front. Phys.* **2020**, *8*, 74. [CrossRef]
8. Simek, M.; Homola, T. Plasma-assisted agriculture: History, presence, and prospects—A review. *Eur. Phys. J. D* **2021**, *75*, 210. [CrossRef]
9. Sonawane, S.K.; Marar, T.; Sonal, P. Non-thermal plasma: An advanced technology for food industry. *Food Sci. Technol. Int.* **2020**, *26*, 727–740. [CrossRef]
10. Misra, N.N.; Schluter, O.; Cullen, P.J. (Eds.) *Cold Plasma in Food and Agriculture*; Elsevier: Amsterdam, The Netherlands, 2016.

11. Morrison, C.F., Jr. Electrosurgical Method and Apparatus for Initiating an Electrical Discharge in an Inert Gas Flow. U.S. Patent 4040426, 9 August 1977.
12. Farin, G.; Grund, K.E. Technology of argon plasma coagulation with particular regard to endoscopic applications. *Endoscop. Surg. Allied Technol.* **1994**, *2*, 71–77.
13. Romero-Mangado, J.; Dey, A.; Diaz-Cartagena, D.C.; Solis-Marcano, N.E.; Lopez-Nieves, M.; Santiago-Garcia, V.; Nordlund, D.; Krishnamurthy, S.; Meyyappan, M.; Koehne, J.E.; et al. Efficacy of atmospheric pressure dielectric barrier discharge for inactivating airborne pathogens. *J. Vac. Sci. Technol.* **2017**, *A35*, 041101. [CrossRef]
14. Kuzminova, A.; Kretková, T.; Kylián, O.; Hanuš, J.; Khalakhan, I.; Prukner, V.; Doležalová, E.; Šimek, M.; Biederman, H. Etching of polymers, proteins and bacterial spores by atmospheric pressure DBD plasma in air. *J. Phys. D Appl. Phys.* **2017**, *50*, 135201. [CrossRef]
15. Babaeva, N.Y.; Kushner, M.J. Dynamics of dielectric barrier discharges over wounded skin. *IEEE Trans. Plasma Sci.* **2011**, *39*, 2964–2965. [CrossRef]
16. Isbary, G.; Zimmermann, J.L.; Shimizu, T.; Li, Y.-F.; Morfill, G.E.; Thomas, H.M.; Steffes, B.; Heinlin, J.; Karrer, S.; Stolz, W. Non-thermal plasma—More than five years of clinical experience. *Clin. Plasma Med.* **2013**, *1*, 19–23. [CrossRef]
17. Fridman, A.; Chirokov, A.; Gutsol, A. Non-thermal atmospheric pressure discharges. *J. Phys. D Appl. Phys.* **2005**, *38*, R1–R24. [CrossRef]
18. Von Woedtkea, T.; Reuter, S.; Masura, K.; Weltmann, K.-D. Plasmas for medicine. *Phys. Rep.* **2013**, *530*, 291–320. [CrossRef]
19. Reuter, S.; Von Woedtke, T.; Weltmann, K.-D. The kINPen—A review on physics and chemistry of the atmospheric pressure plasma jet and its applications. *J. Phys. D Appl. Phys.* **2018**, *51*, 233001. [CrossRef]
20. Foest, R.; Kindel, E.; Lange, H.; Ohl, A.; Stieber, M.; Weltmann, K.D. RF capillary jet—A tool for localized surface treatment. *Contrib. Plasma Phys.* **2007**, *47*, 119–128. [CrossRef]
21. Fricke, K.; Steffen, H.; Von Woedtke, T.; Schroder, K.; Weltmann, K.D. High rate etching of polymers by means of an atmospheric pressure plasma jet. *Plasma Process. Polym.* **2011**, *8*, 51–58. [CrossRef]
22. Jablonowski, L.; Fricke, K.; Matthes, R.; Holtfreter, B.; Schluter, R.; Von Woedtke, T.; Weltmann, K.D.; Kocher, T. Removal of naturally grown human biofilm with an atmospheric pressure plasma jet: An In Vitro study. *J. Biophotonics* **2016**, *10*, 718–726. [CrossRef] [PubMed]
23. Arndt, S.; Schmidt, A.; Karrer, S.; Von Woedtke, T. Comparing two different plasma devices kINPen and Adtec SteriPlas regarding their molecular and cellular effects on wound healing. *Clin. Plasma Med.* **2018**, *9*, 24–33. [CrossRef]
24. Benova, E.; Topalova, Y.; Marinova, P.; Todorova, Y.; Atanasova, M.; Bogdanov, T.; Yotinov, I. Surface-wave-sustained plasma for model biological systems treatment. In Proceedings of the XXXIII International Conference on Phenomena in Ionized Gases (ICPIG), Estoril, Portugal, 9–14 July 2017; Alves, L.L., Tejero-del-Caz, A., Eds.; Instituto de Plasmas e Fusão Nuclear, Instituto Superior Técnico, Universidade de Lisboa: Lisboa, Portugal, 2017. Topic Number: 17. p. 87.
25. Krčma, F.; Tsonev, I.; Smejkalová, K.; Truchlá, D.; Kozáková, Z.; Zhekova, M.; Marinova, P.; Bogdanov, T.; Benova, E. Microwave micro torch generated in argon based mixtures for biomedical applications. *J. Phys. D Appl. Phys.* **2018**, *51*, 414001. [CrossRef]
26. Tsvetkov, V.; Hinkov, A.; Todorov, D.; Benova, E.; Tsonev, I.; Bogdanov, T.; Shishkov, S.; Shishkova, K. Effect of Plasma-Activated Medium and Water on Replication and Extracellular Virions of Herpes Simplex Virus-1. *Plasma Med.* **2019**, *9*, 205–216. [CrossRef]
27. Bogdanov, T.; Tsonev, I.; Traikov, L.L. Microwave plasma torch for wound treatment. *J. Phys. Conf. Ser.* **2020**, *1598*, 012001. [CrossRef]
28. Milusheva, S.; Nacheva, L.; Benova, E.; Marinova, P.; Dimitrova, N.; Georgieva-Hristeva, A. Experiments on Plum pox virus inactivation from micropropagated plum plants through non-thermal plasma treatment. *Plant Prot. Bull.* **2020**, *60*, 83–90.
29. Nedyalkova, S.; Bozhanova, V.; Benova, E.; Marinova, P.; Tsonev, I.; Bogdanov, T.; Koleva, M. Study on the Effect of Cold Plasma on the Germination and Growth of Durum Wheat Seeds Contaminated with Fusarium Graminearum. *Int. J. Innov. Approaches Agric. Res.* **2019**, *3*, 623–635. [CrossRef]
30. Bogdanov, T.; Tsonev, I.; Marinova, P.; Benova, E.; Rusanov, K.; Rusanova, M.; Atanassov, I.; Kozáková, Z.; Krčma, F. Microwave Plasma Torch Generated in Argon for Small Berries Surface Treatment. *Appl. Sci.* **2018**, *8*, 1870. [CrossRef]
31. Moisan, M.; Beaudry, C.; Leprince, P. A new HF device for the production of long plasma columns at a high electron density. *Phys. Lett.* **1974**, *55A*, 125–126. [CrossRef]
32. Moisan, M.; Beaudry, C.; Leprince, P. A small microwave plasma source for long column production without magnetic field. *IEEE Trans. Plasma Sci.* **1975**, *3*, 55–59. [CrossRef]
33. Moisan, M.; Leprince, P.; Beaudry, C.; Bloyet, E. Devices and Methods of Using HF Wave to Energize a Column of Gas Enclosed in an Insulating Casing. U.S. Patent No. 4049940, 20 September 1977.
34. Zakrzewski, Z.; Moisan, M.; Glaude, V.M.M.; Beaudry, C.; Leprince, P. Attenuation of a surface wave in an unmagnetized RF plasma column. *Plasma Phys.* **1977**, *19*, 77–83. [CrossRef]
35. Moisan, M.; Zakrzewski, Z.; Pantel, R. The theory and characteristics of an efficient surface wave launcher (surfatron) producing long plasma columns. *J. Phys. D Appl. Phys.* **1979**, *12*, 219–237. [CrossRef]
36. Zhelyazkov, I.; Atanassov, V. Axial structure of low-pressure high-frequency discharges sustained by travelling electromagnetic surface waves. *Phys. Rep.* **1995**, *255*, 79–201. [CrossRef]

37. Moisan, M.; Nowakowska, H. Contribution of surface-wave (SW) sustained plasma columns to the modeling of RF and microwave discharges with new insight into some of their features. A survey of other types of SW discharges. *Plasma Sources Sci. Technol.* **2018**, *27*, 073001. [CrossRef]

38. Petrova, T.; Benova, E.; Petrov, G.; Zhelyazkov, I. Self-consistent axial modeling of surface-wave-produced discharges at low and intermediate pressures. *Phys. Rev. E* **1999**, *60*, 875–886. [CrossRef] [PubMed]

39. Moisan, M.; Pantel, R.; Hubert, J.; Bloyet, E.; Leprince, P.; Marec, J.; Ricard, A. Production and Applications of Microwave Surface Wave Plasma at Atmospheric Pressure. *J. Microw. Power* **1979**, *14*, 57–61. [CrossRef]

40. Al-Shamma'a, A.I.; Wylie, S.R.; Lucas, J.; Pau, C.F. Design and construction of a 2.45 GHz waveguide-based microwave plasma jet at atmospheric pressure for material processing. *J. Phys. D Appl. Phys.* **2001**, *34*, 2734–2741. [CrossRef]

41. Uhm, H.S.; Kwak, H.S.; Hong, Y.C. Carbon dioxide elimination and regeneration of resources in a microwave plasma torch. *Environ. Pollut.* **2016**, *211*, 191–197. [CrossRef] [PubMed]

42. Uhm, H.S.; Hong, Y.C.; Shin, D.H. Microwave plasma torch and its applications. *Plasma Sources Sci. Technol.* **2006**, *15*, S26–S34. [CrossRef]

43. Castanos-Martiinez, E.; Moisan, M.; Kabouzi, Y. Achieving non-contracted and non-filamentary rare-gas tubular discharges at atmospheric pressure. *J. Phys. D Appl. Phys.* **2009**, *42*, 012003. [CrossRef]

44. Henriques, J.; Tatarova, E.; Ferreira, C.M. Microwave N2–Ar plasma torch. *J. Appl. Phys.* **2011**, *109*, 023301. [CrossRef]

45. Bruggeman, P.; Schram, D.C.; Kong, M.G.; Leys, C.H. Is the Rotational Temperature of OH(A–X) for Discharges in and in Contact with Liquids a Good Diagnostic for Determining the Gas Temperature? *Plasma Process. Polym.* **2009**, *6*, 751–762. [CrossRef]

46. Bruggeman, P.J.; Sadeghi, N.; Schram, D.C.; Linss, V. Gas temperature determination from rotational lines in non-equilibrium plasmas: A review. *Plasma Sources Sci. Technol.* **2014**, *23*, 023001. [CrossRef]

47. Benova, E.; Marinova, P.; Atanasova, M.; Petrova, T.Z. Surface-wave-sustained argon plasma kinetics from intermediate to atmospheric pressure. *J. Phys. D Appl. Phys.* **2018**, *51*, 474004. [CrossRef]

48. Ridenti, M.A.; Spyrou, N.; Amorim, J. The crucial role of molecular ions in the radial contraction of argon microwave-sustained plasma jets at atmospheric pressure. *Chem. Phys. Lett.* **2014**, *595–596*, 83–86. [CrossRef]

49. Ridenti, M.A.; Amorim, J.; Pino, A.D. Causes of plasma column contraction in surface-wave-driven discharges in argon at atmospheric pressure. *Phys. Rev. E* **2018**, *97*, 013201. [CrossRef] [PubMed]

MDPI
St. Alban-Anlage 66
4052 Basel
Switzerland
Tel. +41 61 683 77 34
Fax +41 61 302 89 18
www.mdpi.com

Applied Sciences Editorial Office
E-mail: applsci@mdpi.com
www.mdpi.com/journal/applsci